Springer Theses

Recognizing Outstanding Ph.D. Research

Aims and Scope

The series "Springer Theses" brings together a selection of the very best Ph.D. theses from around the world and across the physical sciences. Nominated and endorsed by two recognized specialists, each published volume has been selected for its scientific excellence and the high impact of its contents for the pertinent field of research. For greater accessibility to non-specialists, the published versions include an extended introduction, as well as a foreword by the student's supervisor explaining the special relevance of the work for the field. As a whole, the series will provide a valuable resource both for newcomers to the research fields described, and for other scientists seeking detailed background information on special questions. Finally, it provides an accredited documentation of the valuable contributions made by today's younger generation of scientists.

Theses are accepted into the series by invited nomination only and must fulfill all of the following criteria

- They must be written in good English.
- The topic should fall within the confines of Chemistry, Physics, Earth Sciences, Engineering and related interdisciplinary fields such as Materials, Nanoscience, Chemical Engineering, Complex Systems and Biophysics.
- The work reported in the thesis must represent a significant scientific advance.
- If the thesis includes previously published material, permission to reproduce this must be gained from the respective copyright holder.
- They must have been examined and passed during the 12 months prior to nomination.
- Each thesis should include a foreword by the supervisor outlining the significance of its content.
- The theses should have a clearly defined structure including an introduction accessible to scientists not expert in that particular field.

More information about this series at http://www.springer.com/series/8790

Snehashis Choudhury

Rational Design of Nanostructured Polymer Electrolytes and Solid–Liquid Interphases for Lithium Batteries

Doctoral Thesis accepted by Cornell University, Ithaca, NY, USA

Springer

Snehashis Choudhury
Department of Chemical Engineering
Stanford University
Stanford, CA, USA

ISSN 2190-5053 ISSN 2190-5061 (electronic)
Springer Theses
ISBN 978-3-030-28942-3 ISBN 978-3-030-28943-0 (eBook)
https://doi.org/10.1007/978-3-030-28943-0

This Springer imprint is published by the registered company Springer Nature Switzerland AG
The registered company address is: Gewerbestrasse 11, 6330 Cham, Switzerland

"The important thing is not to stop questioning. Curiosity has its own reason for existing. One cannot help but be in awe when he contemplates the mysteries of eternity, of life, of the marvelous structure of reality. It is enough if one tries merely to comprehend a little of this mystery every day." – Albert Einstein

Dedicated to my family and my mentees
and to all those who live with a passion

Supervisor's Foreword

I am very pleased to introduce the PhD thesis work of my former student, Dr. Snehashis Choudhury. Snehashis' doctoral thesis focuses on the development of design rules for electrolytes and interphases in electrochemical cells that use reactive metals such as Li and Na as anodes. Such cells are of contemporary interest because they offer substantially higher charge storage capacity per unit mass or volume than state-of-the-art Lithium-ion battery technology. They are scientifically interesting because they present opportunities for chemical and materials engineering science research on transport and ion-matter field interactions on length scales ranging from the few molecule-thick interfacial phases that form spontaneously on and passivate reactive metal anodes to micron length scale phases associated with morphological instability and dendritic electrodeposition at the metal electrode. Ultimately, these physics manifest on macroscopic length scales in the form of metal dendrites that proliferate in the interelectrode space to produce catastrophic battery failure when they bridge the electrodes, short-circuiting the cell.

Until recently, it was taught that only a solid electrolyte with modulus that exceeds that of the metal could be effective in stopping the cascade of instabilities that lead to dendritic deposition. As a consequence, all serious efforts to create practical secondary batteries that utilize Li or Na metal anodes placed priority on what has so far been an insurmountable design objective – find a suitable solid electrolyte that is chemically stable in extended contact with the metal, which has a high-enough mechanical modulus to block dendrites and which enables fast enough ion transport to enable battery operation at temperatures close to ambient. Dr. Choudhury's thesis is a tour de force exploration of clever materials design and metrology strategies that lead to a rather different set of design rules. His work shows that electrolytes formed by tethering Li-ion conducting polyethers to inorganic nanoparticles are able to suppress metal dendrite formation by forming a nanoporous, tortuous network with pore sizes below a theory-defined critical value at which surface tension forces are effective in retarding the morphological instability that produces dendrites. As significant is the discovery that these effects are largely agnostic to the configuration of the polymer-grafted nanoparticle electrolytes – free-standing membranes formed by covalently cross-linking the structures, solvent-free soft glassy electrolytes formed when the tethered chains interpenetrate to jam the material, and nanoparticle-rich interphases formed on the metal by spontaneous absorption of grafted particles have all been shown to be effective. This indicates that the design principle developed in the thesis, stabilizing metal electrodeposition by confining the deposition to nanometer length scales, is quite general.

With an eye towards practical application in batteries where the metal anode is paired with a cathode, the thesis also considers design rules for anode and cathode electrolyte interphase that, respectively, limit reductive and oxidative degradation of polyether-based electrolytes. By means of detailed experimental and computational studies, Dr. Choudhury has shown that purpose-built interphases composed of salts or polymer thin films which promote de-solvation of metal ions are effective in stabilizing reductive and oxidative degradation of polyether electrolytes.

Lynden A. Archer

James A Friend Family Distinguished
Professor of Engineering
David Croll Director of the Cornell
Energy Systems Institute
Deputy Editor, Science Advances
Robert Frederick Smith School of Chemical
and Biomolecular Engineering
Cornell University
Ithaca, NY, USA

Acknowledgments

I have to apologize as my words will not do justice to all the help and support I have received from everyone in my PhD journey and in my life. I would like to firstly thank Prof. Lynden Archer for his help not only in my PhD work but also in my career counseling. His knowledge in a wide spectrum of subjects and his passion for science have always driven me to improve every single day I spent at Cornell. I greatly appreciate the help from my committee members, Prof. Yong Joo and Prof. Geoffrey Coates. I greatly enjoyed the discussions with them on my research.

Perhaps, one of the most satisfying and inspiring experiences at Cornell was my teaching assistant responsibilities. Overall, I served as a TA for four different times in the Chemical Engineering and Physics Department. I want to thank every student in these classes who have inspired me and helped me in discovering my love for teaching. I am grateful to Prof. Julius Lucks, Prof. William Olbricht, and Prof. Chris Alabi for providing me guidance and independence in conducting lectures and recitations in my TA classes.

Thanks to all the former and current members of Archer group. I am greatly thankful to Rajesh for being my mentor in my first year at Cornell as a Master of Engineering student. I have to say, I wouldn't have been in the PhD program without Rajesh's guidance. I am sincerely grateful to Akanksha who have been a great friend, mentor, and collaborator. I will always miss our long conversations about research, people, and life with her while simultaneously working long hours in the lab. Also, in the course of my PhD, I met a great human being and researcher, Zhengyuan, whose dedication towards research was extraordinary. I thank him for making my life so easy in handling different projects and serving as a great partner in everything I worked on in my PhD. Over the 5 years at Cornell, I have worked with several undergraduate and master's students who have been more of my mentor than vice versa. I am thankful to Charles, Dylan, and Sanjuna for their constant support in research and making my experience in the group so memorable.

A big thanks to Himanshu, Prayag, Samanvaya, Sanjuna, Dylan, Ritesh, Pooja, Zhengyuan, Anubhav, Yue, Alex, Kaihang, Nijam, Rohit, Prajwal, Rahul, Mun Sek, and Sampson for their strong support in my research and life. I am greatly obliged to all my collaborators outside Archer group, Prof. Tomas Arias, Prof. Mendoza Cortes, Prof. Ravishankar Sundararaman, Prof. Donald Koch, and Dr. A. Nijamudheen. I am especially grateful to Professor Lena Kourkoutis and Dr. Michael Zachman for their constant support in cryo-electron microscopy. Also, I am thankful to all the staff scientists for their help at Argonne National Lab and Cornell High Energy Synchrotron Facility.

Lastly, and most importantly, I want to thank my parents and my brother and his family for their love and sacrifice all through my life.

Contents

Parts of this thesis have been published in the following journal articles:

- S. Choudhury, Z. Tu, A. Nijamudheen, M. J. Zachman, S. Stalin, Y. Deng, Q. Zhao, D. Vu, L. F. Kourkoutis, J. L. Mendoza-Cortes, and L. A. Archer. Stabilizing Polymer Electrolytes in High-Voltage Lithium Batteries. *Nature Communications*, 10, 3091 (2019). https://doi.org/10.1038/s41467-019-11015-0
- S. Choudhury, D. Vu, A. Warren, M. Tikekar, Z. Tu, and L. A. Archer. Confining Electrodeposition of Metals in Structured Electrolytes. *Proceedings of the National Academy of Sciences*, 115 (26), 6620–6625 (2018)
- S. Choudhury, S. Stalin, Y. Deng, and L. A. Archer. Soft Colloidal Glasses as Solid-State Electrolytes. *Chemistry of Materials*, 30 (17), 5996–6004 (2018)
- S. Choudhury, C. T. Wan, W. I. Al Sadat, Z. Tu, S. Lau, M. J. Zachman, L. F. Kourkoutis, and L. A. Archer. Designer Interphases for the Lithium-Oxygen Electrochemical Cell. *Science Advances*, 3 (4), E1602809 (2017)
- S. Choudhury, S. Wei, Y. Ozhabes, D. Gunceler, M. J. Zachman, Z. Tu, J. H. Shin, P. Nath, A. Agrawal, L. F. Kourkoutis, T. A. Arias, and L. A. Archer. Designing Solid-Liquid Interphases for Sodium Batteries. *Nature Communications*, 8, 898 (2017)
- S. Choudhury, Z. Tu, K. Fawole, S. Stalin, D. Gunceler, R. Sundararaman, and L. A. Archer. Electroless Formation of Hybrid Lithium Anodes for Fast Interfacial Ion Transport. *Angewandte Chemie International Edition*, 56, 1–9 (2017)
- S. Choudhury and L. A. Archer. Lithium Fluoride Additives for Stable Cycling of Lithium Batteries at High Current Densities. *Advanced Electronic Materials*, 2, 1500246 (2016)
- S. Choudhury, A. Agrawal, S. Wei, E. Jeng, and L. A. Archer. Hybrid Hairy Nanoparticle Electrolytes Stabilizing Lithium Metal Batteries. *Chemistry of Materials*, 28 (7), 2147–2157 (2016)
- S. Choudhury, R. Mangal, A. Agrawal, and L. A. Archer. A Highly Reversible Room-Temperature Lithium Metal Battery Based on Cross-Linked Hairy Nanoparticles. *Nature Communications*, 6, 10101 (2015)
- S. Choudhury, S. Stalin, D. Vu, A. Warren, P. Biswal, and L. A. Archer. Solid-State Polymer Electrolytes for High-Performance Lithium Metal Batteries. *Nature Communications* (In press)

Chapter 1
Introduction

1.1 The Lithium Metal Battery

Upsurge in consumer electronics and electrical appliances has necessitated the need for high power, lightweight, long-lasting batteries at low cost. In this regard, lithium-ion batteries have found their place in the commercial world because of their robustness and reversibility [1–3]. Moving forward towards further advancement in energy storage technology, the next-generation batteries are also predicted to be based on lithium metal; however, instead of using a graphitic anode for lithium intercalation, a metallic lithium thin film can serve as anode. Use of the metallic lithium anode can mean a paradigm shift in battery technology. Some of the advantages of a lithium metal battery over state-of-art lithium-ion batteries are as follows: (1) high power density, because rate of intercalation of Li ions into graphite anode is much slower than Li ion plating onto lithium anode; (2) low cost, as use of Li anode avoids the cost of graphite in the anode; (3) lightweight, for not using graphitic anode; (4) high-energy density, use of Li anode provides the liberty of using many cathode materials like sulfur and oxygen that have the higher specific capacity than any currently used cathode; and (5) batteries can be made in variety of form factors.

However, the main bottleneck in the practicality of lithium metal batteries is the safety issue and low reversibility. During the alternate charge and discharge cycles, lithium may get unevenly deposited onto the anode, resulting in growth of dendritic structures having the potential of puncturing the separator and causing internal short circuit and at times explosion. The low reversibility is related with the slow degradation of electrolyte due to parasitic reactions with the anode. Although these problems are also persistent in lithium-ion batteries, they are exacerbated in a lithium metal battery, where the reactive surface of lithium is exposed to the organic electrolyte. Broadly, these issues can be divided in three categories: (1) morphological instability, inevitable dendritic deposition, especially at high current densities; (2) chemical instability, unregulated reactions at electrode-electrolyte interface; and (3) hydrodynamic

S. Choudhury, *Rational Design of Nanostructured Polymer Electrolytes and Solid–Liquid Interphases for Lithium Batteries*, Springer Theses,
https://doi.org/10.1007/978-3-030-28943-0_1

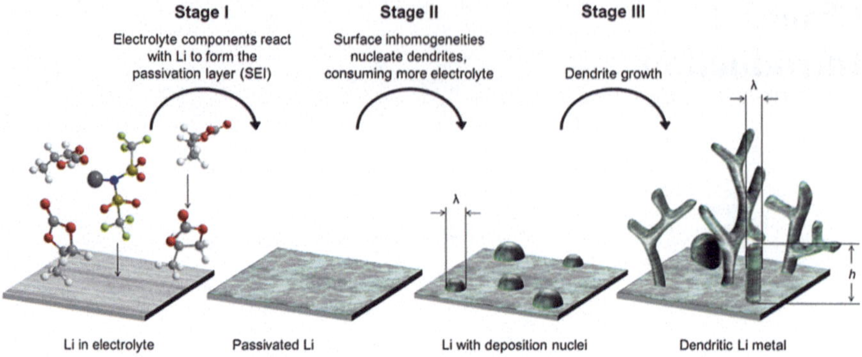

Fig. 1.1 Schematic showing different stages of instabilities during electrodeposition on a lithium metal anode [1]

instability, unstable ion transport that create space charge near the electrode surface. We believe that the origin of these instabilities is a three-stage process [1], as shown in Fig. 1.1. In stage 1, the electrolyte passivates the Li metal electrode by side reactions owing to its high reactivity; however this passive layer so-called solid electrolyte interface (SEI) is nonuniform, creating gradients of surface conductivity. In stage 2, the Li ions selectively electrodeposit on regions having higher conductivity, leading to formation of uneven and sharp nucleates (deposits). These rough deposits increase the surface area of contact between the electrode and electrolyte, thus amplifying the unwanted parasitic reactions resulting in battery-capacity fade over time. Finally, in stage 3, by repeated battery cycling, the lithium ions continue to deposit on the tips of these nucleates due to higher local electric field causing the formation of needlelike dendrites that grow bigger in size to ultimately short-circuiting the cell, which often accompanied with fire and explosions. There has been extensive scientific research in this field for the past four decades to understand the origin of such instabilities in a battery and techniques to mitigate them.

1.2 Design Principles

The three frameworks discussed in the last section all lead to different recommendations for enabling Li anodes. Some common methods include use of additives for stabilizing electrode-electrolyte interface [4–9], high mechanical strength separator for providing compression force on the surface [10–13], nanostructured electrolytes for controlling the length scale of electrodeposition [14–16] and ion transport modification using single-ion conducting electrolytes [17, 18] (as shown in Fig. 1.2). Overall, these strategies have led to significant attention among the researcher to fundamentally understand the transport and thermodynamic aspects of nanostructured electrolytes and solid-liquid interphases. The ongoing efforts in these two broad areas are discussed here.

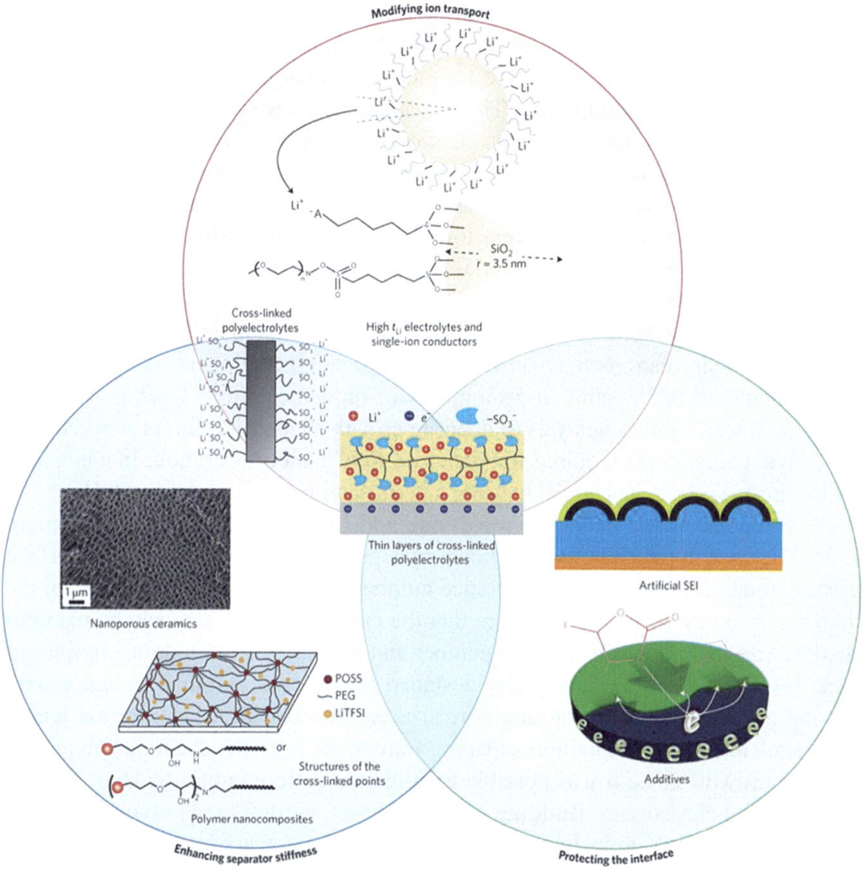

Fig. 1.2 Summary of the different approaches utilized in the literature to mitigate the instabilities associated with electrodeposition of reactive metals (lithium, sodium) [1]

1.2.1 Nanostructured Electrolytes

Solid-state electrolytes have recently gained significant attention because liquid electrolytes pose a safety issue due to their flammable nature. Previous work by Monroe and Newman using solid mechanics showed that dendrite growth can be prevented using a solid electrolyte with modulus twice that of the electrode [19, 20]. In this regard, ceramics have been of primary focus of investigation as candidates for SSE's. Previously, Dudney et al. [21] designed a SSE with LiPON chemistry that demonstrated over 10,000 cycles of galvanostatic charge and discharge in lithium vs. NMO configuration, with minimal capacity fading and close to 100% coulombic efficiency. The long-cycle life performance sets a benchmark for rechargeable battery operation, which to best of our knowledge has not been reported in batteries based on liquid electrolytes. However, there are major shortcomings associated with the LiPON solid electrolyte, and the most notable is the low ionic conductivity ($<10^{-5}$ S cm^{-1}) at room temperature. It is for this reason, the cell reported in the mentioned work is microscale, and similar performance has not been demonstrated in pouch level or even coin-type cell configurations. There has been intensive research in this area for the chemical modification of ceramic electrolytes to improve the conductivity; however the electrochemical stability is seen to deteriorate. Another major disadvantage with ceramic-based solid electrolytes is the poor solid-solid contact between electrode and electrolyte that causes very low interfacial conductivity. Furthermore, the brittleness and high cost of raw material impart a huge challenge in the commercialization of metal batteries with inorganic solid electrolytes.

Recently, nanostructured electrolytes based on alumina membrane [15, 22] and cross-linked hairy nanoparticles [14] have gained significant attention owing to the unprecedented stable cycling and simplicity of implementation. Tikekar et al. [13, 23] using linear stability analysis of dendrite growth analyzed different properties of electrolyte components required for "dendrite-free" battery operation. In a nutshell, the electrode surface was modeled using a perturbation equation: $H_c = L + H_c' e^{\sigma t} e^{ikx}$, where σ represents the dendrite growth rate and k is the inverse deposition length scale. The growth rate was further shown to depend on multiple factors like surface tension, modulus, and anion transference number of the electrolyte as shown in the state diagram in Fig. 1.3. It is evident that the critical nucleate size can be manipulated by varying anion transference number and the electrolyte modulus. In light of these revelations, we systematically designed nanoporous alumina separator with varying pore size. It was seen that pore size correlated with the deposition length scale, such that the high-modulus separator framework restricted the higher than pore dendrite growth. Thus, it was possible to eliminate battery failure by short circuits even in liquid electrolytes. Building on this analysis, we designed polymer composite membrane using cross-linked hairy nanoparticles for a scalable and low-cost

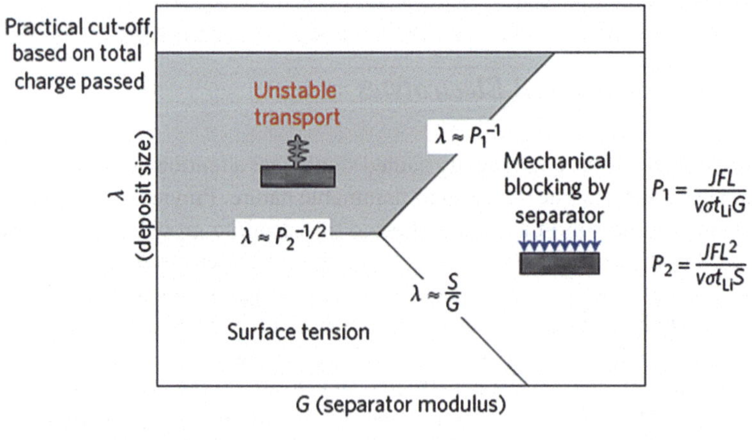

Fig. 1.3 Theoretical state diagram showing how different parameters like surface tension, deposit size, transference number, or modulus affect the deposition stability [1]

solution. The membrane was designed by interlinking polyethylene oxide-grafted silica nanoparticles with polypropylene glycol-based cross-linker. In contrast to most previously reported polymer electrolytes, the cross-linked membrane simultaneously showed good mechanical strength (~1 MPa) and high ionic conductivity at room temperature (~5 mS/cm), which is a consequence of the high cross-linking as node points in these membranes. Direct visualization experiments were performed to understand the effect of pore size on dendrite growth, which showed remarkable agreement with the theoretical predictions. Furthermore, when operated in a battery, the cross-linked membrane showed one of the highest short circuit time compared to similar electrolytes reported in the literature. Unlike previous studies of nanocomposite electrolytes where nanoparticles were used as fillers in polymer solutions, here nanoparticles were linked with polymer matrix using covalent bonding. Consequently, there have several follow-up studies involving synthesis of advanced materials with similar architecture for energy applications [24–28]. Other nanostructure designs for inhibiting dendrite growth include on single-ion conducting electrolytes with silica nanoparticle fillers with sulfonate groups tethered on their surface [18] as well as UV-cross-linked pegylated sulfonic groups [29]. These electrolytes have been reported to have transference number higher than 0.8, while having high ionic conductivity at room temperature. As previously mentioned, the high transference number or fixed anionic species lowers the threshold of instability during electrodeposition. In another set of studies, it was reported that ionic liquids (e.g., imidazolium, piperidinium) when tethered to the silica nanoparticle surface [30, 31], or cross-linked by electric field demonstrated stable cycling in lithium metal batteries [32]. These observations throw light on two different stabilizing mechanisms; ionic liquids act as supporting electrolytes in battery electrolyte, such that these species tend to crowd on specific unstable electrode regions, thereby normalizing the overall electric field. Also, the silica nanoparticles provide mechanical reinforcement and block the continuous reaction of the electrolyte with electrode [33]. A lingering issue in the single-ion conducting and ionic liquid-based structured electrolytes is the limitation in the overall ionic conductivity of the electrolyte. Recently Tu et al. showed that using an ionomer (lithiated nafion)-based coating on the metal anode, one can modify the interfacial ion transport, while the overall conductivity is unaffected as shown in Fig. 1.4 [34].

1.2.2 Solid-Liquid Interphases

Chemical modification of the solid electrolyte interphase (SEI) is an important methodology of eliminating the dendrite forming nucleates. Conventional electrolytes form a relatively thick and insulating SEI layer on the Li metal that is prone to damages during Li ion insertion and desertion, exposing fresh metal to the electrolyte for further degradations. A concept of artificial SEI has been proposed and demonstrated in the literature that takes the advantage of thin-film technology or electrolyte additives to form a desired chemistry at the interface. However, a major

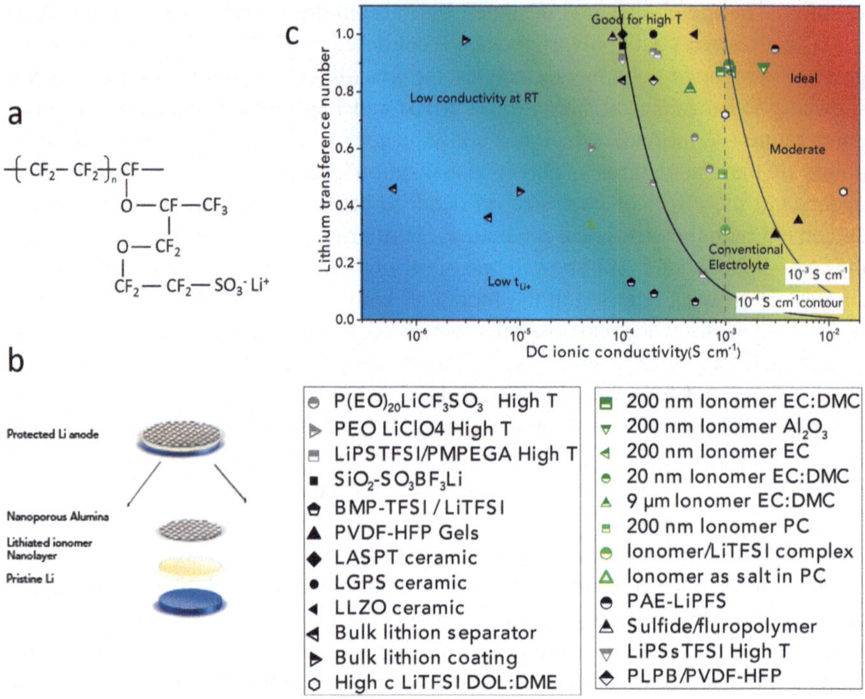

Fig. 1.4 Ionomer coating on lithium metal anode: (**a**) Chemical structure of lithiated nafion; (**b**) battery configuration showing the interfacial layer of lithiated nafion, electrodes, and nanoporous membrane; (**c**) comparison of the conductivity and transference number of various electrolytes and hybrid interfaces reported in the literature [34]

challenge in this research area is the fact that most theoretical and analytical tools are unable to explicitly model the structure-property relationship of the SEI; thus the most approaches render to a somewhat trial-and-error effort to form a stable SEI.

Aurbach et al. [35] realized that no electrolytes are stable when in contact with lithium metal causing the formation of a passivation layer that worsens at high charge and discharge rates due to large volume changes. However, some electrolytes have indeed shown to form a more stable SEI layer than others. For example, 1,3 Dioxalane (DOL) undergoes ring opening reaction at the Li surface to form an elastic surface layer of polydioxalane oligomers that expand and contract with lithium insertion and desertion [34]. Similarly, glyme-based electrolytes like dimethoxyethane (monoglyme), diglyme, and tetraglyme are also good for formation of alkoxy-based SEI (ROLi), which stabilizes electrodeposition in both Li and Na metal batteries [36, 37]. In contrast, propylene carbonate (PC) and other carbonate-based electrolytes form a very unstable, thick, and insulating passivation layer mostly comprising of lithium carbonate (Li_2CO_3) that can yield a maximum efficiency of ~77% in absence of any additives [14]. Thus, it is evident that the chemical composition of the interfacial layer is of outmost importance in determining the degree of coulombic stability. Recent joint density functional theory (JDFT) calculations [38] revealed that surface diffusion barriers for Li are much lower in halide salts (LiBr, LiI, LiF) compared to regular SEI salts like LiOH and Li_2CO_3 [39] (see Fig. 1.5). This means that Li ions can move laterally and rearrange on the interfacial layer before getting deposited as Li atoms, thus having a lesser chance of forming dendritic nucleates. Based on this concept, Lu et al. [40] used LiF additive in the electrolyte that showed remarkable results in terms of dendrite suppression and capacity retention. An LiF-rich SEI layer can also act as a thin-film barrier between the anode and electrolyte, which is recently confirmed by coulombic efficiency measurements in presence of carbonate-based electrolytes resulting in over 10% improvement [6]. In other studies, an interfacial layer of LiF is formed by using fluoride-based additives in the electrolyte. These include hydrogen fluoride (HF) [41] and fluoro-ethylene carbonate (FEC) [42], which stabilizes both Li metal- and silicon-based anodes giving very high capacity for several cycles. Use of dual salt of LiTFSI and LiFSI is also one such technique, where the side reaction between these two salts in presence of lithium metal forms a thin layer of LiF that improves the efficiency of the battery [43]. Electrolytes based on high concentration of LiFSI in DME leads to the same outcome as it is known from computational chemistry that LiFSI decomposes to LiF, which ultimately stabilizes lithium metal deposition [37]. There have been several other additives used in LMBs for improving the battery performance. $LiNO_3$ has been identified in the past as a very promising electrolyte additive as it can improve the coulombic efficiency of a bare electrolyte significantly [44]. A similar behavior is also seen when the lithium anode is protected by a layer of Li_3N, which possess the additional advantage of high Li conductivity [45]. Other additives include vinylene carbonate [7, 8, 46], sultones [9, 47], LiBOB [48, 49], and many others. Some of these additives have additional advantages. For example, alkyl phosphates and fluorine-based additives can act as flame retardants in the batteries [50].

The second approach in lithium metal protection is ex situ formation of a thin film that serves as artificial SEI layer protecting the anode from parasitic reactions with electrolytes and dendrite formation. This is a multidisciplinary field where a broad range of techniques can be utilized to manufacture thin films on lithium. A recent example is the use of atomic layer deposition (ALD) technique for depositing

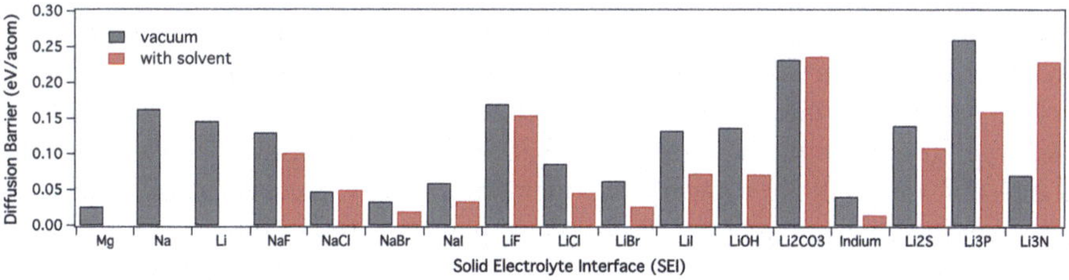

Fig. 1.5 Bar chart showing the surface diffusion barrier for various compounds typically exists in the interfacial layer of lithium or sodium metal battery cycling using liquid as well as solid superionic electrolytes [39]

a monolayer of alumina on lithium surface [51]. An emerging ex situ coating of lithium metal anode includes usage of a secondary metal coating based on tin or indium that can act as a host for the lithium ions to alloy or intercalate before electrodeposition. Such a monolayer works excellently for preventing corrosion of lithium metal from not only the electrolyte but also ambient air [52–54]. In another study, polyacrylonitrile (PAN) was oxidized to enhance ion transport and electrospun to form PAN nanofibers, which was then placed on the electrode surface [55]. This nanostructured network allowed smooth electrodeposition by confining the lithium ion plating inside the network. The other artificial SEI layers include delithiated nafion thin films [34], self-healing viscoelastic polymers [56], etc. In the future, polymer-based artificial SEI layer would be popular because of several reasons, including ease of industrial scale manufacturing, low cost, and minimal reduction in ion conductivity if the polymer has its intrinsic ion transport pathway. In situ cross-linking or network formation with rational choice of polymer chemistry on the lithium metal can also be an effective means of lithium metal protection.

1.3 Outline

The research works reported in this book utilize multiscale transport modeling and experiments to fundamentally understand and to thereby develop rational designs for polymer electrolytes and electrode-electrolyte interphases that overcome the rampant instabilities in metal-based batteries. Overall, the present book is broadly based on the understanding of three important physical concepts regarding electrochemical systems based on metallic electrodes – morphological instabilities in nanostructured electrolytes (Chaps. 2, 3, and 4), chemical instabilities in electrode-electrolyte interfaces (Chaps. 5, 6, 7, and 8), and electrochemical instabilities in polymer electrolytes (Chaps. 9, 10, and 11).

In Chaps. 2, 3, and 4, we design and synthesize novel architectures of polymer-grafted nanoparticles where poly(ethylene oxide) is grafted covalently on the surface of silica nanoparticles. On suspending in a liquid electrolyte, they show simultaneous good conductivity and high mechanical modulus. Furthermore, the silica nanoparticles act as good flame retardant even when suspended in flammable electrolyte solvents. We further discuss on cross-linked polymer electrolytes and their role in preventing dendritic growth in metal batteries. These membranes are synthesized by interlinking hairy nanoparticles with polymer chains. They serve as model nanoporous media because the silica nanoparticles act as barriers to impede dendrite growth, while the interparticle spacing hosts oligomer chains that enable metal ion transport. In a battery, the membranes show excellent performance in symmetric cell cycling as well as when paired with a commercial cathode. Importantly, these studies showed that while the tendency for battery failure by dendrite-induced short circuits can be reduced in polymer electrolytes, the issue of capacity fading as a result of continuous reactions of the metal with liquid electrolyte persists. In Chaps. 5, 6, 7, and 8 we show how multiscale analysis of transport at interfaces enables design of stable solid-liquid interphases for reactive metal batteries. Specifically, we used density functional theory (DFT) calculations to quantify the diffusion energy barrier of ions on different electrodes and interfacial chemistries, and interestingly, we found that the diffusion barrier is significantly lower for metals like magnesium and interfaces based on halides (LiF, LiBr, etc.), compared to bare lithium or sodium electrode of traditional interfaces based on carbonates or oxides. This finding guided the experimental research towards the surface chemistry design of the metal electrodes to enhance the mobility of ions at the electrode-electrolyte interface resulting in stable electrodeposition. Based on these fundamental understandings, in Chaps. 9, 10, and 11, we designed solid and gel polymer electrolytes for lithium metal batteries. However, it was seen that while these electrolytes are successful in stabilizing the negative electrode (lithium metal), they show restricted oxidative stability window for successful operation in a battery with a high-voltage cathode. We embarked on the idea of designing artificial protection layer with ionic polymers on the high-voltage cathode to prevent the electrochemical degradation of the electrolytes. This allowed our electrolytes to be used in a high-voltage (>4 V) battery where a lithium anode is paired against a commercial nickel manganese cobalt oxide (NMC) cathode. We believe this interfacial design strategy can be widely employed by researchers for operating any ether-based solid electrolytes which are well known to degrade at such high potentials.

References

1. Tikekar, M.D., Choudhury, S., Tu, Z., Archer, L.A.: Design principles for electrolytes and interfaces for stable lithium-metal batteries. Nat. Energy. **1**, 16114 (2016)
2. Lin, D., Liu, Y., Cui, Y.: Reviving the lithium metal anode for high-energy batteries. Nat. Nanotechnol. **12**, 194–206 (2017)
3. Armand, M., Tarascon, J.-M.: Building better batteries. Nature. **451**, 652–657 (2008)
4. Zhang, S.S.: A review on electrolyte additives for lithium ion batteries. J. Power Sources. **162**, 1379–1394 (2006)
5. Pieczonka, N.P.W., et al.: Impact of lithium bis(oxalate)borate electrolyte additive on the performance of high-voltage spinel/graphite Li-ion batteries. J. Phys. Chem. C. **117**, 22603–22612 (2013)
6. Choudhury, S., Archer, L.A.: Lithium fluoride additives for stable cycling of lithium batteries at high current densities. Adv. Electron. Mater. **2**, 1500246 (2015)
7. Guo, J., Wen, Z., Wu, M., Jin, J., Liu, Y.: Vinylene carbonate–LiNO3: a hybrid additive in carbonic ester electrolytes for SEI modification on Li metal anode. Electrochem. Commun. **51**, 59–63 (2015)
8. Aurbach, D., et al.: On the use of vinylene carbonate (VC) as an additive to electrolyte solutions for Li-ion batteries. Electrochim. Acta. **47**, 1423–1439 (2002)
9. Li, B., Xu, M., Li, T., Li, W., Hu, S.: Prop-1-ene-1,3-sultone as SEI formation additive in propylene carbonate-based electrolyte for lithium ion batteries. Electrochem. Commun. **17**, 92–95 (2012)
10. Khurana, R., Schaefer, J.L., Archer, L.A., Coates, G.W.: Suppression of lithium dendrite growth using cross-linked polyethylene/poly(ethylene oxide) electrolytes: a new approach for practical lithium-metal polymer batteries. J. Am. Chem. Soc. **136**, 7395–7402 (2014)

11. Bouchet, R., et al.: Single-ion BAB triblock copolymers as efficient electrolytes for lithium-metal batteries. Nat. Mater. **12**, 452–457 (2013)
12. Srivastava, S., Schaefer, J.L., Yang, Z., Tu, Z., Archer, L.A.: 25Th anniversary article: polymer-particle composites: phase stability and applications in electrochemical energy storage. Adv. Mater. **26**, 201–234 (2014)
13. Tikekar, M.D., Archer, L.A., Koch, D.L.: Stability analysis of electrodeposition across a structured electrolyte with immobilized anions. J. Electrochem. Soc. **161**, A847–A855 (2014)
14. Choudhury, S., Mangal, R., Agrawal, A., Archer, L.A.: A highly reversible room-temperature lithium metal battery based on crosslinked hairy nanoparticles. Nat. Commun. **6**, 10101 (2015)
15. Tu, Z., Kambe, Y., Lu, Y., Archer, L.A.: Nanoporous polymer-ceramic composite electrolytes for lithium metal batteries. Adv. Energy Mater. **4**, 1300654 (2014)
16. Tu, Z., Nath, P., Lu, Y., Tikekar, M.D., Archer, L.A.: Nanostructured electrolytes for stable lithium electrodeposition in secondary batteries. Acc. Chem. Res. **48**, 2947–2956 (2015)
17. Lu, Y., et al.: Stable cycling of lithium metal batteries using high transference number electrolytes. Adv. Energy Mater. **5**, 1402073 (2015)
18. Schaefer, J.L., Yanga, D.A., Archer, L.A.: High lithium transference number electrolytes via creation of 3-dimensional, charged, nanoporous networks from dense functionalized nanoparticle composites. Chem. Mater. **25**, 834–839 (2013)
19. Monroe, C., Newman, J.: The impact of elastic deformation on deposition kinetics at lithium/polymer interfaces. J. Electrochem. Soc. **152**, A396 (2005)
20. Monroe, C., Newman, J.: The effect of interfacial deformation on electrodeposition kinetics. J. Electrochem. Soc. **151**, A880 (2004)
21. Li, J., Ma, C., Chi, M., Liang, C., Dudney, N.J.: Solid electrolyte: the key for high-voltage lithium batteries. Adv. Energy Mater. **5**, 1401408 (2015)
22. Tu, Z., et al.: Nanoporous hybrid electrolytes for high-energy batteries based on reactive metal anodes. Adv. Energy Mater. **7**, 1602367 (2017)
23. Tikekar, M.D., Archer, L.A., Koch, D.L.: Stabilizing electrodeposition in elastic solid electrolytes containing immobilized anions. Sci. Adv. **2**, e1600320 (2016)
24. Hu, J., et al.: Flexible organic–inorganic hybrid solid electrolytes formed via thiol–acrylate photopolymerization. Macromolecules. **50**, 1970–1980 (2017)
25. Zhang, J., et al.: Flexible and ion-conducting membrane electrolytes for solid-state lithium batteries: dispersion of garnet nanoparticles in insulating polyethylene oxide. Nano Energy. **28**, 447–454 (2016)
26. Li, Y., Wong, K.W., Dou, Q., Ng, K.M.: A single-ion conducting and shear-thinning polymer electrolyte based on ionic liquid-decorated PMMA nanoparticles for lithium-metal batteries. J. Mater. Chem. A. **4**, 18543–18550 (2016)
27. Lu, Q., et al.: Dendrite-free, high-rate, long-life lithium metal batteries with a 3D cross-linked network polymer electrolyte. Adv. Mater. **29**, 1604460 (2017)
28. Lee, Y.-G., et al.: Dendrite-free lithium deposition for lithium metal anodes with interconnected microsphere protection. Chem. Mater. **29**, 5906–5914 (2017)
29. Ma, L., Nath, P., Tu, Z., Tikekar, M., Archer, L.A.: Highly conductive, sulfonated, UV-cross-linked separators for Li–S batteries. Chem. Mater. **28**, 5147–5154 (2016)
30. Lu, Y., Das, S.K., Moganty, S.S., Archer, L.A.: Ionic liquid-nanoparticle hybrid electrolytes and their application in secondary lithium-metal batteries. Adv. Mater. **24**, 4430–4435 (2012)
31. Wei, S., et al.: A stable room-temperature sodium–sulfur battery. Nat. Commun. **7**, 11722 (2016)
32. Wei, S., et al.: Highly stable sodium batteries enabled by functional ionic polymer membranes. Adv. Mater. **29**, 1605512 (2017)
33. Choudhury, S., Agrawal, A., Wei, S., Jeng, E., Archer, L.A.: Hybrid hairy nanoparticle electrolytes stabilizing lithium metal batteries. Chem. Mater. **28**, 2147–2157 (2016)
34. Tu, Z., et al.: Designing artificial solid-electrolyte interphases for single-ion and high-efficiency transport in batteries. Joule. **1**, 394–406 (2017)

35. Aurbach, D., Zinigrad, E., Cohen, Y., Teller, H.: A short review of failure mechanisms of lithium metal and lithiated graphite anodes in liquid electrolyte solutions. Solid State Ionics. **148**, 405–416 (2002)
36. Seh, Z.W., Sun, J., Sun, Y., Cui, Y.: A highly reversible room-temperature sodium metal anode. ACS Cent. Sci. **1**, 449–455 (2015)
37. Qian, J., et al.: High rate and stable cycling of lithium metal anode. Nat. Commun. **6**, 6362 (2015)
38. Choudhury, S., Wei, S., Ozhabes, Y., Gunceler, D., Nath, P.: Designing solid-liquid interphases for sodium batteries. Nat. Commun. **8**, 898 (2017)
39. Wei, S., Choudhury, S., Tu, Z., Zhang, K., Archer, L.A.: Electrochemical interphases for high-energy storage using reactive metal anodes. Acc. Chem. Res. **51**, 80–88 (2017)
40. Lu, Y., Tu, Z., Archer, L.A.: Stable lithium electrodeposition in liquid and nanoporous solid electrolytes. Nat. Mater. **13**, 961–969 (2014)
41. Kanamura, K.: Electrochemical deposition of very smooth lithium using nonaqueous electrolytes containing HF. J. Electrochem. Soc. **143**, 2187 (1996)
42. Etacheri, V., et al.: Effect of fluoroethylene carbonate (FEC) on the performance and surface chemistry of Si-nanowire li-ion battery anodes. Langmuir. **28**, 965–976 (2012)
43. Miao, R., et al.: Novel dual-salts electrolyte solution for dendrite-free lithium-metal based rechargeable batteries with high cycle reversibility. J. Power Sources. **271**, 291–297 (2014)
44. Li, W., et al.: The synergetic effect of lithium polysulfide and lithium nitrate to prevent lithium dendrite growth. Nat. Commun. **6**, 7436 (2015)
45. Li, Y., et al.: Robust pinhole-free Li3N solid electrolyte grown from molten lithium. ACS Cent. Sci. **4**, 97–104 (2018)
46. Chen, L., Wang, K., Xie, X., Xie, J.: Effect of vinylene carbonate (VC) as electrolyte additive on electrochemical performance of Si film anode for lithium ion batteries. J. Power Sources. **174**, 538–543 (2007)
47. Pires, J., et al.: Role of propane sultone as additive to improve the performance of lithium-rich cathode material at high potential. RSC Adv. **5**, 42088 (2015). https://doi.org/10.1039/C5RA05650K
48. Zhuang, G.V., Xu, K., Jow, T.R., Ross, P.N.: Study of SEI layer formed on graphite anodes in PC/LiBOB electrolyte using IR spectroscopy. Electrochem. Solid-State Lett. **7**, 224–227 (2004)
49. Zhu, Y., Li, Y., Bettge, M., Abraham, D.P.: Electrolyte additive combinations that enhance performance of high-capacity $Li_{1.2}Ni_{0.15}Mn_{0.55}Co_{0.1}O_2$–graphite cells. Electrochim. Acta. **110**, 191–199 (2013)
50. Tan, S., et al.: Tris(hexafluoro-iso-propyl)phosphate as an SEI-forming additive on improving the electrochemical performance of the Li[Li0.2Mn0.56Ni0.16Co0.08]O2 cathode material. J. Electrochem. Soc. **160**, A285–A292 (2013)
51. Kozen, A.C., et al.: Next-generation lithium metal anode engineering via atomic layer deposition. ACS Nano. **9**, 5884–5892 (2015)
52. Choudhury, S., et al.: Electroless formation of hybrid lithium anodes for fast interfacial ion transport. Angew. Chemie Int. Ed. **56**, 13070–13077 (2017)
53. Zhang, T., Liao, K., He, P., Zhou, H.: A self-defense redox mediator for efficient lithium-O2 batteries. Energy Environ. Sci. **9**, 1024–1030 (2015)
54. Tu, Z., et al.: Fast ion transport at solid–solid interfaces in hybrid battery anodes. Nat. Energy. **3**, 310–316 (2018)
55. Liang, Z., et al.: Polymer nanofiber-guided uniform lithium deposition for battery electrodes. Nano Lett. **15**, 2910–2916 (2015)
56. Lopez, J., et al.: Effects of polymer coatings on electrodeposited lithium metal. J. Am. Chem. Soc. **140**, 11735–11744 (2018)

Chapter 2
Hybrid Hairy Nanoparticle Electrolytes Stabilize Lithium Metal Batteries

2.1 Introduction

A rechargeable battery that uses metallic lithium as the anode is among the most sought-after technologies for portable storage of electrical energy. Such batteries are attractive for multiple reasons: first, lithium has the lowest redox potential (−3.04 V vs. standard hydrogen electrode (SHE)); second, lithium has a low gravimetric density (0.534 g/cm^3) and high theoretical capacity (3860 mAh/g); and [1–3] third, because the anode is metallic lithium, the cathode in such lithium metal batteries (LMBs) can be an unlithiated material, such as sulfur, oxygen, or carbon dioxide/oxygen mixtures, which opens up opportunities for batteries with very high specific energies (SE), relative to today's Li-ion technology (e.g., $SE_{Li\text{-}ion} = 0.15$ kWh/kg; $SE_{Li\text{-}S} = 2.5$ kWh/kg; $SE_{Li\text{-}O2/CO2} = 10.5$ kWh/kg). Four decades of focused research aimed at creating LMBs that live up to the promise of this technology has revealed multiple major problems associated with the reactivity of the metal with liquid electrolytes and the stability of Li electrodeposition at the anode during cell recharge.

During normal battery operation, exposed surfaces of a Li metal electrode react with the electrolyte to form a passivation layer loosely termed the solid electrolyte interface (SEI), which ideally would prevent further side reactions between the Li electrode and electrolyte. In practice, the SEI formed on a Li metal surface breaks and reforms due to the expansion and compression of lithium, during plating (charge) and stripping (discharge), resulting in continuous consumption of electrolyte [4]. The lifetime and cyclability of a lithium metal battery is therefore now understood to depend on the ability to create a mechanically and electrochemically robust SEI layer, which also allows for relatively fast transport of Li ions across the electrolyte/electrode interface. After extensive studies, Aurbach et al. [5] in 2002 concluded that none of the aprotic organic electrolyte solvents in current use is compatible with a lithium metal anode because all react with Li to form unstable SEI layers at the high current densities required for practical battery operation.

S. Choudhury, *Rational Design of Nanostructured Polymer Electrolytes and Solid–Liquid Interphases for Lithium Batteries*, Springer Theses,
https://doi.org/10.1007/978-3-030-28943-0_2

Researchers have pursued several approaches for stabilizing the SEI on a Li metal anode. Chemical electrolyte additives, such as vinylene carbonate (VC), which react preferentially with Li metal to form cross-linked polymers have been used as sacrificial agents to create mechanically robust passivating layers on Li that also offer interfacial ionic conductivities comparable to a bulk liquid electrolyte [6–8]. More recently, a larger number of electrolyte additives, including LiBOB [9], $LiNO_3$ [10], LiF [11, 12], and sultones [13, 14], have been shown to produce SEI compositions that not only limit contact between the electrolyte solvent and Li metal but which allow the SEI layer to retain its mechanical integrity during charge and discharge cycles. Methods of introducing such stabilizing components indirectly in the SEI using specific solvents, e.g., fluoroethylene carbonate [15], or salts (e.g., binary LiTFSI-LiFSI [16] or excess LiFSI [1]) in the electrolyte have also been reported. In a recent departure from these approaches, Cui et al. [4] showed that a thin film comprising of hollow carbon nanospheres on the lithium surface provides a robust artificial SEI layer that improved coulombic efficiency of a Li metal anode to over 99%. Similar results have been observed with coatings of boron nitride [17] and polyacrylonitile [18] on Li metal. Problems involving parasitic reaction between Li anodes and liquid electrolytes are exacerbated under conditions of uneven/dendritic electrodeposition of the metal during cell recharge because such deposition increases the contact surface area between the electrolyte and anode, promoting formation of new SEI after every charge cycle. Modifying ion transport in the electrolyte via single ion conducting species [19, 20] or at the electrolyte/electrode interface by means of halide salts [12, 21] has been shown to prevent growth of dendritic structures during charging. Nanostructured separators with high modulus components have also been reported to yield flatter, more compact deposition [7, 12, 22–25].

Weston and Steele [26] in 1982 were to our knowledge among the first to propose using nanocomposites, comprised of nanoparticle fillers in a liquid or polymer-based electrolyte, as electrolytes in rechargeable batteries. Since then there have been myriad studies that have established the benefits of such nanocomposite electrolytes for making portable, leakage-free, non-flammable batteries [24, 27, 28]. High modulus inorganic fillers have also been successfully utilized in the past to suppress rough electrodeposition [24]. In this respect, surface modification of such fillers by special ionically active species like PEO [29], anionic groups [20], and ionic liquids (IL) [30, 31] has been illustrated as particularly effective in stabilizing electrodeposition of Li as a result of ion transport modifications. Korf et al. [32], for example, recently used silica nano-fillers covalently grafted with the ionic liquid 1-methyl-1-propylpiperidinium bis(trifluoromethanesulfone) imide in a liquid propylene carbonate (PC) electrolyte in symmetric Li/Li cells and, interestingly, on postmortem analysis of the Li electrodes following several cycles of plating and stripping found that the regions where nanoparticles adsorb to the Li electrodes exhibited significantly smoother morphology compared to those where nanoparticles are absent. This finding is important because it is counter to what one would expect if the insulating SiO_2 particles are assumed to retard ion transport to the electrode; it also opens up opportunities for more targeted use of nanocomposites as interfacial stabilizers and as building blocks for creating artificial SEI layers on battery electrodes.

Unfortunately, previous reports on nanocomposite electrolytes suggest that their ionic conductivity are too low at high particle volume fractions to realize these beneficial effects under ambient conditions [33, 34]. Very recently we showed that this difficulty can be overcome by deliberately introducing size polydispersity to suspensions of hairy nanoparticles in liquid electrolyte hosts [28]. Herein, we build on these findings to create nanocomposite SEI layers on Li metal electrodes and to carry out an in-depth exploration of their transport and interfacial properties. We find that a SEI layer enriched with a polydisperse, nanoparticle-rich electrolyte imparts novel interfacial and transport behaviors at the electrode/electrolyte interface of a LMB and may also be used to enhance the high-voltage stability of aprotic liquid electrolytes.

2.2 Materials and Methods

2.2.1 Synthesis

Trimethoxysilane functionalized polyethylene glycol methyl ether (PEG, MW~500 g/mol, Gelest chemicals) were grafted to silica nanoparticles with size 10 nm (Ludox, SM-30, Sigma-Aldrich) and 25 nm (Ludox, TM-50, Sigma-Aldrich) using a previously described method [35]. The particles were purified by performing dialysis using a snake skin membrane to remove an unattached PEG chains, followed by repeated centrifugation using ethanol-hexane as solvent-non-solvent to remove any residual PEG from the material. The particles were then dried at 60 °C for 24 h in a convection oven, followed by vacuum drying for 48 h. Thermogravimetric analysis (TGA) was used to estimate the residual inorganic content to compute the grafting densities (number of chains per unit surface area) which were evaluated to be Σ~1.3 chains/nm^2 and Σ~1.5 chains/nm^2 for 10 nm and 25 nm particles, respectively. Hybrid electrolytes were prepared in argon-filled glove box by dispersing the hair nanoparticles in propylene carbonate (PC, Sigma-Aldrich) at various core volume fractions ϕ ranging from 0.2 to 0.4. At each ϕ, the relative fraction of larger particles added was varied which is given by the parameter $x_L = \phi_L/\phi$. 1 M of bis(trifluoromethanesulfonimide) (LiTFSI, Sigma-Aldrich) salt was added to the resultant solution of hybrid silica nanoparticles in PC to create SiO_2-PEG/PC hybrid electrolytes.

2.2.2 Characterization

The dispersion and structure of hybrid particles was determined from transmission electron microscopy (TEM) and small-angle X-ray scattering (SAXS) measurements. For TEM measurements, dilute suspensions of the hybrid electrolyte dissolved in chloroform were dropped on copper grids, with subsequent solvent evaporation and annealing at 70 °C for 48 h.

SAXS measurements were performed at D-1 beamline of Cornell High Energy Synchrotron Source (CHESS) using a point-collimated X-ray beam source. The scattered X-ray intensity from the hybrid electrolyte is measured with the variation in the wave vector q and can be given as a function of particle form factor $P(q,D)$ and structure factor $S(q,\phi,D)$

$$I(q,\phi,D) = \Phi \Delta\rho_e V P(q,D) S(q,\phi,D)$$

where V and $\Delta\rho_e$ are the volume of a single particle core and electron density contrast between the particle and the surrounding medium, respectively. The form factor $P(q,D)$ gives the information about the size and shape of a single particle.

The glass transition temperature of the electrolytes was determined from differential scanning calorimetric (DSC, Q2000 TA instruments) measurements.

2.2.3 *Electrochemical Measurements*

The ionic conductivities of the electrolytes were measured at room temperature using a Novocontrol Broadband Dielectric spectrometer with a frequency range of 0.1–3 × 10^6 zHz. The DC conductivities were obtained from the plateau of real part of the conductivity versus frequency curve [36]. The molar diffusivities were obtained using the Nernst-Einstein equation:

$$D = \frac{\sigma k_B T}{C_0 q^2}$$

where σ (S-m^2/mol) is the D.C. molar ionic conductivity, k_B (J/K) is the Boltzmann constant, T (K) is the temperature, C_0 (#/m^3) is the ion concentration, and q (C) is the charge of the diffusing entity.

The AC impedance measurements were performed on a Solartron electrochemical impedance spectrometer using a symmetric coin cell with lithium metal as the electrodes and SiO_2-PEG/PC as the electrolyte at different ϕ and x_L at ambient temperature. The frequency was varied from 1 MHz to 0.1 Hz at amplitude of 50 mV. Error bars in the final impedances were evaluated as deviations from running each measurement for 12 symmetric cells.

Electrochemical stability voltage window was determined by linear scan voltammetry measurements performed at a scan rate 1 mV/s between −0.2 V and 6.5 V. The stability was further confirmed by performing floating-point test measurements where the voltage was stepped up to 7 V, with each voltage step of 0.5 V maintained for 5 h.

2.2.4 *Analyzing the Coulombic Efficiency*

Li| electrolyte |stainless steel 2032-type coin cells were assembled in an argon-filled glove box. Control liquid electrolyte comprising 1 M LiTFSi-PC with 1 wt% lithium nitrate ($LiNO_3$) and 2 vol% vinylene carbonate and a standard Celgard separator was compared against hybrid nanoparticles dispersed in the same electrolyte with $\phi = 0.3$ and $x_L = 0.5$. In both cases, prior to the measurements, cells were conditioned by cycling them between 0 and 0.5 V for 10 cycles, to ensure the formation of a stable SEI layer on the electrodes as shown previously [7]. To characterize the coulombic efficiency, the conditioned cells were first discharged at a constant current density of 0.25 mA/cm^2 for 2 h to transfer an amount of lithium corresponding to 0.50 mAh/cm^2 of charge from the Li electrode to the stainless steel electrode. The amount of charge recovered in the reverse cycle was then recorded where the cell was charged back to 0.5 V at the same current density, and the fractional recovery between the two cycled was used to determine the coulombic efficiency.

The lithium electrode surface was characterized using a similar procedure for hybrid electrolytes with $\phi = 0.2$ and $x_L = 0.5$ without any additives. The coin cells were disassembled inside the glove box, and the lithium foil was repeatedly washed with pure PC to dissolve any excess of particles on the surface. Scanning electron microscopy (SEM) measurements were performed on the lithium foil to image the electrode surface.

2.2.5 *Cell Lifetime Study*

$LiFePO_4$ slurry was prepared by mixing the active material, super-P carbon, and PVDF binder in 8:1:1 ratio in a ball-mill in presence of NMP solvent. The slurry was casted on an aluminum sheet using a doctor blade to prepared cathode sheets of 0.5 mAh/cm^2 surface capacity. Half-cells were made in glove box with lithium foil as anode, lithium iron phosphate as cathode, and the hybrid binary nanocomposite as electrolyte. The batteries were cycled at constant current between the voltage limits of 2.5 V and 3.8 V.

2.3 Results and Discussion

2.3.1 *Physical Characterization and Ion Transport*

Aggregation and phase separation are among the most important hurdles that have limited application of nanoparticle-polymer composites in many fields of technology [33, 37–41]. Surface functionalization of particles with short polymer chains is a widely practiced technique for creating uniform dispersion of particles in polymers and liquid hosts [42–50]. Here we utilize previous chemistry to densely graft short polyethylene glycol methyl ether (PEGME, MW~500 g/mol) chains to SiO_2 nanoparticles (d_p = 10 nm and 25 nm) and dispersed these particles and their binary mixtures in propylene carbonate (PC)-based liquid electrolytes. The dispersion state of hairy nanoparticles in the resultant hybrid electrolytes was analyzed by observing their transmission electron micrographs (TEM). Figure 2.1a shows that irrespective of the volume ratio (x_L) of larger (d_p = 25 nm) to smaller (d_p = 10 nm) particles, the particles appear as well-dispersed objects in the hybrid electrolytes. This finding is confirmed by analyzing the variation in the scattered intensity $I(q)$ of X-rays from the bulk hybrid electrolytes using small-angle X-ray scattering (SAXS). Figure 2.1b reports typical $I(q)$ vs. q data from SAXS measurements performed on a hybrid electrolyte with fixed volume fraction of nanoparticles ($\phi = 0.3$) but with varying fractions x_L of the larger particles. $I(q)$ is seen to exhibit a plateau for all x_L values in the low q region and a q^{-4} scaling in the high q region. Both of these features are known characteristics of uniformly dispersed, un-aggregated spheres [42, 51].

Previous studies on bi-disperse suspensions of star polymers and hairy nanoparticles have shown that introduction of a larger species to a monodisperse suspension comprised of smaller ones results in a transition from a highly jammed glass to a weakly jammed suspension and, for large enough size ratios, ultimately to a liquid

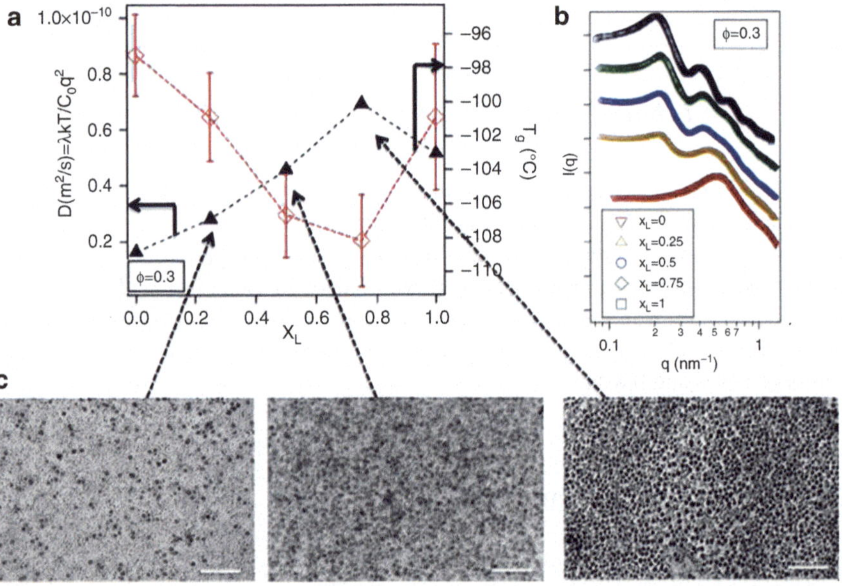

Fig. 2.1 Physical characterization: (**a**) Variation in ionic diffusivity, D (triangles), as calculated from Nernst-Einstein equation, and glass transition temperature, T_g (diamonds), with fraction of larger particles, x_L. (**b**) Intensity, $I(q)$, as a function of wave vector q at $\Phi = 0.3$, as obtained from SAXS measurements for different values of x_L, increasing from bottom to top. (**c**) Transmission electron micrographs (TEM) for bi-disperse hybrid electrolytes at $\Phi = 0.3$ for $x_L = 0.25$, 0.5, and 0.75. The scale bar for all the images is 500 nm

state [52–54]. This transition has also been reported in self-suspended suspensions of bi-disperse hairy particles and is thought to reflect differences in the degree of interpenetration of tethered oligomer chains produced by differences in curvature of the two different nanocore populations [54]. As a consequence, tethered oligomer chains in a bi-disperse blend of hairy particles may acquire higher configurational entropy even at the same overall particle volume fraction. It has been observed previously that lithium-ion transport in electrolytes containing PEO occurs predominantly by local diffusion of polymer chain segments and hopping of Li ions between ether groups on PEO chains [33, 34]. Figure 2.1a reports the diffusivity of lithium ions obtained, via the Nernst-Einstein equation, from ionic conductivity data for a SiO_2-PEGME/PC hybrid electrolyte with $\phi = 0.3$. The figure shows that the diffusivity is a maximum near $x_L = 0.75$. This higher diffusivity can be explained in terms of the lower interpenetration and hence higher mobility of PEGME chains in the binary hairy particle mixtures; the higher mobility of PEGME chains allows Li ions in an electrolyte to migrate at a faster rate, which results in an increase in ionic conductivity and higher ionic diffusivity in the bi-disperse hybrid electrolytes. Figure 2.1a also reports the glass transition temperature, T_g, for SiO_2-PEGME/PC hybrid electrolytes as a function of x_L. Consistent with the explanation of the enhanced diffusivity at $x_L \approx 0.75$, it is seen that T_g exhibits a minimum at x_L close to where the maximum in diffusivity is observed. A reduced T_g for the bi-disperse suspension implies an increased free volume, which is consistent with our hypothesis that the tethered PEGME chains are less constrained in the bi-disperse hybrid electrolytes. Supplementary Fig. 2.1 presents results for electrolytes with $\phi = 0.2$ and 0.4 and shows that the observation of a maximum diffusivity at $x_L \approx 0.75$ and T_g minimum near this value is a generic feature of these materials.

2.3.2 Structural Factor Analysis

Further insight about the structure of the SiO_2-PEGME/PC electrolytes and about the confinement of PEGME chains can be obtained by analyzing the SAXS structure factor S(q) of the materials. Figure 2.2a compares the experimentally obtained S(q) with the calculated structure factor for a binary hard sphere (HS) suspension using the Percus-Yevick (PY) approximation [55]. To account for the fact that a few PEGME chain segments nearest the anchor points on the particles are likely to be completely correlated with their particle substrate [43], the theoretical $S(q)$ approximates the particle size in terms of an effective radius, $a_{\mathrm{eff},i} = a_i + n_{m,i} l_m \cos(\theta/2)$, such that a certain number of monomers, $n_{m,i}$, are assumed to be part of the core. Here i = s,l are for smaller particles and larger particles, respectively; a_i is the original radius of the particle; $l_m = 0.35$ nm is the monomer length for PEO; and $\theta = 68°$ is the bond tetrahedral angle. In analyzing the experimental data using the PY approximation, $n_{m,i}$ was treated as an unknown variable and its value adjusted so that the peak positions of $S(q)$ computed for a binary HS suspension model coincide with those seen in the experimental $S(q)$. It can be observed from Fig. 2.2a that whereas

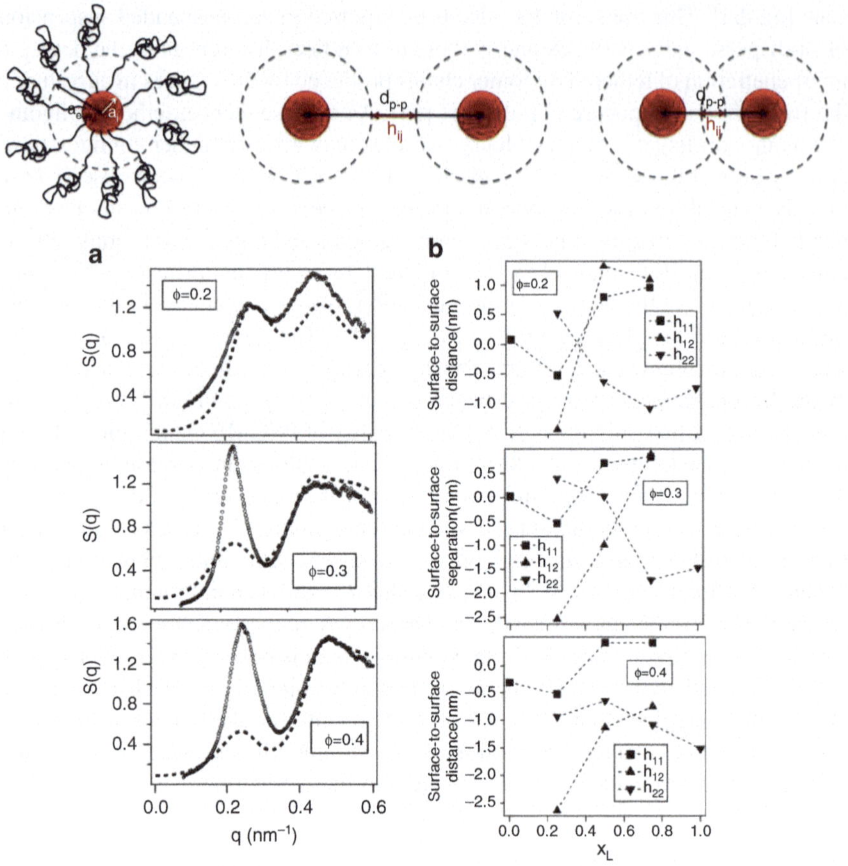

Fig. 2.2 Structure factor: (**a**) Evolution of structure factor, $S(q)$, with the wave vector q for different volume fractions, $\Phi = 0.2$, 0.3, and 0.4 at $x_L = 0.5$. The open symbols are the experimentally obtained $S(q)$ values, and the dashed lines are predictions from binary HS model for the same systems. The schematic above illustrates the effective particle radius, which includes some fraction of monomers as part of the core. (**b**) Surface-to-surface distances, h_{11}, h_{12}, and h_{22}, for the same volume fractions as a function of x_L. The graphics illustrates the two scenarios for surface-surface distances which will be positive when there is no overlap and negative when the polymer chains have significant overlap

the $S(q)$ peak amplitudes and positions estimated from PY-HS analysis are in good agreement with the experimental values for electrolytes with low ϕ, the first $S(q)$ maxima are much higher than the predicted ones in electrolytes with high ϕ. A higher first peak height for $S(q)$ implies a stronger interparticle correlation, which results in the hybrid particles from enhanced interaction of tethered corona chains [35, 56, 57]. Supplementary Fig. 2.2 reports the corresponding results for electrolytes with a range of particle size polydispersities, $x_L = 0$, 0.25, 0.75, and 1. The discrepancy in the calculated and measured peak heights, especially the first peak, is clearly greatest in systems with the largest asymmetry in nanoparticle size. The discrepancy is thought to reflect enhanced interaction between tethered oligomer chains in the hybrid particle electrolytes, which are not included in the PY-HS analysis.

Additional insights about the structure of the electrolytes can be obtained from the PY-HS predictions by resolving the structure factor into its three components, S_{11}, S_{12}, and S_{22}, where 1 and 2 denote smaller and larger particles, respectively (see Supplementary Fig. 2.3). From the first peak positions of each of these components, three different interparticle distances can be obtained, viz., $d_{p\text{-}p,11}$, $d_{p\text{-}p,12}$, and $d_{p\text{-}p,22}$. Figure 2.2b shows the variation in surface-to-surface distance for $\phi = 0.2$, 0.3, and 0.4, as a function of fraction of larger particles x_L. The surface-to-surface distance is calculated by subtracting the effective core diameter from the interparticle distance, i.e., $h_{ij} = d_{p\text{-}p,ij} - 2a_{\text{eff},ij}$, where $_{i,j} = 1$ and 2 and $_i = _j$ when analyzing the distance between the same species. It can be observed that for certain values of ϕ and x_L, h_{ij} is negative which implies that the corona chains overlap or interpenetrate for those regions. It can be seen that as ϕ increases, as expected h_{ij} assumes higher negative values, implying higher degree of interpenetration and thereby confinement for tethered chains at higher volume fractions. For any core volume fraction ϕ, it can also be observed that initial addition of larger particles to a pure monodisperse suspension of smaller ones first decreases the distance between the smaller particles after which the h_{11} value increases with further addition of larger particles. For the same x_L values, however, h_{22} is larger for a smaller fraction of large particles and decreases with further increase in x_L. This implies that for smaller x_L values, the larger particles confine the smaller particles and increases the distance between themselves, and subsequent increase in x_L values leads to an increase in the average distance between smaller particles. This interpretation is confirmed from the h_{12} values, which always increase with increase in x_L, meaning that the distance between a smaller and a larger particle always increases. Also, for the pure monodisperse suspensions of smaller and larger particles, the h_{ij} values are either negative or close to zero, which implies that the tethered corona chains interpenetrate much more in these suspensions, while h_{ij} values for bi-disperse suspensions are either positive or less negative than their monodisperse counterparts, which indicates less interpenetration and hence reduced confinement of corona chain motions in bi-disperse suspension. This observation is in agreement with the trends seen in the diffusivity and T_g values, which confirms the hypothesis that in a bi-disperse suspension, the tethered PEO chains are less confined or have higher mobility which subsequently leads to a higher ionic conductivity values for these hybrid electrolytes as observed previously [28].

2.3.3 *Variation of Interfacial Resistance*

Figure 2.3a–c reports Nyquist plots obtained using electrochemical impedance spectroscopy measurements to evaluate the impedance in bulk electrolyte and at the electrode-electrolyte interface in symmetric Li/Li cells. The experimental plot is

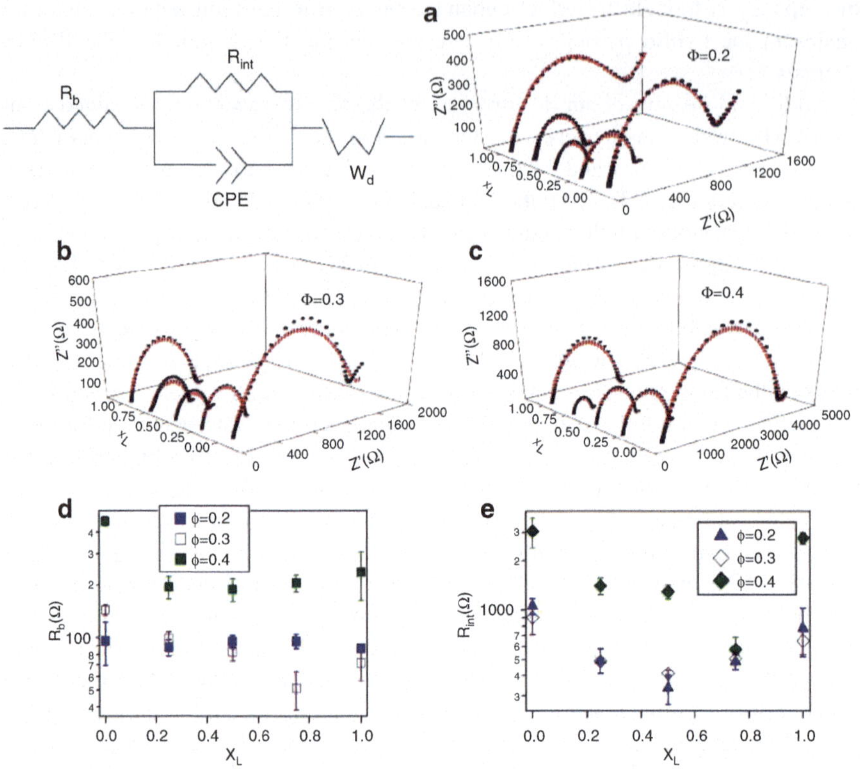

Fig. 2.3 Electrochemical characterization: Impedance spectra for electrolytes with core volume fractions (**a**) $\Phi = 0.2$, (**b**) 0.3, and (**c**) 0.4 measured at different fractions of larger particles added, x_L. The black symbols are experimental data, and the red lines are fit to the equivalent circuit model as shown. (**d**) Bulk resistance, R_b, and (**e**) interfacial resistance, R_i, measured at different core volume fractions as a function of x_L

fitted using the circuit model shown schematically in Fig. 2.3, where R_b corresponds to the resistance in bulk electrolyte, R_{int} represents interfacial resistance to ion migration at the electrode-electrolyte interface, CPE is the constant phase element capacitance near the electrode surface, and W_d is the Warburg diffusion element [23]. Figure 2.3d, e summarizes the magnitude of R_b and R_{int} as a function of x_L at different core volume fractions ϕ, respectively. While R_b does not show significant variation with x_L, R_{int} decreases significantly and exhibits similar trends to those discussed earlier for the ionic diffusivity and T_g. We postulate that the relative invariance in R_b occurs because of adsorption of significant fraction of particles from the bulk electrolyte to the electrode surface, which results in bulk resistance values similar to those of liquid electrolytes with low nanoparticle loadings where ion transport is dominated by diffusion of electrolyte solvent-ion associates. This would then also lead to variations in interfacial resistance similar to trends seen in diffusivity due to the presence of particles.

2.3.4 Surface Characterization of Li Anode

In order to evaluate the hypothesis that surface adsorption of SiO_2-PEGME particles is the source of the weaker dependence of R_b and R_{int} on x_L, we performed scanning electron microscopy (SEM) analysis of the lithium electrodes harvested from symmetric Li/Li cells following galvanostatic cycling at a low current density of 0.03 mA/cm^2 for ten cycles. Prior to SEM analysis, the electrodes were vigorously washed with pure PC to remove any loosely bound material. Figure 2.4a shows a representative SEM image of the Li metal electrode for small and higher magnifications. It is apparent that a dense layer of bi-disperse particles of size 10 nm and 25 nm is adsorbed to the surface of the electrode. The hairy particles appear to form a continuous, protective film on the electrode surface.

It has been observed in a typical lithium-ion battery (LIB) that during charging or discharging, oppositely charged ions accumulate near the surface of respective electrodes [58–60]. The electrochemical capacitance (EC) near the electrode surface can be determined from the relaxation frequency f_0 deduced from the Nyquist plot [61–63] and its relationship, $2\pi f_0 R_{int} C_d = 1$, to the interfacial resistance and capacitance [64]. The calculated capacitance can be written in terms of the dielectric constant at the interface as $C_d = \varepsilon\varepsilon_0 A/l$, where ε is the dielectric permittivity of the material at the interface, ε_0 is the vacuum permittivity, A is the surface area of the electrode, and l is the thickness of the capacitive layer. Figure 2.4b shows the variation in the electrochemical capacitance C_d as a function of x_L for $\phi = 0.3$. It is apparent that hybrid electrolytes based on bi-disperse SiO_2-PEGME nanoparticles display higher interfacial capacitance than the monodisperse particle suspension. This observation can be explained either in terms of an increase in the effective dielectric constant or a reduction in the SEI layer thickness as x_L increases. Based on the earlier observation that the particles form a stable, dense film on the Li anode, we conclude that l is likely to be a fixed number of the order of average particle diameter in the electrolytes. Inserting this value in the above formula allows us to estimate ε and, from it, an apparent Debye screening length ($\lambda = \sqrt{\frac{\varepsilon\varepsilon_o RT}{2F^2 C_0}}$), which is also shown as a function of x_L in Fig. 2.4b for $\phi = 0.3$; while $\phi = 0.2, 0.4$ is shown in Supplementary Fig. 2.4. Here F is Faraday's constant, and C_0 is the molar concentration of salt in the electrolyte. For a pure PC electrolyte containing 1 M salt, $\lambda \approx 0.26$ nm, which is comparable to the highest values obtained in Fig. 2.4b but somewhat larger than the values estimated for the electrolytes with $x_L = 0$ or $x_L = 1$, which is consistent with expectations for the factor of 5–10 difference in the reported dielectric constants for PEG and PC [37, 65–68]. One explanation of the high C_d and λ values apparent at intermediate x_L is that they reflect greater access of PC to the SEI, which would increase ε. It is important to note, however, that even the largest λ values are more than an order of magnitude lower than l, meaning that the electric field gradients in the SEI layer formed by a hairy particle coating of Li are substantially smaller than in a pure electrolyte, which would lower both the diffusive and electroconvective flux of anions to the Li metal surface. The Li^+ transference number

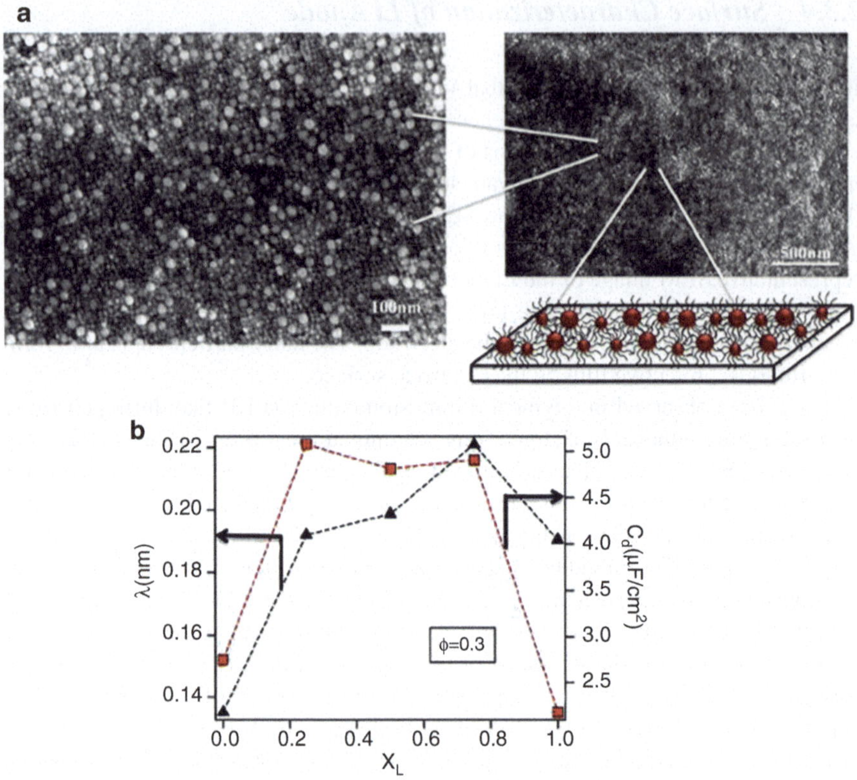

Fig. 2.4 Surface characterization: (**a**) Scanning electron micrographs (SEM) for lithium electrode surface after the symmetric cell was cycled for ten cycles at 0.03 mA/cm^2 with the bi-disperse hybrid electrolyte. The schematic illustrates the electrode surface as observed in SEM, where the particles form protective layer on the electrode surface. (**b**) Electrical double layer, λ (triangles), and electrochemical capacitance per unit area, C_d (squares), for $\phi = 0.3$ as a function of x_L

(t_{Li}^+) was determined by the Bruce-Scrosati method [69, 70] (see Supplementary Fig. 2.5) which was found to be approximately 0.32 for this electrolyte, in agreement with previously reported transference numbers for PEG-based electrolytes [34]. The presence of particles in the SEI therefore does not influence the lithium ion transference number, which is consistent with expectations based on the theoretical predictions of Jacob [71], where it was found that the transference number has a strong dependence of the ratio of pore size to double layer thickness.

2.3.5 *Enhanced Electrochemical Stability of Nanocomposites*

If the Li metal in a LMB is unprotected, the electrolyte continuously reacts with the anode to form insulating products, which increase the interfacial resistance of the battery with time. The interface stability in symmetric cells containing SiO_2-

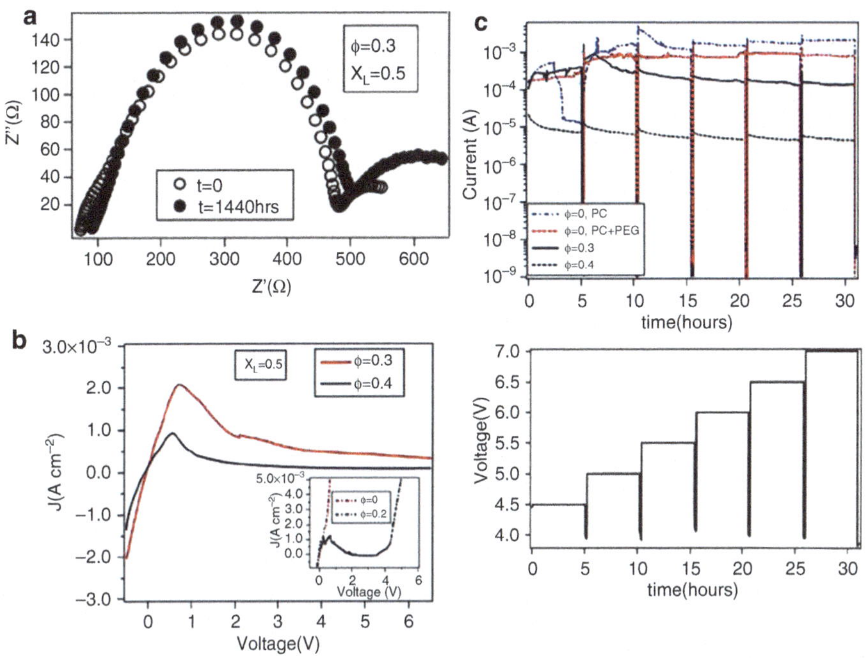

Fig. 2.5 Electrochemical stability: (**a**) Initial impedance spectra (open symbols) for a symmetric cell with hybrid particle electrolyte at $\Phi = 0.3$ and $x_L = 0.5$ compared with the spectra at $t = 1440$ h (closed symbols). (**b**) Variation in faradic current density, J as a function of voltage as measured from Linear sweep voltammetry measurements for $x_L = 0.5$ compared for $\Phi = 0.3$ (red line) and $\Phi = 0.4$ (black line). Inset shows the measurements for neat PC (red line) and $\Phi = 0.2$ (black line). (**c**) Leak current measured as a function of time for different steps of voltage as measured in a floating-point test. Profiles for symmetric cell with hybrid electrolyte at $\Phi = 0.3$ (black solid line) and $\Phi = 0.4$ (black dashed line) are compare against neat electrolytes with PC (blue line) and with PEG and PC (red line) blended in the same ratio as that for hybrid particle electrolyte with $\Phi = 0.3$

PEGME/PC hybrid electrolyte with $\phi = 0.3$ and $x_L = 0.5$ were evaluated using impedance measurements as a function of time for a period of 2 months. It can be seen from Fig. 2.5a that the time-resolved impedance plots overlay well onto each other, indicating that the particle-rich SEI layer imparts long-term enhanced chemical stability to Li metal/electrolyte interface. This enhanced stability is consistent with the existence of a passivating particle film on the Li anode as deduced from the earlier SEM experiments. Figure 2.5b reports results from linear sweep voltammetry and electrochemical floating-point measurements on lithium cells containing the hybrid electrolytes. The linear sweep voltammogram reveals that at higher particle volume fractions (e.g., $\phi = 0.3$ and $\phi = 0.4$) and for $x_L = 0.5$, the hybrid electrolytes are stable up to 6.5 V vs. Li/Li^+. In contrast, electrolytes containing lower fractions $\phi = 0.2$ of particles (inset of Fig. 2.5b) become unstable at around 4 V. The initial peak at 0.2 V in all the measurements is the cathodic peak associated with the Li/Li^+ reaction [7, 12]. Its presence for all of the materials studied means that this process is not compromised in the hybrid electrolytes. Similar trends were also observed in the floating-point tests in Fig. 2.5c, where hybrid electrolytes containing higher nanoparticle volume fractions are seen to manifest negligible leakage currents even after exposure to voltages as high as 7.5 V for 5 h. In contrast, the particle-free PC electrolyte or a blend of untethered PEG in PC already exhibits high leakage currents at 5 V. Together, these results clearly show that SiO_2-PEGME/PC hybrid electrolytes with high SiO_2 are exceptionally stable, in agreement with previous reports [72].

2.3.6 Analyzing Galvanostatic Performance

As a final assessment of the ability of a nanoparticle-rich SEI to stabilize Li metal anodes against reaction with liquid electrolytes, we analyzed the coulombic efficiency obtained in galvanostatic measurements employing Li/stainless steel electrodes. In these experiments, a fixed amount of Li^+ ions are stripped from lithium metal and deposited to the stainless steel substrate and plated back in successive runs. By measuring the capacity ratio for the strip and plate segments of the cycle, the coulombic efficiency (CE) of the cell can be quantified. It has been reported that PC-based electrolytes spontaneously form an insulating and unstable SEI layer on Li, yielding a low CE of ~70% [73]. It has also been shown that the CE for PC can be improved by employing VC and $LiNO_3$ as electrolyte additives [6, 7]. To study the effect of particles on CE of Li metal electrodes, we utilized a PC-based electrolyte reinforced with 1% $LiNO_3$ and 2% VC as a control. It can be seen from Fig. 2.6a that although both the control and hybrid PC electrolytes exhibit high CE, the control electrolyte is unable to maintain the improvements beyond ~25 cycles, implying that the VC and $LiNO_3$ are consumed. In contrast, the SiO_2-PEGME/PC hybrid electrolyte displays high CE for at least 80 cycles. In the control experiment, despite use of additives, the PC electrolyte decomposes over cycles due to its low stability window and uneven electrodeposition which exposes higher effective surface area of the electrode for the electrolyte to react; thus the additive fails to maintain high-coulombic efficiency over ~25 cycles. On the other hand, as previously inferred from electrochemical stability tests and electron microscopy, the SiO_2-PEGDME form a protective monolayer on the electrode surface that prevents constant exposure of liquid electrolyte with lithium surface and breakdown of the additive-rich SEI. This synergistic effect of the additives and SiO_2-PEGDME helps in maintaining high-coulombic efficiency large number of cycles compared to the control without particles. Figure 2.6b shows the room-temperature performance of the binary hybrid nanocomposite electrolytes in a Li/$LiFePO_4$ half-cell in which a SiO_2-PEGME/PC hybrid with $\phi = 0.3$ and $x_L = 0.5$ is employed. The figure reports the voltage profiles (voltage versus capacity) for the 1st, 10th, and 50th cycles of charge and discharge. The low efficiency in the first cycle is associated with irreversible reactions that help in forming a stable passivation layer on the Li anode. The results in Fig. 2.6b shows that even after 50 cycles, the cells maintain a high discharge capacity of ~130 mAh/g at C/3 rate.

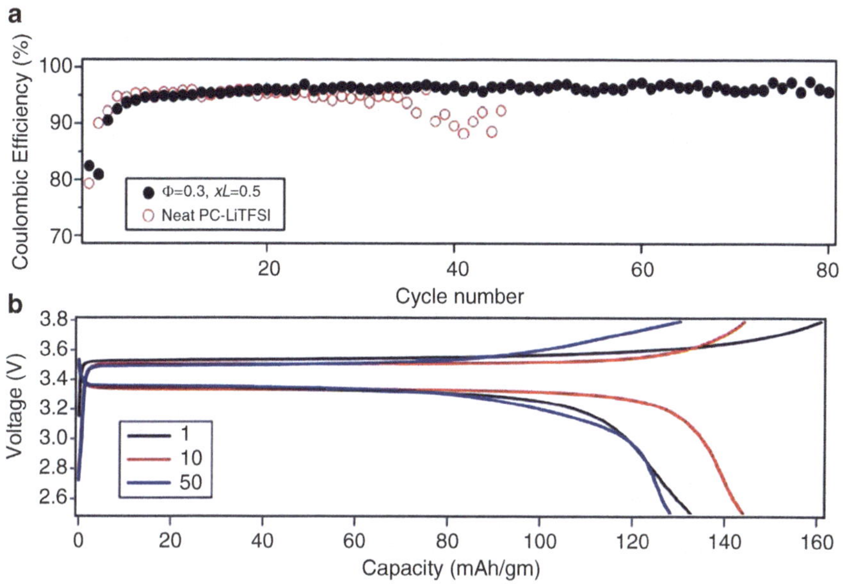

Fig. 2.6 Galvanostatic performance: (**a**) Comparison of coulombic efficiency as a function of cycle number for neat PC electrolyte and hybrid particle electrolyte with $\Phi = 0.3$ and $x_L = 0.5$ for Li|electrolyte|stainless steel configuration. Measurements were performed at a current density of 0.25 mA/cm^2. (**b**) Charge-discharge profiles for Li/LFP half-cell containing hybrid particle electrolyte with $\Phi = 0.3$ and $x_L = 0.5$ at a fixed rate of C/3 in the initial, 10th, and 50th cycle

2.4 Conclusion

In conclusion, we have studied hybrid electrolytes created by blending a low volatility carbonate (PC) with a bi-disperse mixture of SiO_2 PEGDME nanoparticles. We show that hybrid electrolytes based on bi-disperse hairy nanoparticles can be designed to provide multiple attractive features, including exceptional high-voltage stability (>7 V) over extended times; protection of a Li metal anode that allows stable long-term cycling of the anode at high-coulombic efficiency; and low bulk and interfacial resistance at room temperature that enables stable cycling of Li/$LiFePO_4$ half-cells at a C/3 rate.

The origin of enhanced ion mobility in bi-disperse hybrid electrolytes is shown by means of DSC measurements, where is is seen that in the bidispersed electrolytes, the tethered PEG chains exhibit reduction in T_g and, in SAXS analysis, increased mobility of tethered PEGME chains due to lower levels of chain interpenetration is observed. By means of interfacial impedance measurements, it was further shown that the interfacial mobility of hybrid electrolytes at a Li metal electrode is a much stronger function of polydispersity than the bulk resistance. We hypothesize that this difference arises from the spontaneous adsorption of particles to the high-energy Li metal electrode to form a dense protective film. Postmortem analysis of the Li electrode surface following galvanostatic cycling shows that a dense particle-rich film accumulates at the Li metal surface, which increases the interfacial capacitance and appears to be very effective in limiting access of liquid electrolyte to the Li metal surface without compromising interfacial transport of Li ions. In contrast to control, particle-free electrolytes, hybrid electrolytes based on bidisperse hairy nanoparticles are shown to enable Li/stainless steel cells with high-coulombic efficiency for at least 80 cycles and Li/$LiFePO_4$ cells with high discharge capacity in extended cycling at room temperature.

Appendix: Supplementary Information

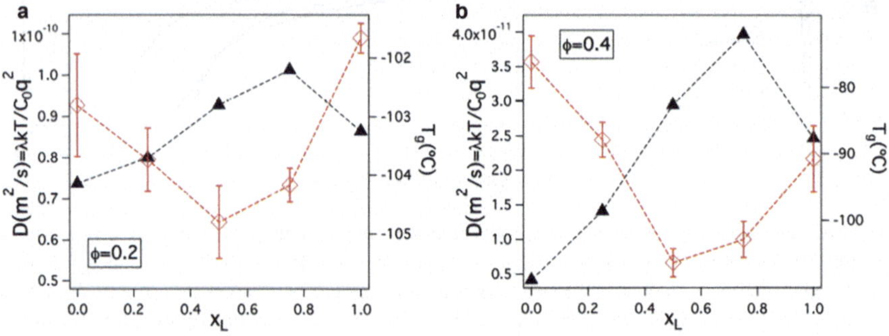

Supplementary Fig. 2.1 Variation of glass transition temperature (open symbols) and diffusivity (close symbols) as a function of fraction of larger particles x_L, for core volume fractions (a) $\phi = 0.2$ and (b) $\phi = 0.4$. The dashed lines are guide to the eye

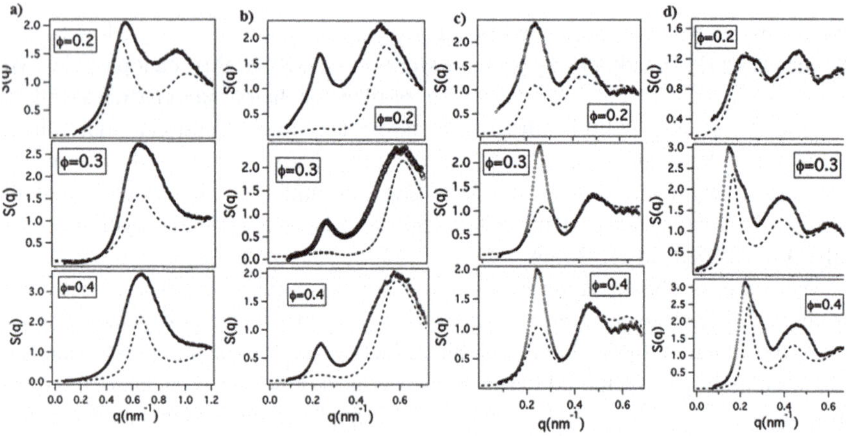

Supplementary Fig. 2.2 Evolution of structure factor S(q) as a function of the wave vector q at volume fractions $\Phi = 0.2$, 0.3, and 0.4 for (**a**) $x_L = 0$, (**b**) $x_L = 0.25$, (**c**) $x_L = 0.75$, and (**d**) $x_L = 1$. The open symbols are experimental values, and the dashed lines are fit to experimental $S(q)$ using binary hard sphere model

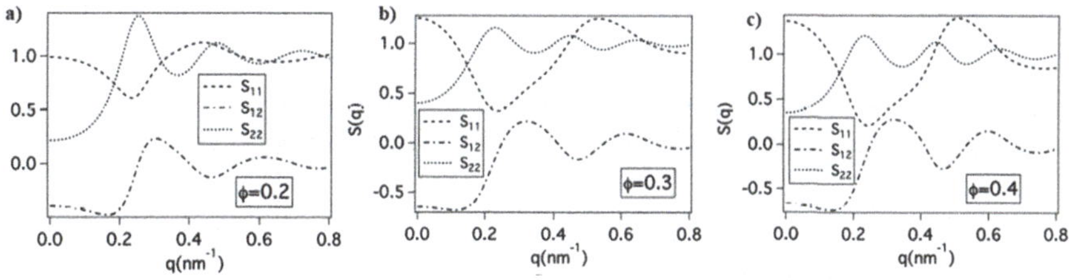

Supplementary Fig. 2.3 Variation in three components of structure factor, viz., S_{11}, S_{12}, and S_{22}, as estimated from binary hard sphere model predictions. Here, 1 and 2 indicate smaller and larger particles, respectively. The structure factors are shown for $x_L = 0.5$ for different volume fractions (**a**) $\Phi = 0.2$, (**b**) $\Phi = 0.3$, and (**c**) $\Phi = 0.4$

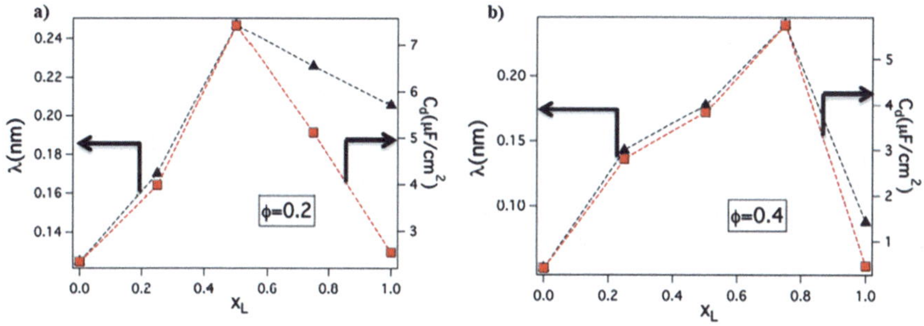

Supplementary Fig. 2.4 Electrical double layer (square symbols) and electrochemical capacitance (triangle symbols) for different values of x_L at (**a**) $\Phi = 0.2$ and (**b**) $\Phi = 0.4$

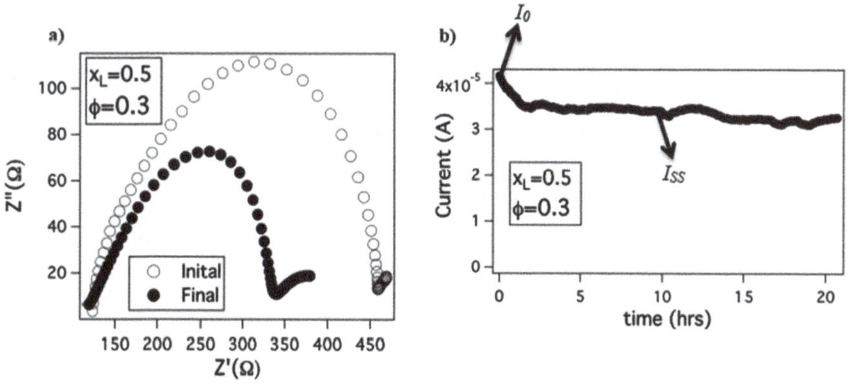

Supplementary Fig. 2.5 (**a**) Electrochemical impedance measurements for a symmetric lithium cell before polarization (open symbols) and after steady state is reached (closed symbols). R_0 and R_{ss} were determined by fitting these impedances to the circuit model mentioned in the main text. (**b**) Current decay for the cell undergoing polarization at 20 mV potential. I_0 and I_{SS} were used in the calculations as observed in the plot

References

1. Qian, J., Henderson, W.A., Xu, W., Bhattacharya, P., Engelhard, M., Borodin, O., Zhang, J.-G.: High rate and stable cycling of lithium metal anode. Nat. Commun. **6**, 6362 (2015)
2. Xu, W., Wang, J., Ding, F., Chen, X., Nasybulin, E., Zhang, Y., Zhang, J.-G.: Lithium metal anodes for rechargeable batteries. Energy Environ. Sci. **7**, 513 (2014)
3. Tu, Z., Nath, P., Lu, Y., Tikekar, M.D., Archer, L.A., et al.: Acc. Chem. Res. **48**(11), 2947–2956 (2015)., acs.accounts.5b00427
4. Zheng, G., Lee, S.W., Liang, Z., Lee, H.-W., Yan, K., Yao, H., Wang, H., Li, W., Chu, S., Cui, Y.: Interconnected hollow carbon nanospheres for stable lithium metal anodes. Nat. Nanotechnol. **9**, 618–623 (2014)
5. Aurbach, D., Zinigrad, E., Cohen, Y., Teller, H.: A short review of failure mechanisms of lithium metal and lithiated graphite anodes in liquid electrolyte solutions. Solid State Ionics. **148**, 405–416 (2002)
6. Guo, J., Wen, Z., Wu, M., Jin, J., Liu, Y.: Vinylene carbonate–LiNO3: a hybrid additive in carbonic ester electrolytes for SEI modification on Li metal anode. Electrochem. Commun. **51**, 59–63 (2015)
7. Choudhury, S., Mangal, R., Agrawal, A., Archer, L.A.: A highly reversible room-temperature lithium metal battery based on crosslinked hairy nanoparticles. Nat. Commun. **6**, 10101 (2015)
8. Chen, L., Wang, K., Xie, X., Xie, J.: Effect of Vinylene Carbonate (VC) as electrolyte additive on electrochemical performance of Si film anode for lithium ion batteries. J. Power Sources. **174**, 538–543 (2007)
9. Xu, K., Zhang, S., Jow, T.R.: LiBOB as additive in LiPF[sub 6]-based lithium ion electrolytes. Electrochem. Solid-State Lett. **8**, A365 (2005)
10. Li, W., Yao, H., Yan, K., Zheng, G., Liang, Z., Chiang, Y.-M., Cui, Y.: The synergetic effect of lithium polysulfide and lithium nitrate to prevent lithium dendrite growth. Nat. Commun. **6**, 7436 (2015)
11. Choudhury, S., Archer, L.A.: Lithium fluoride additives for stable cycling of lithium batteries at high current densities. Adv. Electron. Mater. **2**, 1500246 (2015)
12. Lu, Y., Tu, Z., Archer, L.A.: Stable lithium electrodeposition in liquid and nanoporous solid electrolytes. Nat. Mater. **13**, 961–969 (2014)
13. Pires, J., Timperman, L., Castets, A., Santos-pena, J., Dumont, E., Levasseur, S., Dedryvere, R., Tessier, C., Anouti, M.: Role of propane sultone as additive to improve the performance of lithium-rich cathode material at high potential. RSC Adv. **5**(52), 42088–42094 (2015)
14. Li, B., Xu, M., Li, T., Li, W., Hu, S.: Prop-1-Ene-1,3-sultone as SEI formation additive in propylene carbonate-based electrolyte for lithium ion batteries. Electrochem. Commun. **17**, 92–95 (2012)
15. Etacheri, V., Haik, O., Goffer, Y., Roberts, G.A., Stefan, I.C., Fasching, R., Aurbach, D.: Effect of Fluoroethylene Carbonate (FEC) on the performance and surface chemistry of Si-nanowire Li-ion battery anodes. Langmuir. **28**, 965–976 (2012)
16. Miao, R., Yang, J., Feng, X., Jia, H., Wang, J., Nuli, Y.: Novel dual-salts electrolyte solution for dendrite-free lithium-metal based rechargeable batteries with high cycle reversibility. J. Power Sources. **271**, 291–297 (2014)
17. Luo, W., Zhou, L., Fu, K., Yang, Z., Wan, J., Manno, M., Yao, Y., Zhu, H., Yang, B., Hu, L.: A thermally conductive separator for stable Li metal anodes. Nano Lett. **15**(9), 6149–6154 (2015)
18. Liang, Z., Zheng, G., Liu, C., Liu, N., Li, W., Yan, K., Yao, H., Hsu, P.-C., Chu, S., Cui, Y.: Polymer nanofiber-guided uniform lithium deposition for battery electrodes. Nano Lett. **15**(5), 2910–2916 (2015)
19. Lu, Y., Tikekar, M., Mohanty, R., Hendrickson, K., Ma, L., Archer, L.A.: Stable cycling of lithium metal batteries using high transference number electrolytes. Adv. Energy Mater. **5**, 1402073 (2015)

20. Schaefer, J.L., Yanga, D.A., Archer, L.A.: High lithium transference number electrolytes via creation of 3-dimensional, charged, nanoporous networks from dense functionalized nanoparticle composites. Chem. Mater. **25**, 834–839 (2013)
21. Ozhabes, Y., Gunceler, D., Arias, T.A.: Stability and surface diffusion at lithium-electrolyte interphases with connections to dendrite suppression. arXiv. **1504**(05799), 1–7 (2015)
22. Khurana, R., Schaefer, J.L., Archer, L.A., Coates, G.W.: Suppression of lithium dendrite growth using cross-linked polyethylene/poly(ethylene oxide) electrolytes: a new approach for practical lithium-metal polymer batteries. J. Am. Chem. Soc. **136**, 7395–7402 (2014)
23. Tu, Z., Kambe, Y., Lu, Y., Archer, L.A.: Nanoporous polymer-ceramic composite electrolytes for lithium metal batteries. Adv. Energy Mater. **4**, 1300654 (2014)
24. Gurevitch, I., Buonsanti, R., Teran, A.A., Gludovatz, B., Ritchie, R.O., Cabana, J., Balsara, N.P.: Nanocomposites of titanium dioxide and polystyrene-poly(ethylene oxide) block copolymer as solid-state electrolytes for lithium metal batteries. J. Electrochem. Soc. **160**, A1611–A1617 (2013)
25. Bouchet, R., Maria, S., Meziane, R., Aboulaich, A., Lienafa, L., Bonnet, J., Phan, T.N.T., Bertin, D., Gigmes, D., Devaux, D., et al.: Efficient electrolytes for lithium-metal batteries. Nat. Mater. **12**, 452–457 (2013)
26. Weston, J.E., Steele, B.C.H.: Effects of inert fillers on the mechanical and electrochemical properties of lithium salt-poly (ethylene oxide) polymer electrolytes _ i b. Solid State Ionics. **7**, 75–79 (1982)
27. Croce, F., Scrosati, B.: Nanocomposite lithium ion conducting membranes. Ann. N.Y. Acad Sci. **984**, 194–207 (2003)
28. Agrawal, A., Choudhury, S., Archer, L.A.: A highly conductive, non-flammable polymer-nanoparticle hybrid electrolyte. RSC Adv. **5**, 20800–20809 (2015)
29. Srivastava, S., Schaefer, J.L., Yang, Z., Tu, Z., Archer, L.A.: 25Th anniversary article: polymer-particle composites: phase stability and applications in electrochemical energy storage. Adv. Mater. **26**, 201–234 (2014)
30. Lu, Y., Korf, K., Kambe, Y., Tu, Z., Archer, L.A.: Ionic-liquid-nanoparticle hybrid electrolytes: applications in lithium metal batteries. Angew. Chemie. **126**, 498–502 (2014)
31. Lu, Y., Das, S.K., Moganty, S.S., Archer, L.A.: Ionic liquid-nanoparticle hybrid electrolytes and their application in secondary lithium-metal batteries. Adv. Mater. **24**, 4430–4435 (2012)
32. Korf, K.S., Lu, Y., Kambe, Y., Archer, L.A.: Piperidinium tethered nanoparticle-hybrid electrolyte for lithium metal batteries. J. Mater. Chem. A. **2**, 11866–11873 (2014)
33. Nugent, J.L., Moganty, S.S., Archer, L.A.: Nanoscale organic hybrid electrolytes. Adv. Mater. **22**, 3677–3680 (2010)
34. Schaefer, J.L., Moganty, S.S., Yanga, D.A., Archer, L.A.: Nanoporous hybrid electrolytes. J. Mater. Chem. **21**, 10094 (2011)
35. Choudhury, S., Agrawal, A., Kim, S.A., Archer, L.A.: Self-suspended suspensions of covalently grafted hairy nanoparticles. Langmuir. **31**, 3222–3231 (2015)
36. Jonscher, A.K.: The “universal” dielectric response. Nature. **267**, 673–679 (1977)
37. Zhang, Q., Archer, L.A.: Poly(ethylene oxide)/silica nanocomposites: structure and rheology. Langmuir. **18**, 10435–10442 (2002)
38. Moganty, S.S., Jayaprakash, N., Nugent, J.L., Shen, J., Archer, L.A.: Ionic-liquid-tethered nanoparticles: hybrid electrolytes. Angew. Chemie. **122**, 9344–9347 (2010)
39. Hussain, F.: Review article: polymer-matrix nanocomposites, processing, manufacturing, and application: an overview. J. Compos. Mater. **40**, 1511–1575 (2006)
40. Palmqvist, A.E.C.: Synthesis of ordered mesoporous materials using surfactant liquid crystals or micellar solutions. Curr. Opin. Colloid Interface Sci. **8**, 145–155 (2003)
41. Balazs, A.C., Emrick, T., Russell, T.P.: Nanoparticle polymer composites: meet two small worlds. Science. **314**, 1107–1110 (2013)
42. Mangal, R., Srivastava, S., Archer, L.A.: Phase stability and dynamics of entangled polymer–nanoparticle composites. Nat. Commun. **6**, 1–9 (2015)

43. Green, D.L., Mewis, J., Engineering, C., Uni, V., Way, E., Charlottes, V.: Connecting the wetting and rheological behaviors of poly (dimethylsiloxane)-grafted silica spheres in poly (dimethylsiloxane) melts. Langmuir. **22**, 9546–9553 (2006)
44. Srivastava, S., Agarwal, P., Archer, L.A.: Tethered nanoparticle-polymer composites: phase stability and curvature. Langmuir. **28**, 6276–6281 (2012)
45. McEwan, M., Green, D.: Rheological impacts of particle softness on wetted polymer-grafted silica nanoparticles in polymer melts. Soft Matter. **5**, 1705 (2009)
46. Dutta, N., Green, D.: Nanoparticle stability in semidilute and concentrated polymer solutions. Langmuir. **24**, 5260–5269 (2008)
47. Lindenblatt, G., Schärtl, W., Pakula, T., Schmidt, M.: Structure and dynamics of hairy spherical colloids in a matrix of nonentangled linear chains. Macromolecules. **34**, 1730–1736 (2001)
48. Oh, H., Green, P.F.: Polymer chain dynamics and glass transition in athermal polymer/nanoparticle mixtures. Nat. Mater. **8**, 139–143 (2009)
49. Meng, D., Kumar, S.K., Lane, J.M.D., Grest, G.S.: Effective interactions between grafted nanoparticles in a polymer matrix. Soft Matter. **8**, 5002 (2012)
50. Ohno, K., Morinaga, T., Takeno, S., Tsujii, Y., Fukuda, T.: Suspensions of silica particles grafted with concentrated polymer brush: effects of graft chain length on brush layer thickness and colloidal crystallization. Macromolecules. **40**, 9143–9150 (2007)
51. Glatter, O., Kratky, O.: Small angle x-ray scattering. Academic Press, New York (1982)
52. Zaccarelli, E., Mayer, C., Asteriadi, A., Likos, C., Sciortino, F., Roovers, J., Iatrou, H., Hadjichristidis, N., Tartaglia, P., Löwen, H., et al.: Tailoring the flow of soft glasses by soft additives. Phys. Rev. Lett. **95**, 268301 (2005)
53. Mayer, C., Zaccarelli, E., Stiakakis, E., Likos, C.N., Sciortino, F., Munam, A., Gauthier, M., Hadjichristidis, N., Iatrou, H., Tartaglia, P., et al.: Asymmetric caging in soft colloidal mixtures. Nat. Mater. **7**, 780–784 (2008)
54. Agrawal, A., Yu, H.-Y., Srivastava, S., Choudhury, S., Narayanan, S., Archer, L.: Dynamics and yielding of binary self-suspended nanoparticle fluids. Soft Matter. **11**, 5224–5234 (2015)
55. Ashcroft, N.W., Langreth, D.C.: Structure of binary liquid mixtures. I. Phys. Rev. **16**, 685–692 (1967)
56. Srivastava, S., Shin, J.H., Archer, L.A.: Structure and rheology of nanoparticle–polymer suspensions. Soft Matter. **8**, 4097 (2012)
57. Yu, H.-Y., Srivastava, S., Archer, L.A., Koch, D.L.: Structure factor of blends of solvent-free nanoparticle-organic hybrid materials: density-functional theory and small angle X-ray scattering. Soft Matter. **10**, 9120–9135 (2014)
58. Qu, D., Qu, D., Shi, H., Shi, H.: Studies of activated carbons used in double-layer capacitors. J. Power Sources. **74**, 99–107 (1998)
59. Shi, H.: Activated carbons and double layer capacitance. Electrochim. Acta. **41**, 1633–1639 (1996)
60. Choi, N.S., Chen, Z., Freunberger, S.A., Ji, X., Sun, Y.K., Amine, K., Yushin, G., Nazar, L.F., Cho, J., Bruce, P.G.: Challenges facing lithium batteries and electrical double-layer capacitors. Angew. Chemie Int. Ed. **51**, 9994–10024 (2012)
61. Flandrois, S., Simon, B.: Carbon materials for lithium-ion rechargeable batteries. Carbon N. Y. **37**, 165–180 (1999)
62. Shiraishi, S., Kurihara, H., Tsubota, H., Oya, A., Soneda, Y., Yamada, Y.: Electric double layer capacitance of highly porous carbon derived from lithium metal and polytetrafluoroethylene. Electrochem. Solid-State Lett. **4**, A5–A8 (2001)
63. Largeot, C., Portet, C., Chmiola, J., Taberna, P.L., Gogotsi, Y., Simon, P.: Relation between the ion size and pore size for an electric double-layer capacitor. J. Am. Chem. Soc. **130**, 2730–2731 (2008)
64. Lanfredi, S., Rodrigues, A.C.M.: Impedance spectroscopy study of the electrical conductivity and dielectric constant. J. Appl. Phys. **86**, 2215 (1999)
65. Payne, R., Theodorou, I.E.: Dielectric properties and relaxation in ethylene carbonate and propylene carbonate. J. Phys. Chem. **76**, 2892–2900 (1972)

66. Simeral, L., Ameyib, R.L., Amey, L.: Dielectric properties of liquid propylene carbonate. J. Phys. Chem. **74**, 1968–1971 (1970)
67. Sengwa, R.J., Kaur, K., Chaudhary, R.: Dielectric properties of low molecular weight poly(ethylene glycol)s. Polym. Int. **608**, 599–608 (2000)
68. Schneider, U., Lunkenheimer, P., Brand, R., Loidl, A.: Broadband dielectric spectroscopy on glass-forming propylene carbonate. Phys. Rev. E. Stat. Phys. Plasmas. Fluids. Relat. Interdiscip. Topics. **59**, 6924–6936 (1999)
69. Bruce, P.: Conductivity and transference number measurements on polymer electrolytes. Solid State Ionics. **28-30**, 918–922 (1988)
70. Appetecchi, G.B.: A new class of advanced polymer electrolytes and their relevance in plastic-like, rechargeable lithium batteries. J. Electrochem. Soc. **143**, 6 (1996)
71. Jorne, J.: Transference number approaching unity in nanocomposite electrolytes. Nano Lett. **6**, 2973–2976 (2006)
72. Capuano, F., Croce, F., Scrosati, B.: Composite polymer electrolytes. J. Electrochem. Soc. **203**, 197–203 (2003)
73. Ding, F., Xu, W., Graff, G.L., Zhang, J., Sushko, M.L., Chen, X., Shao, Y., Engelhard, M.H., Nie, Z., Xiao, J., et al.: Dendrite-free lithium deposition via self-healing electrostatic shield mechanism. J. Am. Chem. Soc. **135**, 4450–4456 (2013)

Chapter 3
A Highly Reversible Room-Temperature Lithium Metal Battery Based on Cross-Linked Hairy Nanoparticles

3.1 Introduction

The search for portable, high-capacity, and safe electrical energy storage technologies remains one of the paramount motivators for materials research. High-voltage cathodes, high-energy anodes, and highly conductive but stable electrolytes for lithium-ion batteries (LIBs) have received a lopsided share of the attention by researchers because of their multiple attractive features, including high energy density, light weight, high operating voltage and minimal memory effects [1–3]. Secondary lithium metal batteries (LMBs), wherein metallic lithium serves as the anode, are an attractive alternative to LIBs but are known to have a serious problem associated with dendrite-induced short circuits [4]. During repeated cycles of charge and discharge, the uneven deposition of Li ion on this metal leads to the formation of ramified structures, which grow unstably, puncture the separator, and ultimately cause cell failure by short circuiting the anode and cathode. Over the years, a growing number of studies have explored electrolyte and separator platforms to suppress the dendrite growth in an effort to enable LMBs [5, 6]. Recent efforts have focused on stabilizing the surface of the Li anode using electrolyte additives [7–9], hybrid ionic liquid nanostructures [10, 11], or a high-modulus separator [12–14], which can also provide a means of applying compression forces to stabilize the anode during deposition. Infusion of a nanoporous ceramic or polymer membrane separator with a liquid electrolyte that can facilitate Li-ion transfer, without compromising the mechanical properties of the nanoporous membrane, provides a more straightforward route towards mechanically strong, room-temperature electrolytes/separators that prevent dendrite growth [15, 16].

Solid or gel polymer electrolytes have been researched extensively for their ability to enable batteries in various form factors that are leakage-free, flexible, yet safer [12–14, 17–21]. However, these gel electrolyte systems have consistently underperformed in terms of the ionic conductivity requirements for room-temperature

S. Choudhury, *Rational Design of Nanostructured Polymer Electrolytes and Solid–Liquid Interphases for Lithium Batteries*, Springer Theses,
https://doi.org/10.1007/978-3-030-28943-0_3

operation of advanced batteries [22]. In a block copolymer solid electrolyte system, the ratio of the hard nonconducting phase to the soft conducting phase determines the mechanical strength of the system. It has been shown, for instance, that in poly(styrene)-poly(ethylene oxide) (PS-PEO) electrolytes, a PS-PEO molar ratio of around unity provides a good balance between mechanical strength and ionic conductivity [23]. However, the abundance of the nonconducting, reinforcing PS phase still results in low bulk conductivity relative to liquid electrolytes, necessitating elevated temperature battery operation, which is a limitation for many consumer-based applications. Nanocomposite electrolytes comprised of liquid or polymeric electrolytes reinforced with nanoparticle fillers can achieve higher modulus at lower reinforcing material content, which potentially offers multiple straightforward paths towards electrolytes with high-modulus and acceptable room-temperature ionic conductivity [24–28]. Uniform dispersion of the fillers in the polymer host is understood to be a prerequisite to prevent particle agglomeration and local inhomogeneity in the electrolyte medium. Unfortunately, strong attractive van der Waals and depletion forces on the particles result in particle aggregation and phase separation [29, 30]. Several recent studies have shown that various physical and chemical modifications of nanoparticle-polymer interactions can lead to dramatic improvements in phase stability and electrolyte properties of such systems [29, 31–34].

A strategy towards a hybrid electrolyte platform, which can provide high ionic conductivity and attractive mechanical properties, is designing a cross-linked polymer web in which hairy nanoparticles serve as cross-linkers. A perhaps obvious benefit of this design is that chemistry introduced on the surface of the precursor particles can be presented in the pores of the cross-linked material to selectively pass Li ions while hindering the unstable dendrite growth. Here, we report the first realization of this concept and show that an electrolyte with the proposed design combining the advantages of a nanocomposite and solid polymer electrolyte. In the systems reported herein, a hydrophobic polymer provides a porous conductive pathway for Li ions, while a short hydrophilic oligomer tethered to the nanoparticle surface constrains the network providing structure and mechanical strength. The nanoparticle-induced cross-linking of the polymer prevents particle aggregation, which is known to compromise flexibility and elasticity of nanocomposites.

3.2 Methods

3.2.1 Materials

Ludox SM30 colloidal silica ($d = 10 \pm 2$ nm), poly(propylene glycol)-toluene-2,4-diisocyanate, propylene carbonate, Bis(trifluoromethane)sulfonimide lithium salt, and ethylene carbonate/diethylene carbonate (1v:1v) with lithium hexafluorophosphate were all purchased from Sigma-Aldrich. Hydroxy-terminated poly(ethylene oxide)-silane was obtained from Gelest. All the chemicals were used as received in appropriate conditions.

3.2.2 Nanoparticle-Polymer Cross-Link Synthesis and Composite Electrolyte Preparation

Cross-linked nanoparticle-polymer composites (CNPCs) were synthesized using a two-step process. In the first step, colloidal silica was grafted with hydroxy-poly(ethylene oxide)-silane [500 Da] in water by the reaction of silanol groups of the PEO chain and OH- groups on the silica particle. A predetermined amount of the Ludox SM30 and a large excess of the hydroxy-PEO-silane was first heated at 80 °C for a period of 2 days, before centrifuging the thus obtained mixture in a chloroform-hexane mixed solvent five times at a speed of 8500 rpm for 10 min in order to remove all unlinked PEO from the hairy nanoparticles. The recovered OH-PEO-SiO_2 hairy particle suspension was rigorously dried and saved for the next reaction step. In the second step, poly(propylene glycol)-toluene-2,4-diisocyanate and the OH-PEO-SiO_2 nanoparticles were combined in a 4:1 ratio by weight and dissolved anhydrous chloroform using a vortex mixer. The obtained solution was casted onto a rectangular Teflon mold of desired size. The mixture was allowed to cross-link overnight under ambient conditions to produce CNPC films/membranes of desired thicknesses. The resultant membranes were characterized using TGA, DSC, and FTIR. In order to use the synthesized CNPC membranes as battery electrolytes, they were soaked for a period of 2 days in conventional liquid electrolytes in an argon-filled glove box.

3.2.3 TEM and Small Angle X-Ray Scattering

The nanoscale structure of the as prepared CNPC membranes was characterized using a FEI T12 Spirit TEM. For these measurements, a thin layer of the precursor material was casted onto the TEM grid, allowing it to cross-link on the grid. Before TEM imaging, coated grids were dried overnight using the same procedure used to prepare the freestanding CNPC membranes. Small angle X-ray scattering (SAXS) measurements were performed at the Cornell High Energy Synchrotron Source (CHESS) on a strip of the CNPC film using a point-collimated X-ray source at room temperature.

3.2.4 Mechanical Properties

Mechanical properties of CNPC membranes and CNPC electrolytes were analyzed using dynamic mechanical analysis (DMA Q800) at room temperature. A fixed strain of 0.1% was applied on the strip over a range of frequency of 1–100 Hz, in order to obtain the storage modulus of the samples. A similar experiment was done on the silica-PPO suspension using an Anton Paar MCR 501 shear rheometer in order to characterize the effect of cross-linking on mechanical properties of the materials.

3.2.5 Electrochemical Characterization

The ionic conductivity of CNPC electrolytes was measured as a function of temperature using a Novocontrol N40 Broadband Dielectric instrument. The d.c. conductivity was obtained from the plateau, high-frequency region of the a.c. conductivity versus frequency data. Impedance spectroscopy measurements were performed using a Solartron frequency analyzer. A two-electrode, symmetric Li|Li cell in which the CNPC electrolyte was sandwiched between lithium foils was studied in a frequency range from 10^{-3} Hz to 10^{7} Hz. Linear sweep and cyclic voltammetry were performed at a sweep rate of 1 mV/s between −0.2 V and 6.5 V using Li/CNPC electrolyte/stainless steel cells to quantify the voltage stability window of the electrolytes.

3.2.6 Cell Lifetime and Failure Studies

Symmetric 2032 type Li|Li coin cells containing either liquid of CNPC electrolytes were prepared inside an argon-filled glove box. To facilitate comparisons, all cells used in this component of the study also included a standard Celgard separator, with a diameter of 6.4 mm. The cells were evaluated using both galvanostatic (strip-plate) cycling and in step current (polarization) modes over a range of current density using a Neware CT-3008 battery tester. In the "strip-plate" experiments, the cells were charged and discharged at a predetermined current density for a period of 3 h each. Failure was deduced either from a sudden drop of the resultant voltage waveform or when the waveform exhibited new harmonics and displayed irregular time dependence. In the polarization experiments, the symmetric cells were constantly charged at a particular current density until the cell failed. In this case failure was determined either by a sudden drop in the resultant voltage profile or by the appearance of irregular transients in the voltage profile. Postmortem characterization of the lithium surface after cycling was used to corroborate our conclusions about failure from the electrical response of the cells and for assessing the surface structure of Li in cells that showed no electrical signatures of failure. For this purpose, cells were dissembled in the glove box, and the Li foil was removed and was washed repeatedly in pure PC to remove the electrolyte, before performing SEM (LEO155FESEM) analysis.

3.2.7 Measuring the Coulombic Efficiency

Li| electrolyte |Stainless steel 2032-type coin cells were assembled in an argon-filled glove box. Control experiments using liquid electrolytes comprised of 1 M LiTFSI-PC with 1% (w.t.) $LiNO_3$ and 2% (vol.) vinylene carbonate additives, and a standard Celgard separator were compared to measurements using CNPC electrolytes created by soaking CNPC membranes in these same liquid electrolytes. In both cases, prior to the measurements, cells were conditioned by cycling them between 0 and 0.5 V for 10 cycles, following a procedure reported in literature [44] thought to enable formation of a stable SEI layer on the electrodes. To characterize the columbic efficiency, the conditioned cells were first discharged at a constant current density of 0.25 mA/cm^2 for 2 h to transfer an amount of lithium corresponding to 0.50mAh/cm^2 of charge from the Li electrode to the stainless steel electrode. The amount of charge recovered in the reverse cycle when the cell is charged back to 0.5 V at the same current density was recorded and the fractional recovery used to determine the coulombic efficiency.

3.2.8 Half-Cell Testing

Both LTO and LFP cathodes were prepared by mixing the active materials with Super P carbon and PVDF binder in the ratio of 8:1:1, by weight, in the presence of NMP, and casting a layer of the resultant slurry onto aluminum foil. Next, the cathode was dried first at room temperature and then at 70 °C overnight under vacuum. The surface capacity of both films was maintained constant as 0.5mAh/cm^2. Li|CNPC electrolyte|LTO and Li|CNPC electrolyte|LFP 2032-type coin cells were assembled under argon environment in a glove box.

3.3 Results

3.3.1 Synthesis and Physical Characterization of Cross-Linked Membrane

Figure 3.1a is a schematic of the reaction steps involved in the synthesis of our cross-linked hybrid electrolyte material. First, short-chained oligomers, poly(ethylene oxide)-500 Da, are grafted onto silica particles to form nanoparticle organic hybrids with reactive end groups. The PEO-tethered silica particles are next linked with difunctional poly(propylene oxide)-2000 Da (PPO) to form the cross-linked nanoparticle-polymer composite (CNPC) in a desired macroscopic shape. The materials used in the present study achieve high levels of cross-linking by two-reaction mechanisms: dimerization between isocyanate end groups in PPO and reaction between isocyanate group in PPO and end-functional hydroxyl groups on the hairy nanoparticles to form stable urethane linkages. The CNPC membrane (Fig. 3.1c) is mechanically tough and has a "rubbery" texture. In order to use it as a composite electrolyte, the membrane was soaked in a liquid electrolyte comprised of propylene carbonate (PC) containing 1 M lithium bis(trifluoromethanesulfonyl)

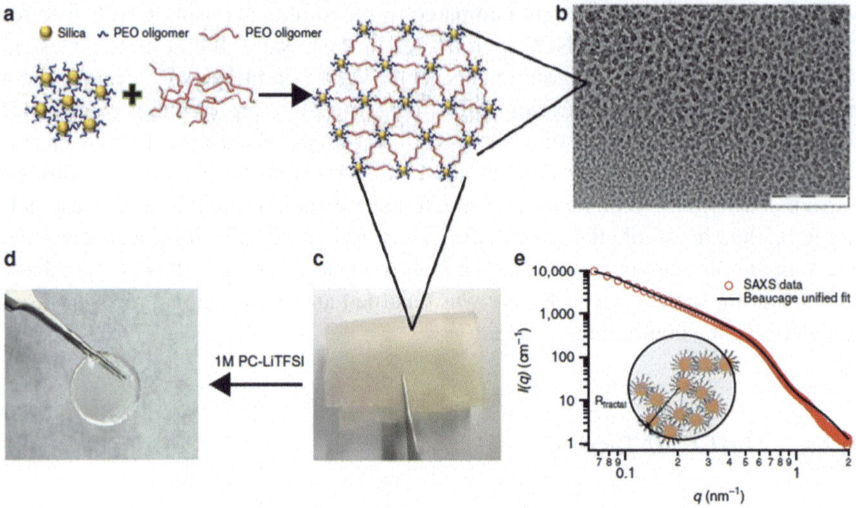

Fig. 3.1 From hybrid nanostructure to macroscale membrane: clockwise. (**a**), The first row shows the scheme of reaction involved in the synthesis of the freestanding cross-linked nanoparticle-polymer composite. Silica tethered with hydroxy-terminated polyethylene oxide is reacted with polypropylene oxide-diisocyanate with a weight ratio of 1:4 at room temperature in a Teflon mold of appropriate shape. (**b**), TEM image of the cross-linked membrane. The black circles are silica particle dispersed in the polymer matrix. (**c**), The photograph of a freestanding membrane is shown. Physically it has a rubbery texture and is transparent. (**d**), Image of the wet polymer gel is shown which is obtained by longtime soaking of the prepared membrane in unimolar electrolyte liquid. (**e**), Scattered intensity $I(q)$ vs. wave vector q profile obtained from SAXS measurements. Solid line represents the fit for Beaucage unified model to the data. The fit in the low q regime shows a power-law scattering ($I(q) \sim q^{-\alpha}$) with an exponent (α) of around 2, indicating the presence of mass fractals

imide (LiTFSI) salt for a period of 2 days. This step yields a material with a swollen, gel-like appearance (shown in Fig. 3.1d) but with little loss of the solid nanocomposite material's mechanical strength. In contrast, it is seen that on soaking with electrolyte, the un-cross-linked components separate out in the liquid, leaving behind a transparent film with no observable particle aggregates. For brevity, the CNPC membranes infiltrated with liquid electrolytes are hereafter referred to as CNPC electrolytes.

The reaction progress can be facilely mapped using Fourier transform infrared spectroscopy (FTIR) as illustrated in Supplementary Fig. 3.1. It is evident that the -NCO bond at (2270 cm^{-1}) present in the neat PPO-diisocyanate polymer is consumed in the reaction and at same time generates a urethane linkage, evident by the -NH vibration (3288 cm^{-1}) and -C=O bond peak (1680 cm^{-1}). TGA analysis of these materials (Supplementary Fig. 3.2) shows that the freestanding cross-linked film contains 6% silica by weight and in the final CNPC electrolyte material, the silica content is 2% (Supplementary Table 3.2); thus there is a preponderance of ion-conducting entity in the material. Significantly, because the silica particles are nano sized (~10 nm) and have multiple chains attached to their surface (~55 chains/ particle), each particle provides a large number of node points for cross-linking the PPO polymer. Thus, even a low particle content in the precursor material is expected to enable extensive cross-linking via particle nodes, which should lead to dramatic improvements in mechanical properties without compromising room-temperature ionic conductivity of the liquid electrolyte hosted by a CNPC membrane.

The synthesized CNPC membranes were characterized using differential scanning calorimetry (DSC) measurements (Supplementary Fig. 3.3), which reveal a low glass transition temperature (T_g) of −63 °C for neat PPO polymer, after cross-linking, T_g increases to −42 °C. The dispersion and arrangement of nanoparticles in the cross-linked polymer matrix is important for the targeted application as nano-structured electrolytes. Figure 3.1b shows a 2D image of a thin layer of the material obtained from transmission electron microscopy (TEM). The image shows that the particles organize in disordered interconnected, stringlike phases to produce tortuous pores with an average diameter of 24 ± 2 nm estimated from the TEM micrograph (see Supplementary Fig. 3.4). Small angle X-ray scattering (SAXS) has been used previously to study particle packing and polymer-nanoparticle interactions in bulk nanocomposite-based electrolyte materials [35–37]. Figure 3.1e shows the experimental intensity $I(q)$ as a function of wave number (q) obtained from SAXS measurements on CNPCs. The solid line through the experimental I(q) data was obtained using the Beaucage unified model (see Supporting Information). The power-law scattering regime $I(q) \sim q^{-\alpha}$, with exponent $\alpha \approx$ 2observed at low q, indicates that the particles in the materials are organized in the form of mass fractals with sizes in the range 50–70 nm. The shoulder at $q = q_s \approx$ 0.6nm-1 implies that the scattering originates from particles with average diameter $d = 2\pi/q_s \approx$ 10.5nm, which is close to the average diameter of the SiO_2 nanocores used for synthesizing the materials. Finally, the absence of any additional structure contribution (maxima) in I(q) in the intermediate and high q regime means that the primary particles are reasonably far apart. The interparticle distance can be theoretically calculated assuming random packing of hard spheres in suspension, $d_{p\text{-}p} = d(0.63/\phi)^{1/3}$, where d is the nanoparticle diameter and ϕ is the volume fraction of particles in the polymer network. This analysis yields an interparticle spacing of approximately 27 nm, which is comparable to the particle separation of 24 ± 2 nm deduced from analysis of the TEM image in Fig. 3.1b (see Supplementary Fig. 3.4). Thus, the TEM and SAXS results imply that in a freestanding CNPC membrane, randomly distributed particles organize in stringlike phases to form tortuous path for ion and mass transport through the materials. Such nanostructured materials have been hypothesized, largely without proof, to be excellent candidates to hinder dendrite-induced short circuit in lithium batteries [13, 14, 20].

3.3.2 *Mechanical and Electrochemical Properties of Cross-Linked Membrane*

Figure 3.2a reports the storage modulus as a function of frequency for a freestanding CNPC film before and after soaking in liquid electrolyte. The storage modulus of a suspension of PPO in silica is also shown in the figure. Clearly, cross-linking increments the modulus by several orders of magnitude. The mechanical strength of the material is about five orders of magnitude higher than a neat PEO-based electrolyte and is comparable to the elastic shear modulus reported for dry PS-PEO block copolymer-based electrolytes [23, 38]. Additionally, it is seen that even after soaking in a liquid electrolyte, the elastic modulus of the CNPC material remains high; it is higher than those recently reported for cross-linked PEO-PE-PEO solid polymer electrolytes [14]. This latter feature we attribute to the large number of cross-

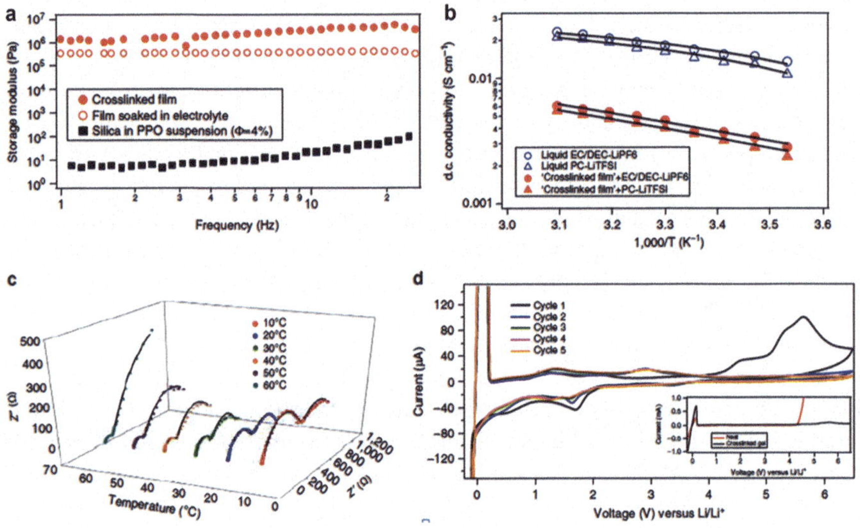

Fig. 3.2 Good mechanical and electrochemical properties: (**a**), storage modulus (Pa) as a function of frequency, comparing the dry polymer network and gel electrolyte with a suspension of silica in PPO polymer with same volume fraction. The modulus of cross-linked gel is more than five orders of magnitude higher than the suspension. (**b**), Conductivity as a function of inverse absolute temperature. The markers show measured values, while the continuous lines are fitted VFT curves. The cross-linked gel shows conductivity of an order of magnitude less than the intrinsic value of electrolyte. (**c**), Impedance spectroscopy results for a symmetric lithium battery, with cross-linked gel electrolyte, are plotted against temperature, where the *x*-axis denotes temperature, *y*-axis denotes real impedance (ohms), and *z*-axis denotes imaginary impedance (ohms). (**d**), Cyclic voltammetry of a cell with configuration Li/cross-linked gel/stainless steel is shown for five cycles, where *x*-axis represents voltage and *y*-axis, current. The inset compares the stability of the cross-linked gel with neat electrolyte of 1 M PC-LiTFSI

links made to a single particle and the short length and stiffness of the polymer fragments used as cross-linkers. Figure 3.2b reports the d.c. conductivity in Arrhenius plot for CNPC electrolytes obtained by soaking the synthesized CNPC membranes in two widely studied liquid electrolytes, PC-LiTFSI and EC/DEC-$LiPF_6$. In comparison with the neat (no nanoparticles) liquid electrolytes, the conductivity is lower, by about one order of magnitude, but is still high enough at room temperature for most lithium battery applications. This high conductivity is attributed to the low glass transition temperature of PPO and the low fraction of non-conductive nanoparticles needed to produce materials with high mechanical strength. The conductivity versus temperature data is well described by the Vogel-Fulcher-Tammann (VFT) relationship $\sigma = A \exp(-B/(T-T_O))$, where A is the pre-exponential factor corresponding to conductivity at infinite temperature, B is the activation energy, and T_O is the reference temperature. The respective coefficients for the different systems are tabulated in Supplementary Table 3.1.

Figure 3.2c reports temperature-dependent impedance spectra for CNPC electrolytes obtained by soaking the membranes in PC-1MLiTFSI electrolyte. The measurements were performed in symmetric (Li|Li), two-electrode cells. The solid lines through the data were obtained using an equivalent circuit model previously reported for nanocomposite based electrolyte systems [39–41]. The bulk and interfacial resistance deduced from the Nyquist plots are plotted as a function of temperature in Supplementary Fig. 3.5. It is evident that both resistances decrease with temperature and the bulk resistance is always lower than the interfacial resistance. This implies that ion transport through the tortuous materials is in reality easier relative to ion transfer across the electrolyte-electrode interface. It is apparent nonetheless that the bulk and interfacial resistances (Supplementary Fig. 3.5) exhibit similar dependence on temperature, indicating that both have similar activation energy. It can therefore be concluded that the CNPC electrolytes make good contact with the Li metal, as just the polymer itself appears to limit ion transport across the interface. The electrochemical stability for the CNPC electrolyte was analyzed at room temperature using cyclic voltammetry in a prototype cell with the following configuration, Li|CNPC+PC-1MLiTFSI|stainless steel. Measurements were performed between −0.2 V and 6.5 V, and a scan rate of 1 mV/s was employed. Figure 3.2d shows the current as a function of voltage for the CNPC electrolyte at different cycle numbers, while the inset shows the equivalent first cycle results obtained using a PC-1MLiTFSI liquid electrolyte. The significant current peaks at ~ −0.2 V vs. Li^+/Li and + 0.2 V vs. Li^+/Li indicate that plating and stripping of Li ions onto/from stainless steel [42]. As previously reported [43], a passivation layer is formed on the Li metal surface at 4.1 V vs. Li^+/Li, indicated by the current peak, which disappears in subsequent cycles. The CNPC electrolyte also shows a high-stability window, with cathodic stability of at least 5 V vs. Li^+/Li, which is superior to what is observed for a PC-LiTFSI liquid electrolyte.

3.3.3 Analyzing Stability of Lithium Electrodeposition Using Cross-Linked Membranes

Figure 3.3a, d report results from a so-called strip-plate test in which symmetric Li|Li cells are charged and discharged sequentially for a period of 3 h. The voltage profiles for liquid electrolytes are shown in Supplementary Fig. 3.6. It is clear that cells containing the CNPC electrolyte exhibit remarkably high cycling stability in comparison with standard cells with the same liquid electrolyte infused in the CNPC membrane. At a current density of 0.20 mA/cm^2, the cells show stable voltage profiles for more than 500 h. In comparison, a control symmetric cell with the same electrolyte and without the cross-linked membranes fails within 60 h (see Supplementary Fig. 3.6). At a higher current density of 1 mA/cm^2, the cells run for more than 120 h, whereas the voltage profile for the control cell is already unstable from the first cycle of the test. Figure 3.3 reports the morphology of the lithium anode surface using scanning electron microscopy (SEM) after 100 h of cycling for both types of cells discussed above, in comparison the SEM image of an uncycled (pristine) lithium is shown in Supplementary Fig. 3.7. At a current density of 0.20 mA/cm^2, the anode surface is smooth with only sporadic patches of rough deposits (Fig. 3.3b) for cells based on the CNPC electrolyte. In contrast, clear sharp dendrites are formed on the surface of the anode cycled in the control liquid electrolyte (Fig. 3.3c). For cells cycled at the higher current density of 1 mA/cm^2, the SEM images show that after 100 h, a dendritic structure is visibly present on the anode surface cycled in both the CNPC and control liquid electrolytes; however, the control cells display comparatively sharper and much larger dendrites.

In an LMB, electrolytes are also prone to degradation as a result of parasitic chemical and electrochemical reactions with the reactive lithium metal anode. Uneven electrodeposition exacerbates this failure mode by creating fresh surface for additional reactions. The coulombic efficiency is an important parameter that allows these effects to be quantified and tracked from cycle to cycle; it can therefore be considered a surrogate measure of the stability of the solid electrolyte interface (SEI) on the lithium metal anode. Several recent studies have considered methods for improving the stability of the SEI layer using protective films on Li anodes [44, 45]. Procedures ranging from deployment of sacrificial, surface-reactive additives in liquid electrolytes [7, 9, 46–48], to use of electrolytes saturated with salt [49] have been reported to be effective in stabilizing lithium metal.

We performed cycling studies in which a fixed amount of Li is transferred from a lithium metal electrode onto a stainless steel substrate and quantify the recovery in the reverse process when Li is stripped from the substrate to determine the coulombic efficiency. Understanding that surface protection of lithium is required for meaningful coulombic efficiencies in LMBs, the control measurements were performed using liquid electrolytes comprised of PC containing 1 wt% $LiNO_3$ and 2 wt% vinylene carbonate (VC) as an additive, which are reported to form a stable SEI on Li [48]. In contrast to the symmetric cell studies discussed in the previous section where reasonable cycling was possible in liquid electrolytes in cells using O-ring style separators, we discovered that the poor adhesion between deposited

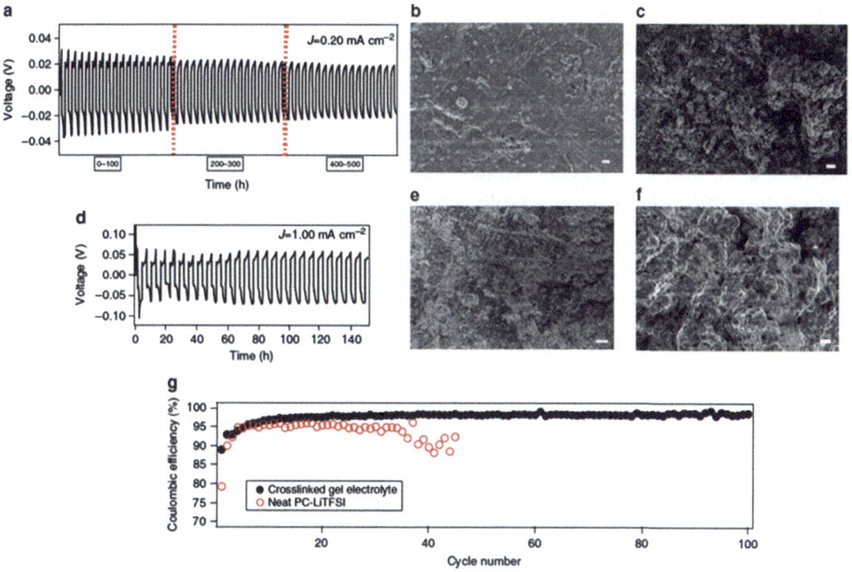

Fig. 3.3 Inhibiting dendrites and enabling smooth electrodeposition: (**a**), plate-strip cycles for a symmetric cell comprising of cross-linked gel electrolyte are shown at a current density of 0.20 mA/cm^2; it shows stable performance for over 500 h. (**b**), The morphology of lithium metal anode after 100 h of cycling of with the cross-linked gel electrolyte is shown, where very smooth surface is observed. (**c**), Similarly, the morphology of lithium metal anode neat electrolyte is shown, where dendritic structures of lithium are seen to have formed. (**d**), Voltage vs. time plots at 1.00 mA/cm^2 for cross-linked gel-based cells are shown; the cells show no sign of short circuit for at least 120 h. (**e**), The postmortem analysis shows scatters of lithium structures. (**f**), The lithium surface after 100 h of cycling shows dense dendrites that are capable of shorting the cells. All white scale bars measure 20 microns. (**g**), Electrochemical tests with Li| electrolyte| stainless steel configuration at a current density of 0.25 mA/cm^2 comparing coulombic efficiency as a function of cycle numbers for pristine and cross-linked gel electrolytes with 2%(vol.) V.C. and 1%(wt.) $LiNO_3$ additive

lithium and the stainless steel counter electrode caused lithium to separate and become electrically disconnected from the stainless steel substrate, resulting in poor columbic efficiencies. Replacing the O-rings with a conventional Celgard separator eliminated this problem.

The coulombic efficiency obtained in this manner is plotted in Fig. 3.3g as a function of cycle number at a current density of 0.25 mA/cm^2 for the control liquid electrolyte and for CNPC electrolytes (CNPC membranes soaked in the control liquid electrolyte). A typical voltage-capacity curve measured at the 40th cycle of these measurements is provided in the Supplementary Fig. 3.8. PC is known to form a very unstable SEI layer, yielding a low coulombic efficiency (~75%) [50]; however, it is evident that the additives improve the efficiency to over 90% both in liquid PC and in the CNPC electrolytes. Nevertheless, it is seen from Fig. 3.3g that the neat electrolyte fails to sustain high coulombic efficiency beyond 45 cycles, while the CNPC electrolytes maintain high coulombic efficiency for at least 100 cycles. This can be rationalized by our earlier observation of smoother electrodeposition in the latter electrolytes, which helps in preventing the rupture and reformation of SEI layer in successive charge and discharge cycles.

Polarization experiments at a fixed current density provide a more aggressive protocol for evaluating stability of lithium electrodeposition. Figure 3.4a compares the cell lifetimes obtained using strip-plate and polarization measurements. A typical voltage vs. time curve is shown in the Supplementary Fig. 3.9. It is again apparent that the CNPC electrolytes enable cells that deliver over two orders of magnitude improvement in lifetime. The short circuit times (T_{SC}) deduced for both the polarization test and strip-plate cycling studies can be fitted with a power-law function of the form $T_{SC} = A.J^{-K}$, where J is the current density and A is a transport coefficient. The power-law exponents for the strip-plate and polarization measurements are, respectively, 1.84 and 1.35, consistent with previous reports [14, 43].

Figure 3.4a compares the short circuit time from our plate-strip and polarization experiments with results from a wide variety of literature studies. For example, Rosso et al. [51] studied the stability of lithium electrodeposition in high molecular weight poly(ethylene oxide) containing lithium salt at 90 °C. The figure shows that at equivalent current densities, the cross-linked nanocomposite electrolyte-based cells register substantially higher short circuit times in comparison with those obtained in the polymer electrolytes studied by Rosso et al. Figure 3.4a also reports results from the work of Liu et al. [39, 40, 52] who performed electrochemical experiments in visualization cells in which surface-modified silica-PEO-based nanocomposite electrolytes, solid polymer electrolytes comprised of PEOxLiTFSI repeating units, and combinations of the two were used to stabilize deposition in lithium metal batteries at elevated temperature; from Sannier et al. [53] who performed symmetric cell studies using gel polymer membranes based on PEO and PVdF-HFP polymers; and from Khurana et al. [14] using cross-linked PEO-PE-PEO polymers. It is important to note that all of these experiments were performed at elevated operating temperature between 60 °C and 90 °C to overcome the poor room-temperature ionic conductivity of the electrolyte materials. It is apparent from the figure that the CNPC electrolytes deliver good or better performance than all of these other studies with the important difference that the experiments the CNPC electrolytes were all performed at room temperature! Finally, Fig. 3.4a reports results from simple liquid electrolytes (here termed neat electrolytes) and other state-of-the-art nanocomposite and polymer-based electrolytes reported previously to be effective inhibitors of dendrite growth in LMBs operated at room temperature. Comparison of these results with those obtained in the present study also shows that cells that utilize CNPC electrolytes exhibit superior ability to stabilize electrodeposition of Li than any of the other room-temperature electrolyte/separator materials [11, 54]. On the basis of these results, we therefore conclude that the CNPC electrolytes reported in the present study are promising candidates towards the goal of dendrite-free room-temperature LMBs.

To further characterize the galvanostatic performance of our electrolyte materials, LMBs were constructed in which metallic lithium was paired with lithium titanate oxide (LTO) and lithium iron phosphate (LFP) as the cathode. For these studies, CNPC membranes were soaked in the commonly used EC: DEC (1v: 1v) electrolyte solvent mixture containing 1 M $LiPF_6$ salt. Again the CNPC electrolytes exhibited

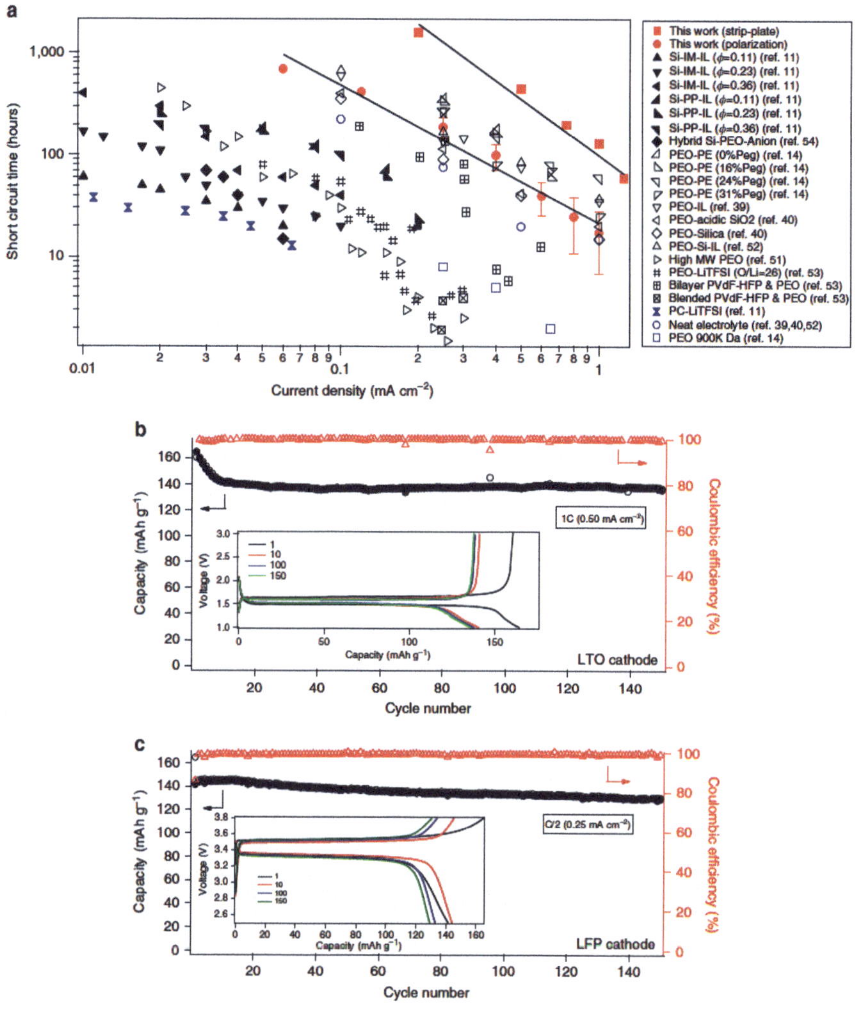

Fig. 3.4 High short circuit time and good cyclic stability: (**a**), short circuit time of cross-linked gel electrolytes compared with other state-of-the-art battery performance. Red squares and red circles indicate the Tsc for strip-plate test and polarization test in this work, respectively. The black filled symbols represent polarization tests done at room temperature, while the open symbols represent elevated temperature experiments. Black closed triangles represent silica tethered with imidazolium (Si-IM-IL) and piperidinium ionic liquid (Si-PP-IL) at various volume fractions of silica, indicated in parenthesis [11]. Black closed diamonds indicate anion-tethered hybrid silica nanocomposites [54]. The high temperature data include cross-linked PE-PEO solid polymer with different plasticizer contents given in parenthesis [14]. Other data points are PVdF-HFP/PEO composite [53], high molecular weight polymer [51], silica-polymer composite [40], polymer with ionic liquid [39], as well as their combination [52]. The blue symbols indicate neat/pristine electrolyte systems. (**b**), Cycling performance for LiI cross-linked gell LTO is shown at 1C (0.50 mA/cm^2). The inset shows the voltage profiles of the same. (**c**), Cyclic performance for LiI cross-linked gell LFP at C-rate of C/2 (0.25 mA/cm^2) is shown, with inset showing voltage profiles. In the last two figures, closed black symbols indicate discharge capacity, and open black indicate charge capacity. The red triangles denote coulombic efficiency

mS/cm-level room-temperature ionic conductivities, allowing the batteries to be operated at room temperature. For practical applications, it is important for a battery to have high capacity for large number of cycles, even at high current densities. Figure 3.4b reports the cycling performance of Li|CNPC+EC:DEC1MLiPF_6|LTO cells at a current density of 0.5 mA/cm^2. The batteries retain high capacity for at least 150 cycles. Similar capacity retention is observed for the LTO based cell using the CNPC electolytes, when cycled at higher current density of 1mA/cm^2 (as seen in Supplementary Fig. 3.10). Figure 3.4c and Supplementary Fig. 3.11 report the performance of battery where we paired metallic lithium with a LFP cathode in the CNPC electrolyte. At a current density of 0.25 mA/cm^2, the battery retains a capacity of over 120mAh/cm^2 upto 150 cycles. The inset of the figures shows the voltage profiles of the respective cells exhibit clear plateaus and low IR losses. These results clearly show that the CNPC electrolytes have good compatibility with lithium metal and with conventional high-performance intercalating cathodes and anodes. This makes them promising candidate materials for LMBs in which a metallic lithium anode is paired with conventional intercalating cathodes.

3.4 Discussion

Our findings provide a novel pathway towards nanostructured membranes in which chemistry introduced on the surface of nanoparticles can be internalized in the pores for regulating ion and mass transport. Here we illustrate the approach using hairy nanoparticles that can be cross-linked with rigid polymers to create ion-conducting membranes with good mechanical properties. Because each particle is functionalized with up to 55 reactive groups, this allows one to achieve highly cross-linked materials at very low particle contents. This makes it possible to create electrolyte membranes with both high mechanical modulus ($G_N = 1$ MPa) and high, liquid-like ionic conductivity ($\sigma_0 = 5$mS/cm) at ambient temperature, where traditional high-modulus electrolyte systems fail to maintain high conductivity. The cross-linked nanoparticle-polymer composite (CNPC) electrolytes are shown to be exceptionally effective in promoting smooth, dendrite-free electrodeposition of lithium metal at intermediate current densities. Comparison of the lithium metal anode lifetimes achievable in the new materials with those reported for other polymer, block copolymer, cross-linked polymers, and polymer-nanoparticle composites shows that CNPC systems are the most promising for room-temperature LMB systems. Further, it is shown that these materials work efficiently in LMBs based on low-voltage LTO and intermediate voltage LFP cathodes, where they can be reversibly cycled with high discharge capacity for over 150 cycles, at high current densities.

Acknowledgments This work was supported by the National Science Foundation, Award No. DMR–1006323 and by Award No. KUS–C1018–02, made by King Abdullah University of Science and Technology (KAUST). Small angle X-ray scattering facilities available through the Cornell High Energy Synchrotron Source (CHESS) were used in the study. CHESS is supported by the NSF and NIH/NIGMS via NSF award DMR-1332208.

Appendix: Supplementary Information

Supplementary Figures

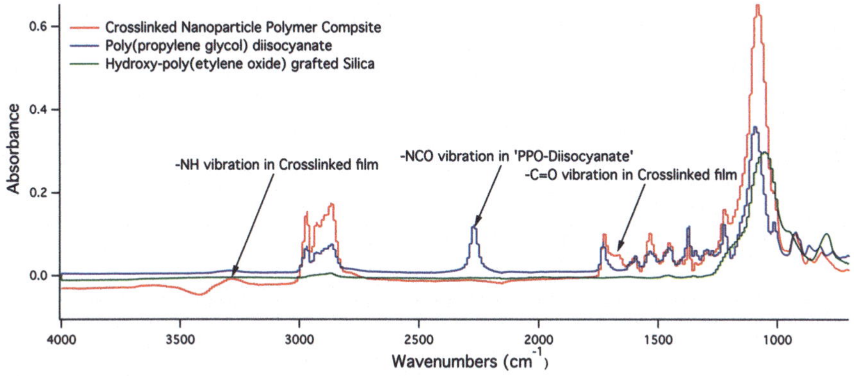

Supplementary Fig. 3.1 Fourier transform infrared spectroscopy (FTIR) characterization of the reaction scheme: the reaction process is confirmed by the IR-peak transition involved in urethane reaction. The –NCO peak of the PPO diisocyanate disappears, while characteristic peaks of –NH vibration and shift in the –C=O peak appear because of the urethane bond formation in the final product

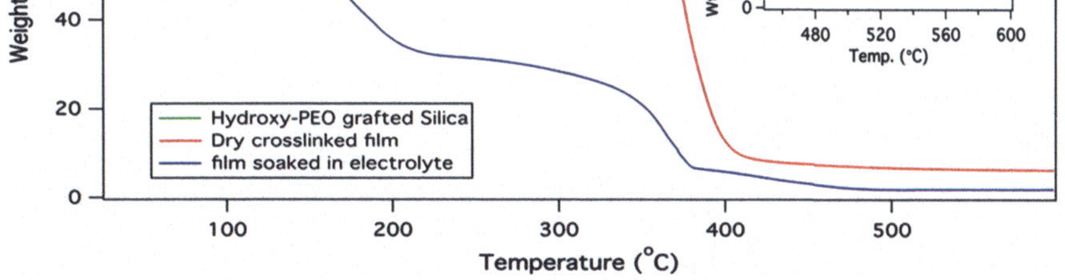

Supplementary Fig. 3.2 TGA analysis: the initial silica content in the hairy nanoparticle (Si-PEO) was 83%, which reduces to 6% silica in the final product of cross-linked film. Thus, the nonconducting entity in the entire material is remarkably low, in spite of having decent mechanical strength. Further soaking the CNPC in 1 M electrolytes comprising of PC-LiTFSI reduces the silica content close to 2%, as shown in the inset of the figure. Also, it is evident that the film is remarkably stable, such that there is degradation of the product up to 120 °C, where PC starts to degrade

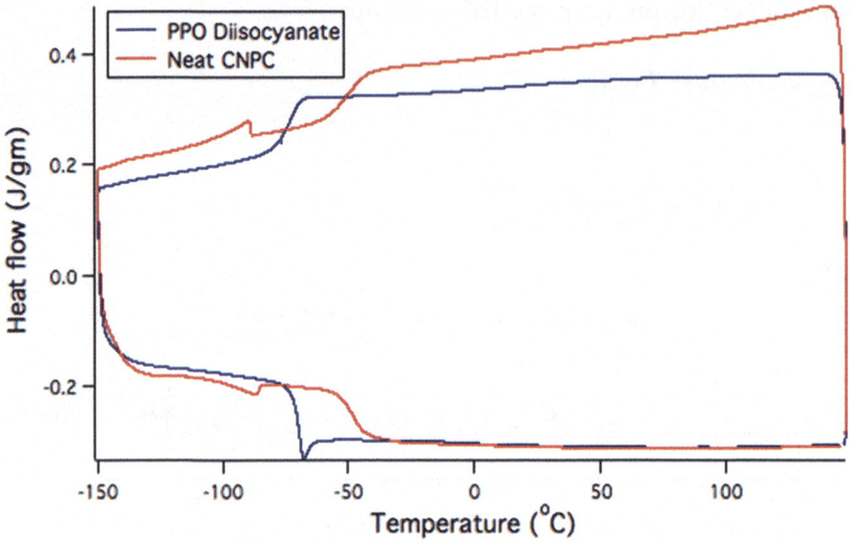

Supplementary Fig. 3.3 DSC characterization: the glass transition temperature of the PPO polymer increases from −63 °C to −42 °C due to the reduction of free volume as the chains are constricted due to the cross-linking. However, the cross-linked film is still in amorphous state at room temperature where it is used as electrolyte in battery systems. The amorphous nature of the polymer membrane enables higher ionic conductivity compared to other high MW PEO based electrolyte at room temperature

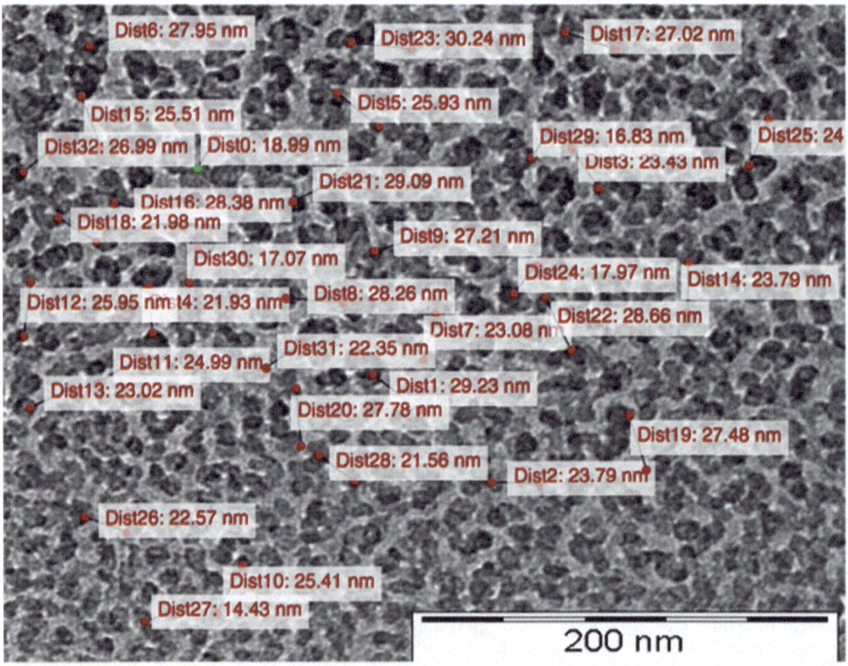

Supplementary Fig. 3.4 TEM analysis: the interparticle spacing in the cross-linked polymer is estimated by graphical analysis of TEM micrograph

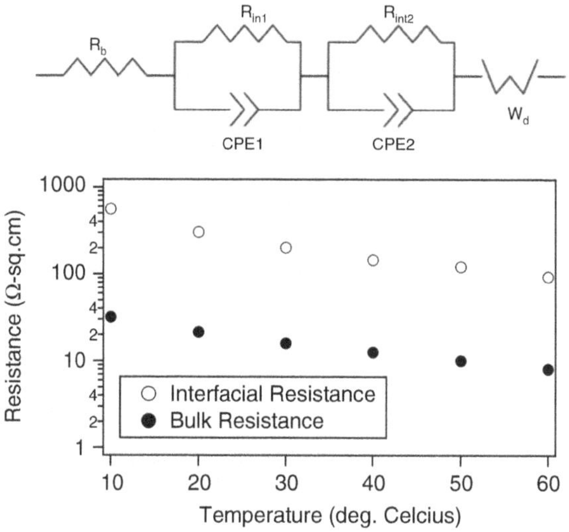

Supplementary Fig. 3.5 Equivalent electric circuit for the impedance spectroscopy results: the Nyquist plot obtained by the measurement of the impedance at a wide range of frequency can be fitted by an equivalent circuit shown above. The bulk resistance and interfacial resistance obtained are plotted against temperature; and it is seen that the interfacial resistance is always higher than the bulk resistance, which indicates interface limited ion transfer

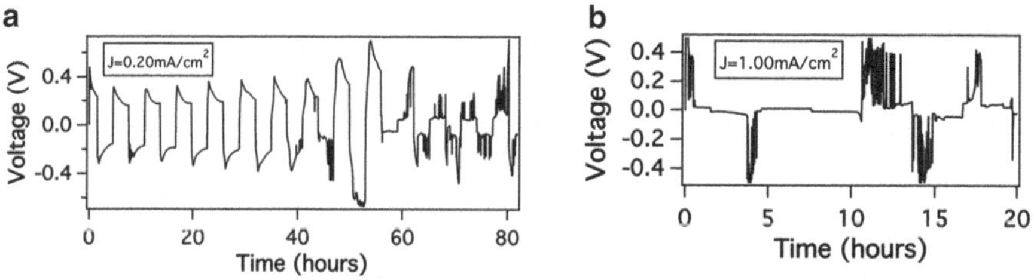

Supplementary Fig. 3.6 Strip-plate measurement of a symmetric lithium cell without CNPC separator: the voltage profile of a control cell (Li/PC + LiTFSI/Li) is plotted against time. (**a**), It is seen that at current density of 0.20 mA/cm^2 about 55 h of charging-discharging, the voltage profile gets distorted and the voltage range drops down which is a signature of dendrite-induced short circuit. (**b**), At current density of 1.00 mA/cm^2, the voltage profile is unstable even at the start of cycle

Supplementary Fig. 3.7 SEM image of pristine lithium: surface of lithium is presented in order to compare the changes in surface morphology after cycling using neat and cross-linked gel-based electrolytes. (Scale bar is 10 microns)

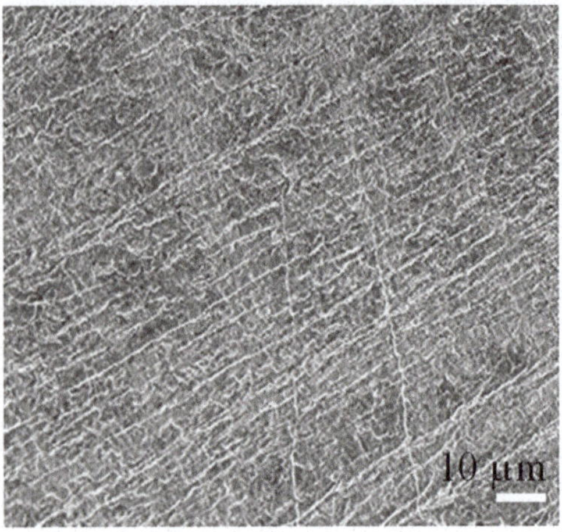

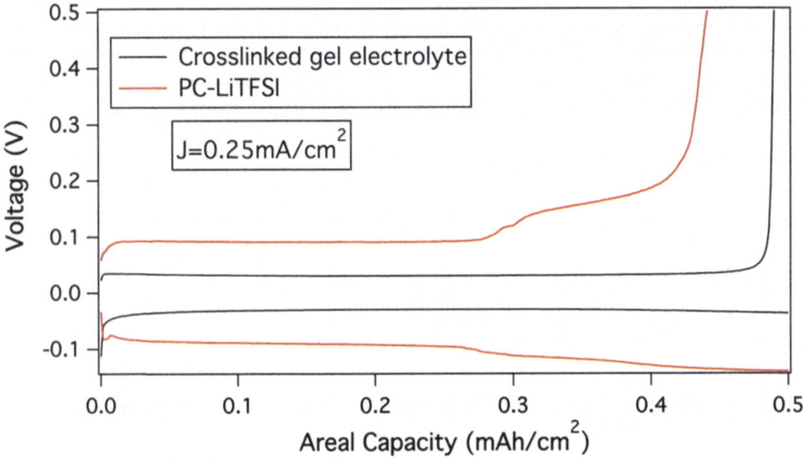

Supplementary Fig. 3.8 Coulombic efficiency test using Li| electrolyte| stainless steel configuration showing 40th cycle: batteries with cross-linked gel electrolyte and pristine PC-LiTFSI were cycled at 0.25 mA/cm^2. It is seen that at the 40th cycle, the neat electrolyte exhibits deposition at a much lower voltage and also low coulombic efficiency compared to the cross-linked gel electrolyte

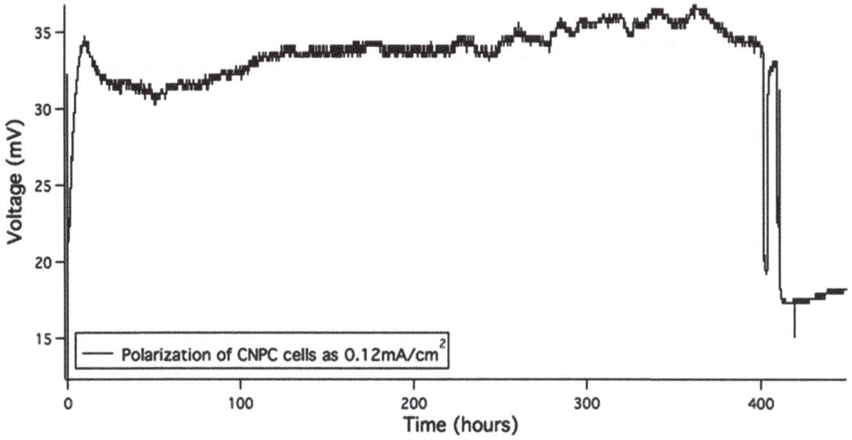

Supplementary Fig. 3.9 Polarization curve of symmetric lithium cell with cross-linked gel electrolyte: a symmetric lithium cell consisting of cross-linked gel electrolyte is charged constantly at a current density of 0.12 mA/cm^2. It is seen that the cell successfully depositing Li ion onto the anode surface for about 400 h before failing is indicated by drop in the voltage profile

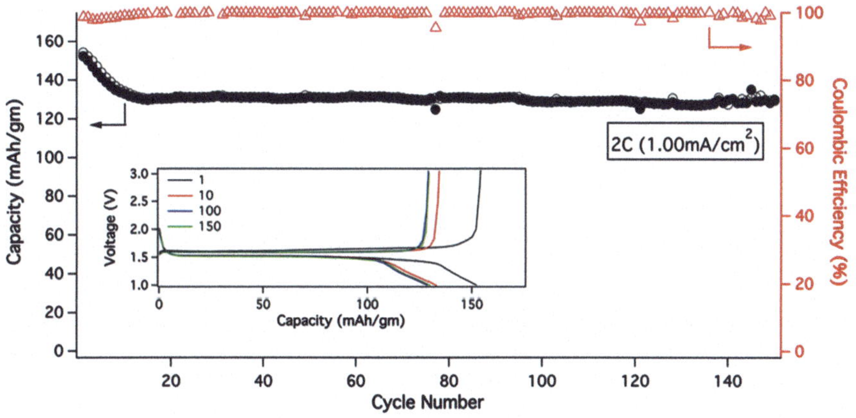

Supplementary Fig. 3.10 Cycling performance using LTO cathode: it is seen that using the cross-linked gel electrolyte, a LTO-based battery cycles well for over 150 cycles at a high current density of 1 mA/cm^2

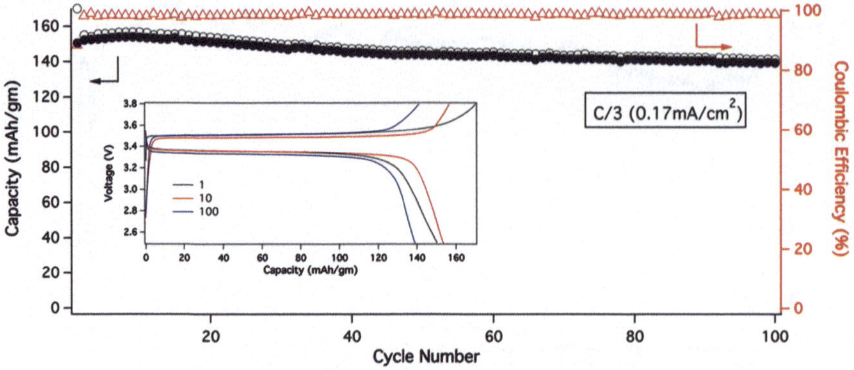

Supplementary Fig. 3.11 Cycling performance of $LiFePO_4$ battery using cross-linked gel electrolyte: it is seen that at a C-rate of C/3, the battery cycles with minimum fade for at least 100 cycles

Supplementary Tables

Supplementary Table 3.1 Content of different components at successive synthesis stage: the weight percent of the silica, PEO, PPO, and electrolyte is given in the table. It is seen that ultimately in the cross-linked gel, electrolyte contains as low as 2% silica, still having a relatively high mechanical modulus

Type of product	Si-PEO	Si-PEO-PPO	Si-PEO-PPO-electrolyte
Component	Wt. %	Wt. %	Wt. %
Silica	83	6	2
PEO	17	1.3	0.4
PPO	–	92.7	30.9
Electrolyte	–	–	66.7

Supplementary Table 3.2 VFT parameters of the different electrolyte configuration: $\sigma = A\exp(-B/(T-T_O))$, where A is the pre-exponential factor corresponding to conductivity at infinite temperature, B is the activation energy, and T_O is the reference temperature

Parameters	A (S/cm)	B (K)	T_O (K)
LiPF6-EC/DEC	0.045	56.9	237
LiTFSI-PC	0.044	61	239
Cross-linked film-LiPF6-EC-DEC	0.239	870.5	84.3
Cross-linked film-LiTFSI-PC	0.237	891.5	85

Supplementary Methods

The size of clusters was obtained by fitting the SAXS data with Beaucage unified Eq. (3.1) as shown below.

$$I(q) = A\exp\left(-\frac{q^2R_p^2}{3}\right) + B\,\mathrm{erf}\left(\frac{\left(\frac{qR_p}{\sqrt{6}}\right)^3}{q}\right)^{p_1} + \sum_i C_i \exp\left(-\frac{q^2R_p^2}{3}\right)\mathrm{erf}\left(\frac{\left(\frac{qR_{c,i}}{\sqrt{6}}\right)^3}{q}\right)^{p_{2,i}} \tag{3.1}$$

First two terms (Guinier and power-law) contribute to the scattering for spheres in a dilute suspension with radius $a = \sqrt{\frac{5}{3}}R_p \sim$ 4–5 nm and power-law exponent $p_1(\sim 4)$. A and B are the Guinier and Porod scaling factors. The last term contributes to the scattering from the fractal objects in the low q regime with $R_{\mathrm{fractal}} = \sqrt{\frac{5}{3}}R_c \sim$ 51–72 nm with a power exponent $p_2 \sim 2$, indicating the fractals to be mass fractals. Since the low q regime has only the power-law scattering, no Guinier term has been included suggesting that the R_{fractal} obtained from the fitting will be the lower bound as exact dimension cannot be determined. C is the prefactor for the power-law scattering in the low q. Absence of any additional structure contribution in intermediate to high q suggests that the particles are reasonably far apart.

Supplementary Reference

Beaucage, G.: Small-angle scattering from polymeric mass fractals of arbitrary mass fractal dimension. J. Appl. Crystallogr. **29**(2), 134–146 (1996)

References

1. Armand, M., Tarascon, J.-M.: Building better batteries. Nature. **451**, 652–657 (2008)
2. Dresselhaus, M.S., Thomas, I.L.: Alternative energy technologies. Nature. **414**, 332–337 (2001)
3. Dunn, B., Kamath, H., Tarascon, J.-M.: Electrical energy storage for the grid: a battery of choices. Science. **334**, 928–935 (2011)
4. Tarascon, J.M., Armand, M.: Issues and challenges facing rechargeable lithium batteries. Nature. **414**, 359–367 (2001)
5. Kim, H., Jeong, G., Kim, Y.U., Kim, J.H., Parke, C.M., Sohn, H.J.: Metallic anodes for next generation secondary batteries. Chem. Soc. Rev. **42**, 9011–9034 (2013)
6. Xu, W., et al.: Lithium metal anodes for rechargeable batteries. Energy Environ. Sci. **7**, 513 (2014)
7. Pieczonka, N.P.W., et al.: Impact of lithium bis(oxalate)borate electrolyte additive on the performance of high-voltage spinel/graphite Li-ion batteries. J. Phys. Chem. C. **117**, 22603–22612 (2013)
8. Li, B., Xu, M., Li, T., Li, W., Hu, S.: Prop-1-ene-1,3-sultone as SEI formation additive in propylene carbonate-based electrolyte for lithium ion batteries. Electrochem. Commun. **17**, 92–95 (2012)

9. Aurbach, D., et al.: On the use of vinylene carbonate (VC) as an additive to electrolyte solutions for Li-ion batteries. Electrochim. Acta. **47**, 1423–1439 (2002)
10. Lu, Y., Das, S.K., Moganty, S.S., Archer, L.A.: Ionic liquid-nanoparticle hybrid electrolytes and their application in secondary lithium-metal batteries. Adv. Mater. **24**, 4430–4435 (2012)
11. Lu, Y., Korf, K., Kambe, Y., Tu, Z., Archer, L.A.: Ionic-liquid-nanoparticle hybrid electrolytes: applications in lithium metal batteries. Angew. Chem. **126**, 498–502 (2014)
12. Bouchet, R., et al.: Efficient electrolytes for lithium-metal batteries. Nat. Mater. **12**, 452–457 (2013)
13. Gurevitch, I., et al.: Nanocomposites of titanium dioxide and polystyrene-poly(ethylene oxide) block copolymer as solid-state electrolytes for lithium metal batteries. J. Electrochem. Soc. **160**, A1611–A1617 (2013)
14. Khurana, R., Schaefer, J.L., Archer, L.A., Coates, G.W.: Suppression of lithium dendrite growth using cross-linked polyethylene/poly(ethylene oxide) electrolytes: a new approach for practical lithium-metal polymer batteries. J. Am. Chem. Soc. **136**, 7395–7402 (2014)
15. Tu, Z., Kambe, Y., Lu, Y., Archer, L.A.: Nanoporous polymer-ceramic composite electrolytes for lithium metal batteries. Adv. Energy Mater. **4**, n/a–n/a (2014)
16. Tikekar, M.D., Archer, L.A., Koch, D.L.: Stability analysis of electrodeposition across a structured electrolyte with immobilized anions. J. Electrochem. Soc. **161**, A847–A855 (2014)
17. Fuller, J., Breda, A.C., Carlin, R.T.: Ionic liquid-polymer gel electrolytes. J. Electrochem. Soc. **144**, 8–11 (1997)
18. Zhang, J., Sun, B., Huang, X., Chen, S., Wang, G.: Honeycomb-like porous gel polymer electrolyte membrane for lithium ion batteries with enhanced safety. Sci. Rep. **4**, 6007 (2014)
19. Stephan, A.M.: Review on gel polymer electrolytes for lithium batteries. Eur. Polym. J. **42**, 21–42 (2006)
20. Cheng, X.-B., Peng, H.-J., Huang, J.-Q., Wei, F., Zhang, Q.: Dendrite-free nanostructured anode: entrapment of lithium in a 3D fibrous matrix for ultra-stable lithium-sulfur batteries. Small. **10**, 4257–4263 (2014)
21. Hallinan, D.T., Mullin, S.A., Stone, G.M., Balsara, N.P.: Lithium metal stability in batteries with block copolymer electrolytes. J. Electrochem. Soc. **160**, A464–A470 (2013)
22. Hallinan, D.T., Balsara, N.P.: Polymer electrolytes. Annu. Rev. Mater. Res. **43**, 503–525 (2013)
23. Singh, M., et al.: Effect of molecular weight on the mechanical and electrical properties of block copolymer electrolytes. Macromolecules. **40**, 4578–4585 (2007)
24. Croce, F., Appetecchi, G.B., Persi, L., Scrosati, B.: Nanocomposite polymer electrolytes for lithium batteries. Nature. **394**, 456–458 (1998)
25. Tang, C., Hackenberg, K., Fu, Q., Ajayan, P.M., Ardebili, H.: High ion conducting polymer nanocomposite electrolytes using hybrid nanofillers. Nano Lett. **12**, 1152–1156 (2012)
26. Bertasi, F., et al.: Single-ion-conducting nanocomposite polymer electrolytes for lithium batteries based on lithiated-fluorinated-iron oxide and poly(ethylene glycol) 400. Electrochim. Acta. **175**, 113 (2015)
27. Agrawal, A., Choudhury, S., Archer, L.A.: A highly conductive, non-flammable polymer-nanoparticle hybrid electrolyte. RSC Adv. **5**, 20800 (2015)
28. Croce, F., Sacchetti, S., Scrosati, B.: Advanced, lithium batteries based on high-performance composite polymer electrolytes. J. Power Sources. **162**, 685–689 (2006)
29. Balazs, A.C., Emrick, T., Russell, T.P.: Nanoparticle polymer composites: where two small worlds meet. Science. **314**, 1107–1110 (2006)
30. Krishnamoorti, R.: Strategies for dispersing nanoparticles in polymers. MRS Bull. **32**, 341 (2007)
31. Smith, G.D., Bedrov, D.: Dispersing nanoparticles in a polymer matrix: are long, dense polymer tethers really necessary? Langmuir. **25**, 11239–11243 (2009)
32. Chandran, S., Begam, N., Padmanabhan, V., Basu, J.K.: Confinement enhances dispersion in nanoparticle-polymer blend films. Nat. Commun. **5**, 3697 (2014)

33. Patra, T.K., Singh, J.K.: Polymer directed aggregation and dispersion of anisotropic nanoparticles. Soft Matter. **10**, 1823–1830 (2014)
34. Srivastava, S., Agarwal, P., Archer, L.A.: Tethered nanoparticle-polymer composites: phase stability and curvature. Langmuir. **28**, 6276–6281 (2012)
35. Litschauer, M., Peterlik, H., Neouze, M.-A.: Nanoparticles/ionic linkers of different lengths: short-range order evidenced by small-angle X-ray scattering. J. Phys. Chem. C. **113**, 6547–6552 (2009)
36. Litschauer, M., Puchberger, M., Peterlik, H., Neouze, M.-A.: Anion metathesis in ionic silica nanoparticle networks. J. Mater. Chem. **20**, 1269 (2010)
37. Moganty, S.S., et al.: Ionic liquid-tethered nanoparticle suspensions: a novel class of ionogels. Chem. Mater. **24**, 1386–1392 (2012)
38. Stone, G.M., et al.: Resolution of the modulus versus adhesion dilemma in solid polymer electrolytes for rechargeable lithium metal batteries. J. Electrochem. Soc. **159**, A222–A227 (2012)
39. Liu, S., et al.: Lithium dendrite formation in Li/poly(ethylene oxide)–lithium Bis(trifluoromethanesulfonyl)imide and N-methyl-N-propylpiperidinium Bis(trifluoromethanesulfonyl)imide/Li cells. J. Electrochem. Soc. **157**, A1092 (2010)
40. Liu, S., et al.: Effect of nano-silica filler in polymer electrolyte on Li dendrite formation in Li/poly(ethylene oxide)–Li(CF3SO2)2N/Li. J. Power Sources. **195**, 6847–6853 (2010)
41. Schaefer, J.L., Moganty, S.S., Yanga, D.A., Archer, L.A.: Nanoporous hybrid electrolytes. J. Mater. Chem. **21**, 10094 (2011)
42. Georén, P., Lindbergh, G.: On the use of voltammetric methods to determine electrochemical stability limits for lithium battery electrolytes. J. Power Sources. **124**, 213–220 (2003)
43. Lu, Y., Tu, Z., Archer, L.A.: Stable lithium electrodeposition in liquid and nanoporous solid electrolytes. Nat. Mater. **13**, 961–969 (2014)
44. Zheng, G., et al.: Interconnected hollow carbon nanospheres for stable lithium metal anodes. Nat. Nanotechnol. **9**, 618–623 (2014)
45. Luo, W., et al.: A thermally conductive separator for stable Li metal anodes. Nano Lett. **15**(9), 6149–6154 (2015)
46. Li, W., et al.: The synergetic effect of lithium polysulfide and lithium nitrate to prevent lithium dendrite growth. Nat. Commun. **6**, 7436 (2015)
47. Pires, J., et al.: Role of propane sultone as additive to improve the performance of lithium-rich cathode material at high potential. RSC Adv. **5**, 42088–42094 (2015)
48. Guo, J., Wen, Z., Wu, M., Jin, J., Liu, Y.: Vinylene carbonate–LiNO3: a hybrid additive in carbonic ester electrolytes for SEI modification on Li metal anode. Electrochem. Commun. **51**, 59–63 (2015)
49. Qian, J., et al.: High rate and stable cycling of lithium metal anode. Nat. Commun. **6**, 6362 (2015)
50. Ding, F., et al.: Dendrite-free lithium deposition via self-healing electrostatic shield mechanism. J. Am. Chem. Soc. **135**, 4450–4456 (2013)
51. Rosso, M., Gobron, T., Brissot, C., Chazalviel, J.-N., Lascaud, S.: Onset of dendritic growth in lithium/polymer cells. J. Power Sources. **97-98**, 804–806 (2001)
52. Liu, S., et al.: Effect of co-doping nano-silica filler and N-methyl-N-propylpiperidinium bis(trifluoromethanesulfonyl)imide into polymer electrolyte on Li dendrite formation in Li/poly(ethylene oxide)-Li(CF3SO2)2N/Li. J. Power Sources. **196**, 7681–7686 (2011)
53. Sannier, L., Bouchet, R., Rosso, M., Tarascon, J.M.: Evaluation of GPE performances in lithium metal battery technology by means of simple polarization tests. J. Power Sources. **158**, 564–570 (2006)
54. Schaefer, J.L., Yanga, D.A., Archer, L.A.: High lithium transference number electrolytes via creation of 3-dimensional, charged, nanoporous networks from dense functionalized nanoparticle composites. Chem. Mater. **25**, 834–839 (2013)

Chapter 4
Confining Electrodeposition of Metals in Structured Electrolytes

4.1 Significance

Electrochemical cells based on lithium and sodium metal anodes are considered among the most versatile platforms for high-energy electrical energy storage. Unfortunately, unstable dendritic metal deposition at currents below the diffusion limit leads to premature cell failure by internal short circuits. In a volatile electrolyte, the ohmic heat generated by these shorts poses significant safety risks. The manuscript reports that electrodeposition of metals can be stabilized by confining ion transport to length scales below a few hundred nanometers. It is also shown that dendrite growth can be arrested in electrolytes with mechanical moduli well below that of the metal. This finding contradicts current orthodoxy, which holds that solid-state electrolytes with moduli higher than the metal are required for preventing dendrite growth.

4.2 Introduction

Rechargeable electrochemical cells based on alkali metal anodes provide opportunities to substantially push the energy storage limits of batteries. Such cells achieve this feat both by increasing the amount of electrical energy that can be stored on a mass or volume basis at the anode and by enabling more flexible material choices for the cathode [1–6]. Parasitic reactions between the chemically reactive anode [7–11] and proliferation of rough, dendritic/mossy electrodeposition at the metal anode during battery recharge have been reported to reduce stability of the cells by increasing the likelihood of failure by internal short circuits and by increasing the surface area of reactive metals in contact with electrolytes [12, 13].

S. Choudhury, *Rational Design of Nanostructured Polymer Electrolytes and Solid–Liquid Interphases for Lithium Batteries*, Springer Theses,
https://doi.org/10.1007/978-3-030-28943-0_4

The stability analysis of electrodeposition reported by Monroe and Newman [14, 15] (MN) is to our knowledge the first to employ continuum modeling of ion transport processes and mechanics in bulk electrolytes to predict morphological transitions at a metal anode. This work showed that marginal stability is achieved at any current density in an electrolyte with mechanical modulus about twice that of the metal electrode. One study using block copolymer electrolytes based on hard polystyrene and ion conductive poly(ethylene oxide) segments to systematically adjust the electrolyte modulus provided early support for this prediction [16–19]. This finding, however, stands in contrast to a large and growing body of recent experimental reports which show that rough deposition of metals can be inhibited in gel-like electrolytes [20, 21], cross-linked polymer electrolytes [18, 22, 23], and ion-conducting polymer electrolytes [24–26], all with mechanical moduli three or more orders of magnitude below that of the metal electrode. Recently, Zhao et al. [27] showed that a composite polymer electrolyte with immobilized ions and ceramic fillers can indeed inhibit dendrites at ambient conditions. Even more dramatic are reports which show that dendrite-induced short circuits in lithium and sodium batteries can be delayed or even eliminated at low and moderate current densities in liquid electrolytes in which a fixed fraction of anions is maintained at the interface [28–33] or in liquid electrolytes in which halide ions are present at the electrode-electrolyte interface [11, 34–37]. Essentially nothing is known about the working mechanisms by which such low-modulus materials prevent metal dendrite proliferation at either low or high current densities, relative to the diffusion limiting current $J*$, and a variety of qualitative arguments have been proposed to explain their working mechanism.

The present study is motivated by recent theoretical work by Tikekar et al. [7, 38], which extended the MN analysis to consider the effect of local stress-transport coupling with ion migration profiles and the impact such coupling has on stability of metal electrodeposition. This analysis shows that at current densities below $J*$, stable electrodeposition of metals can be achieved under a broad range of conditions, including in electrolytes with shear moduli as low as 0.1 MPa (i.e., more than four orders of magnitude lower than that of the metal electrode). Another conclusion from the analysis is that electrolytes created by hosting simple liquids in porous materials should be able to prevent rough electrodeposition of any metal when the average pore size of the host falls below a certain critical electrolyte-dependent value. This prediction is important because it means that solid-state electrolytes are not required to stabilize electrodeposition at metal anodes in batteries; it is also testable. Here, we design structured electrolytes based on cross-linked nanoparticle-polymer hybrid membranes in which the effective pore size can be facilely manipulated by changing the volume fraction of particles in the precursor material. We also use these electrolytes to systematically investigate the effect of electrolyte network structure and mechanical properties on the stability of metal deposition at currents close to the critical current density, $J*$. The study takes advantage of an optical visualization technique that allows time-dependent changes in morphology of the metal-electrolyte interface to be directly imaged. Used in combination with the analysis of Tikekar et al. [38], these efforts are shown to lead to a comprehensive understanding of how and why structured electrolytes that do not meet the MN modulus criteria are able to prevent metal dendrite proliferation.

4.3 Materials and Methods

4.3.1 Materials

Lithium rods, ethylene carbonate, dimethylene carbonate, lithium hexafluorophosphate, anhydrous copper chloride, dimethyl sulfoxide, glass fibers Ludox SM30 colloidal silica ($d = 10 \pm 2$ nm), and poly(propylene glycol)-toluene-2,4-diisocyante were all purchased from Sigma-Aldrich. Hydroxy-terminated poly (ethylene oxide)-silane was obtained from Gelest. Copper foil was obtained from Alfa Aesar. All the chemicals were used as received in after rigorous drying in a ~0 ppm water level and <5 ppm oxygen glove box.

4.3.2 Linear Stability Analysis

The details regarding the linear stability analysis are presented in a previous paper by Tikekar et al. [38]

4.3.3 Cross-Linked Hairy Nanoparticles Synthesis

The reaction procedure for cross-linking between the grafted silica nanoparticles and functionalized PPO polymer is given in a previous work [18]. Different cross-linked nanoparticles with varying pore sizes were synthesized by adjusting the ratio between the silica nanoparticles and the PPO polymer as given in Supplementary Table 4.1. The ratio between the liquid electrolyte and the cross-linked composite was kept constant at 2:1 by weight.

4.3.4 Dielectric Spectroscopy

The dielectric spectroscopy measurements were done using a Novocontrol N40 Broadband Dielectric instrument. The cross-linked hairy nanoparticles as well as neat PPO (without liquid electrolyte) were casted at the center of a Teflon O-ring. The measurements were done in a frequency range from 10^{-3} to 10^{7} Hz at various temperatures.

4.3.5 Transmission Electron Microscopy

The TEM measurements were performed to understand the nanostructure of the cross-linked hairy nanoparticles using a FEI T12 Spirit TEM. A thin layer of the materials was casted onto the TEM grid using chloroform as the solvent, and it was allowed to cross-link in the TEM grid. Before the measurements, the samples were dried overnight at 60 °C.

4.3.6 Scanning Electron Microscopy

SEM analysis was done using the LEO155FESEM instrument. In this experiment, symmetric lithium coin cell was constructed using the cross-linked electrolyte (r.c.p. = 20 nm). "Plate-strip" measurements were performed at a current density of 0.1 mA/cm^2 for 100 h, with each cycle comprising of 3 h; thereafter the lithium electrode was extracted and dried in the antechamber of glove box overnight.

4.3.7 Mechanical Properties

The cross-linking process was studied mechanically by casting the precursor materials in the rheometer with parallel-plate geometry of diameter 25 mm using a ARES rheometer. Specifically, for time sweep measurements, a constant strain amplitude of 5% and fixed frequency of 10 Hz was applied until the storage modulus reached a fixed value. The frequency sweep measurements were performed at a fixed strain amplitude of 5% between 1 Hz and 100 Hz for all materials. All the measurements were done without any addition of liquid electrolytes and at the same temperature value of 60 °C.

4.3.8 Direct Visualization Experiments

The visualization experiment was carried out for understanding the electrodeposition process with various forms of electrolytes mentioned in the manuscript. In all the measurements, the ratio between the weight of the composite material and liquid electrolyte (1 M $LiPF_6$ EC/DMC) was kept to be constant at 1:2. The details of the visualization cell for lithium deposition measurements are presented in a previous paper [10]. The dendrite growth was done using a semiautomatic analysis in MATLAB, where the sizes of the focused dendrites are measured over time.

4.4 Results

Figure 4.1a shows the synthetic route used to create cross-linked hairy nanoparticle-based electrolytes. Briefly, silica nanoparticles (diameter ~10 nm) are first grafted in aqueous solution with an alkoxy silane-terminated oligomeric polyethylene glycol (PEO-OH, M_w = 500 g/mol). As shown in our previous work [18, 39, 40], the approach produces SiO_2 nanoparticles densely grafted ($\sum \approx 1$ chain/nm^2; from thermogravimetric analysis (TGA)) with PEO chains bearing a reactive terminal hydroxyl group. The resultant SiO_2–PEO-OH nanoparticles are subsequently used as multifunctional node points for cross-linking poly(propylene oxide) (PPO, M_w = 2000 g/mol) chains functionalized with isocyanate groups at both ends. A straightforward calculation shows that on average, there are ~300 PEO-OH molecules anchored to a single SiO_2 nanoparticle, meaning that materials with a high degree of cross-linking are achieved. Additionally, because the PEO chains are short and the PPO linkers are amorphous at ambient temperatures, the cross-linked materials exist as freestanding membranes with high flexibility and toughness for silica content of 6.5% (by weight) (see Fig. 4.1b). The large number of grafting sites available on each SiO_2 nanoparticle makes it possible to introduce different functionalities, including ionic species [41] or flame retardants such as phosphates [42] in the materials. Cross-linking these nanostructures provides a facile route to nanoporous membranes in which the surface chemistry of the pores can be manipulated for regulating the ion transport characteristics or for preventing thermal runway in lithium batteries. Furthermore, by varying the volume fraction of particles, it is possible

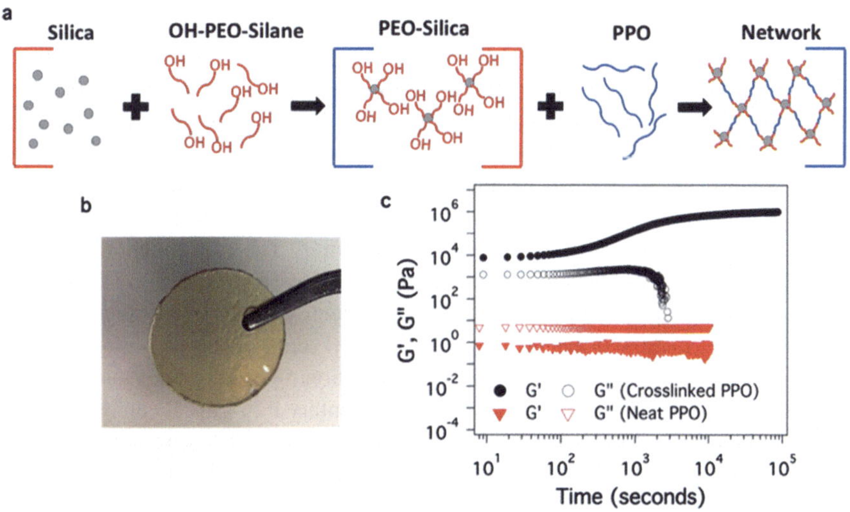

Fig. 4.1 Synthesis cross-linked hairy nanoparticles: (**a**) Schematic showing synthesis procedure of the cross-linked nanoparticles; (**b**) photograph of the freestanding membrane with 6.8% (by weight) silica content; (**c**) time sweep measurements under oscillatory shear at strain 1% and frequency of 10 Hz at 60 °C

to systematically change the configuration of the particle-tethered PEO and PPO chains in the cross-linked materials to create membranes with vastly different pore sizes and mechanical properties.

It is possible to monitor the progress of the cross-linking reaction by measuring the time-dependent development of elasticity in the materials by means of dynamic shear rheological measurements (Fig. 4.1c). Specifically, we employed oscillatory shear to continuously measure the storage and loss modulus of the initial slurry of PEO-grafted nanoparticles and PPO-diisocyanate chains at 60 °C using a small-amplitude shear strain of 1% and a fixed oscillatory shear frequency of 10 Hz. It is observed that the storage modulus increases over time indicative of the cross-linking process, which reaches its maximum value in about 24 h. It is important to note that although the chemical reaction between an isocyanate and hydroxyl group is very fast, the overall cross-linking reaction kinetics are under diffusion control as it becomes progressively harder for PPO chains tethered to one particle to bond to a neighboring site on another particle to increase network elasticity. It can also be seen from Fig. 4.1c that the loss modulus decreases and rolls over to very low values at long times, consistent with the gradual loss of dissipation processes as the network becomes more fully cross-linked. These effects are most dramatically demonstrated by comparison of the storage and loss moduli for the unlinked PPO polymer. It is seen that whereas the loss modulus dominates storage for the unlinked, liquid-like material, the storage modulus dominates as cross-linking nears completion. It is remarkable that at just 6.8 wt% of SiO_2 in the materials, there is increment of over five orders of magnitude in modulus.

More in-depth knowledge of the configuration and dynamics of the linker polymer chains is possible from dielectric relaxation measurements. It is known that polymers such as poly(propylene oxide) and cis-1,4-polyisoprene possess type-A dipole that facilitates alignment of the polymer chain end-to-end vector in the direction of an imposed electric field; thus one can quantify the end-to-end chain relaxation from the dielectric loss spectrum [43, 44]. Additionally, from the magnitude of the loss peak, one can calculate the dielectric strength, which for flexible, amorphous chains can be used to estimate the relative degree of chain stretch [45]. Supplementary Fig. 4.1a, b shows the loss permittivity of the neat PPO and cross-linked PPO at various frequencies and temperatures. The experimental data was fitted using Havriliak-Negami (H-N) equation: $\varepsilon^*(\omega) = \varepsilon_\infty + \Delta\varepsilon/[(1 + (i\omega\tau)^\alpha)^\beta]$, where α and β are associated with broadening dielectric peaks, τ represents the H-N relaxation time, ω is the frequency, ε_∞ is the dielectric constant (ε') at high-frequency limit, and $\Delta\varepsilon$ is the dielectric strength. The relaxation times at various temperatures obtained from the H-N fits are shown in Supplementary Fig. 4.1c. It can be seen that the relaxation time for all temperatures follows Vogel-Fulcher-Tammann (VFT) model, represented by the continuous lines in the figure. The VFT model is given as: $\tau = A \exp(-B/k(T - T_0))$, where A is the prefactor, B is the activation energy, and T_o is the reference temperature, shown in Supplementary Table 4.1. It can be seen that the activation energy of the relaxation dynamics of PPO polymer increases dramatically from *9.8 kJ/mole* to *20 kJ/mole* as a result of cross-linking, which is consistent with expectations based on the molecular weight of the polymers and high levels of cross-linking achieved [18]. Comparison of the dielectric strength for the untethered and tethered PPO indicates that the linkers are also highly stretched (shown in Supplementary Fig. 4.1d). For random coil type-A dielectric polymer chains, the dielectric strength $\Delta\varepsilon$ is proportional to the mean squared end-to-end vector of the polymer chain. The results in Supplementary Fig. 4.1d show that $\Delta\varepsilon$ measured using the cross-linked membranes is approximately 250 times larger than the corresponding value for the linear unlinked polymer. Considering that the molecular weight of the PPO chains is just 2 kDa, the large increases in $\Delta\varepsilon$ also imply that dipoles on the particle tethered chains are also highly correlated.

The cross-linked hairy nanoparticles provide an opportunity to understand the underlying physics of metal electrodeposition in heterogeneous environment, because these materials serve as model nanoporous structures such that the silica nanoparticles act as non-conductive physical barriers and the polymeric chains in the interparticle gaps as pathways for ion migration. In this context, the volume fraction of silica nanoparticles in the entire composite was varied, and the corresponding interparticle distance was estimated by assuming random closed packing (r.c.p.) arrangement of spherical nanoparticles: $d_{p-p} = d\left[\left(0.63/\Phi\right)^{1/3} - 1\right]$, as reported in Supplementary Table 4.2 for different weight fraction of silica content. We utilized transmission electron microscope (TEM) to analyze the spatial arrangement of the nanoparticles in the polymer composite as shown in Supplementary Fig. 4.2a for r.c.p. pore sizes of 20 nm, 100 nm, and 500 nm. The interparticle distance of the silica nanoparticles was further measured from the TEM images, and the corresponding variation was fitted to a *normal distribution* function obtained from the respective mean and standard deviations and plotted as inset of the images. The average interparticle distances measured from TEM are observed to be similar to r.c.p. pore sizes (Fig. 4.2a). However, the variance of the distribution is seen to increase with decreasing silica volume fractions (Supplementary Fig. 4.3). We further performed frequency sweep measurements at a constant amplitude of 5% for different cross-linked PPO as well as neat PPO polymer. The storage and loss modulus obtained from the measurements are plotted in Supplementary Fig. 4.2b, c, respectively. The results show that the composite materials at all cross-linking densities have higher G' than G", indicating elastic behavior that is unseen at such low-core volume fractions in colloidal suspensions. The elastic modulus is seen to progressively increase with increasing core volume fraction; the associated cage volume (*kT/G'*) is reported in Fig. 4.2a. Membranes with r.c.p. pore size 20 nm show the highest shear modulus of approximately 1 MPa, while in the uncross-linked state, the materials are simple Newtonian liquid. The higher variance of spatial distribution of silica nanoparticles and reminiscence of elasticity at low-core fractions indicate that the particles form hierarchical stringlike mass fractals due to the fixed length of the cross-linker (PPO), as shown in Fig. 4.2b, also confirmed using small-angle X-ray scattering in our previous work [18].

Next, we systematically study the morphology of the lithium-electrolyte interface in the cross-linked hairy nanoparticles soaked with liquid electrolyte using direct visualization of the lithium. For the conditions used in the experiments, the diffusion-limited current density for the different membranes was calculated from

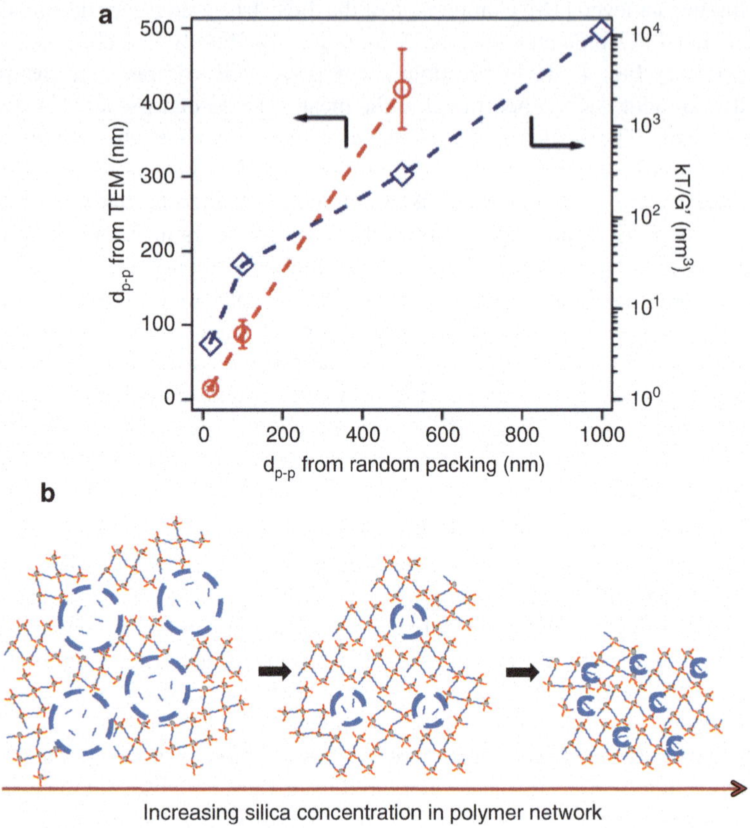

Fig. 4.2 Characterization of pore architecture in the cross-linked structure: (**a**) Comparing the pore sizes obtained from TEM as well as cage sizes obtained from frequency sweep measurements with the random closed packing pore sizes; (**b**) schematic showing the changes in pore structures with the variation of silica nanoparticle volume fractions. In all the sub-parts, the sample names are characterized by the random closed packing interparticle distance between silica nanoparticles in the composites

the measured conductivity and transference number: $J* = 2zc_oFD_{app}(t_aL)^{-1}$. Here, z is the charge number of cation, F is the Faraday constant, c_o is salt metal anode during electrodeposition in an optical cell. The cell comprises of a lithium metal and stainless steel electrodes separated by a chamber containing electrolyte of 1 M $LiPF_6$-EC/DMC [10]. concentration, t_a is the anion transference number, and L is the interelectrode spacing (1 mm) and D_{app} is the diffusion coefficient. Using the Bruce-Vincent method [33], the transference number (see Supplementary Fig. 4.4) was determined to be 0.38 for the cross-linked nanoparticles (r.c.p. = 20 nm). The variation of conductivity and critical current density ($J*$) for the cross-linked nanoparticles at different pore sizes are reported in Supplementary Fig. 4.5. It is seen that $J*$ varies in the range ~5 to 9 mA/cm^2 for the materials studied. Using the previously reported values of Li^+ transference number (0.36) and lithium ion diffu-

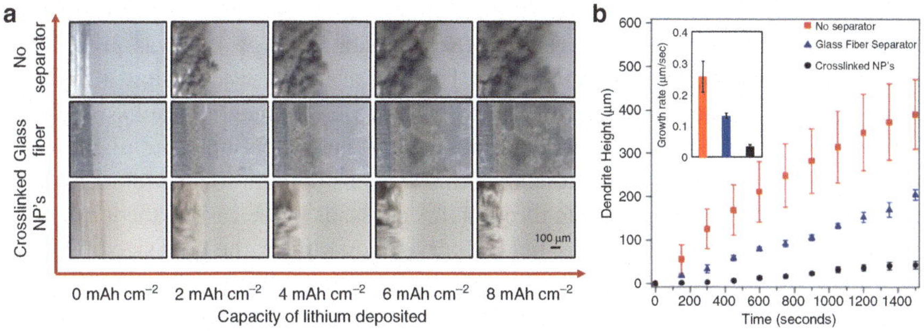

Fig. 4.3 Stability of electrodeposition by direct visualization of electrode: (**a**) Snapshots of the electrode and electrolyte in every 15 min during charging at the rate of 8 mA cm^{-2}; (**b**) height of dendrite at various points of the electrode for the initial 1500 s; the inset compares the growth rate by assuming a linear growth for electrodeposition at 8 mA cm^{-2}

sivity (3*10^{-6} cm^2/sec) for 1 M EC:DMC-LiPF$_6$ electrolyte, the diffusion-limited current density can be estimated to be 9 mA/cm^2 for the same interelectrode spacing [46, 47].

Figure 4.3a reports snapshots of the negative electrode at different stages of electrodeposition for the cases without any separator, with a commercial glass fiber separator and cross-linked hairy nanoparticles (r.c.p. pore size = 20 nm) at a constant current density of 8 mA/cm^2, which is 0.9 times the $J*$ calculated for liquid electrolyte. Through visual inspection, it is obvious that Li electrodeposition in the absence of a separator is uneven, mossy, and blackened, which is indicative of the side reactions between electrode and electrolyte as well as porous nature of the deposited layers. At first appearance, the deposits are already quite large, at least 100 μm in diameter, which is much larger than typical Li dendrite nucleate sizes assumed in the literature [6]. Electrodeposition studies based on a commercial glass fiber separator show that deposition is relatively compact until a capacity of about 2 mAh/cm^2 (i.e., approximately 10 μm/cm^2 Li) is transferred, thereafter dendritic structures develop and proliferate, ultimately piercing the separator. This sequence is associated with the short circuiting phenomena described in the Introduction section and can lead to catastrophic cell failure by fire, explosion, or both. In comparison to the cases with and without separator, the deposition pattern with the cross-linked hairy nanoparticles displays uniform deposition for entire 1 h of the charging process.

By tracking the advancing front of the Li electrode over time, it is possible to quantify the dendrite growth as a function of time (see Fig. 4.3b). The apparent growth rate obtained from the initial time through the assumption of linearity is shown in the inset. The larger-sized error bars for the neat electrolyte case are a reflection of the unevenness of electrodeposition. The apparent growth rate of dendrite in the commercial glass fiber separator is calculated to be ~0.13 μm/sec, while for the cross-linked hairy nanoparticles, it is estimated to be 0.03 μm/sec which is modestly larger than the theoretical rate (0.011 μm/sec) for perfectly smooth electrodeposition at 8 mAh/cm^2.

This idea that one could control the stability of electrodeposition by manipulating the nanostructure of an electrolyte (Supplementary Fig. 4.6) was previously discussed in the theoretical analysis by Tikekar et al. [7, 38] and has also been adapted in an ad hoc manner to explain previous experimental results [48, 49]. Deposition in membranes, with pores larger than the critical nucleate diameter at which electrodeposition is unstable, is in this analysis thought to lead to unchecked growth of dendrites that penetrate through this pores, as in the case of the glass fiber separator. In contrast, in the cross-linked hairy nanoparticles, the pore size of 20 nm is evidently much smaller than the nucleate size leading to a significant retardation of dendrite growth. To determine the effect of SEI formation and chemical instability on the topological features observed during the electrodeposition process, we performed postmortem analysis of lithium electrode after repeated charge and discharge cycling in a coin cell. Specifically, we utilized a symmetric lithium battery with the cross-linked nanoparticle electrolyte (20 nm) and operated the battery at a current of 0.1 mA/cm^2 for 100 h, with each half cycle comprising of 3 h, as shown in Supplementary Fig. 4.7. We intentionally chose low current densities for the operation to isolate the effect of chemical and not morphological instabilities. The SEM image of the extracted lithium metal is seen to be smooth and unperturbed (Supplementary Fig. 4.8). This provides evidence that the observed evolution of electrode surface in visualization experiment is dominant by dendritic growth of lithium and not due to parasitic reactions at the interface. One could make these tentative ideas more concrete by interrogating the evolution of the interface in cross-linked electrolytes in which the average pore size is systematically varied. Supplementary Fig. 4.9 shows the snapshots of deposition at same current density of 8 mA/cm^2, and Supplementary Fig. 4.10 reports the analytical results from such studies in which the morphology of the Li electrolyte interface is characterized in cross-linked nanoparticles with varying pore size (or, weight fraction of SiO_2 in the composite). The liquid electrolyte weight was maintained to be same in all cases reported in this work, which is twice as that of the composite weight. The results show that when the effective pore size of the membrane is 1000 nm, dendrite growth is unimpeded and rapid. However, at high volume fractions of silica nanoparticles corresponding to interparticle distance of order 500 nm, Supplementary Fig. 4.10 shows two regimes of dendrite growth, wherein the growth rate at the initial time is low, followed by an accelerated rate of dendrite propagation at longer time. The results in Supplementary Fig. 4.9 also show that for the membranes with 500 nm pores, dendrites with "needlelike" morphology are visible at 4 mAh/cm^2, in contrast to the mossy-blackened-deposited lithium apparent when the pores are larger and in neat electrolytes. Furthermore, it is seen that in membranes with effective pore size of around 100 nm, the deposition is similarly restricted as for the 20 nm case, implying that there is a threshold pore size between 500 nm and 100 nm where the resistance to dendrite proliferation is large enough to prevent growth.

Figure 4.4a, b summarizes the growth rate deduced from experiments as a function of average membrane pore size and elastic modulus, respectively, at 8 mA/cm^2 as well as for varying current densities such that J/J^* remains constant at 0.9 (see Supplementary Fig. 4.5). Optical snapshots of the initial deposition and the

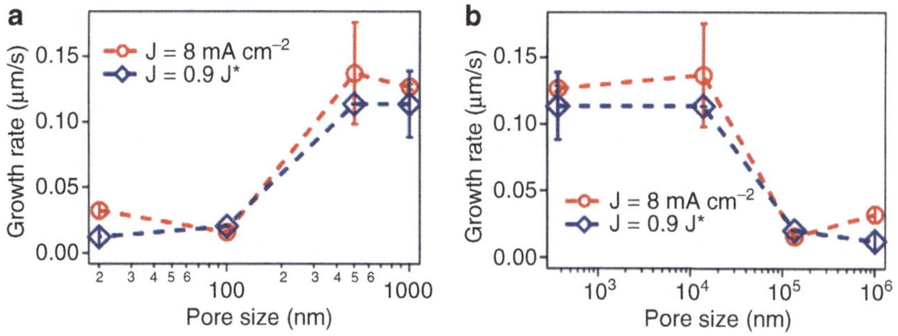

Fig. 4.4 Analyzing pore size dependence on dendrite growth: (**a**, **b**) Growth rate at different pore sizes and corresponding elastic modulus, respectively, for the cross-linked hairy nanoparticles at current density of 8 mA/cm^2 as well as for variable current densities matching 0.9 times of the limiting current densities measured for respective samples

corresponding current densities (at $J/J* = 0.9$) are reported in Supplementary Fig. 4.11. The experimental results reveal a sudden transition from stable deposition to dendrite growth between 100 to 500 nm pore size in agreement with our hypothesis that electrodeposition stability is function of electrolyte length scale. Also, the shear modulus measured from rheology doesn't have a linear relation with growth rate that can be rationalized by the fact that the cross-linked membrane is comprised of building blocks having different inherent modulus (SiO_2 and polymer). As the deposition takes place in the interparticle gap, the growing "dendrites" only experience the modulus of the liquid electrolyte and polymers; thus the length scale of the polymeric phase plays a greater role than the overall modulus of the material. We also analyze the growth rate of dendrites as a function of pore size obtained from TEM micrograph analysis as well as the pore volume calculated from plateau modulus as kT/G'. As shown in Supplementary Fig. 4.12, a similar nonlinear behavior is observed.

Figure 4.5a plots the perturbation growth rate versus wavelength deduced from the analysis of Tikekar et al. [38] It is seen that the growth rate is negative for small wavelengths and goes through a maximum at a particular wavelength that decreases as the current density approaches critical current density, $J*$. The first of these results can be shown to arise from the greater influence of surface tension in suppression growth of small dendrite nucleates, while the presence of the maximum reflects the effect of the physical (macroscopic) dimensions of the simulation cell in restricting nucleate size below certain limits. As the current density increases, the growth rate of the fastest (most unstable) mode increases, and its wavelength decreases because the interfacial tension has to compete with progressively more aggressive electroconvection to stop the roughening. The fastest growing mode can therefore be thought of as representative of the typical size of dendrites that one would observe in an experiment under similar conditions. A porous material would restrict the size of the growing dendrite, since the deposition can only happen on length scales smaller than the pore size. In Fig. 4.5b, we show the calculated growth rate for the fastest growing mode as a function of the critical wavelength. For very large pores, the pore size is predicted to have no effect on the dendrite growth, and

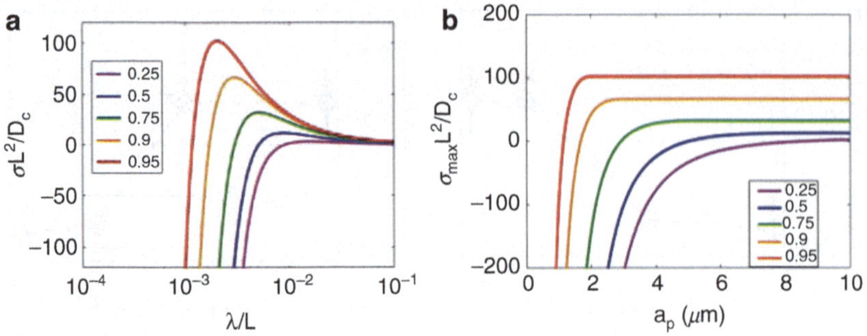

Fig. 4.5 Linear stability analysis of dendrite growth: (**a**) Normalized growth rate of the deposited nucleate at different normalized wave numbers obtained by linear stability analysis; (**b**) maximum growth at different theoretical pore sizes at various effective currents. The different lines in parts (**a**) and (**b**) represent different values of *J*/*J*∗

the growth occurs more or less at the same rate as in the absence of the porous material. For smaller pores however, the pore diameter restricts the sizes of the dendrites that can form in them, causing the growth rate of the fastest mode to plummet. Below a certain size, there can be no unstable deposition, as surface tension adequately competes with current density driven dendritic growth at all possible length scales. This critical pore size decreases with current density due to the requirement of a higher contribution from surface tension as seen in Supplementary Fig. 4.13. The inset in Fig. 4.13 plots the analogous growth rate deduced from experiments as a function of membrane pore size for a fixed current density, *J*/*J*∗ *~0.9*. The experimental results show a similar trend as calculated from the linear stability analysis, wherein there is sudden transition from stable deposition to dendrite growth. The transition point from the theoretical prediction corresponding to non-zero growth is ~1.6 μm, while the experiments indicate the crossover between 100 and 500 nm pore size as previously mentioned in Fig. 4.4a. Considering the variety of sources of error in comparing the experiments and theory, including the tortuosity of the pores in the cross-linked materials, the polydispersity of the particle size, uncertainty in the value of the Li transference number under the conditions of the experiments [47], and the deviation from linearity in calculation of initial growth rate, the agreement between the theoretical and observed transition is fair.

4.5 Conclusion

In conclusion, we utilize a cross-linked network by covalently linking polymer-grafted nanoparticles, where varying the volume fraction of the nanoparticles in the composite can conveniently change the effective pore size. It was shown that these cross-linked nanoparticles when used as gel electrolytes showed stable electrodeposition in contrast to mossy and dendritic morphology for cases without separator and with glass fibers, respectively. We performed visualization experiments for different pore sizes determined by volume fractions to demonstrate that below 500 nm, the deposition is smooth and compact, while dendritic for other cases. These experimental observations were rationalized using linear stability analysis of dendrite growth at different perturbation length scale representing pore diameter of the electrolytes. It was understood that below the limiting current density, the electrodeposition can be stabilized by implementing a separator with pore size lower than the length scale of the most unstable nucleate. Overall, this work will provide guidelines for designing solid-state electrolytes and separators for metal-based batteries.

Acknowledgments This work was supported by the National Science Foundation, Division of Materials Research, through Award No. DMR–1609125.

Appendix: Supplementary Information

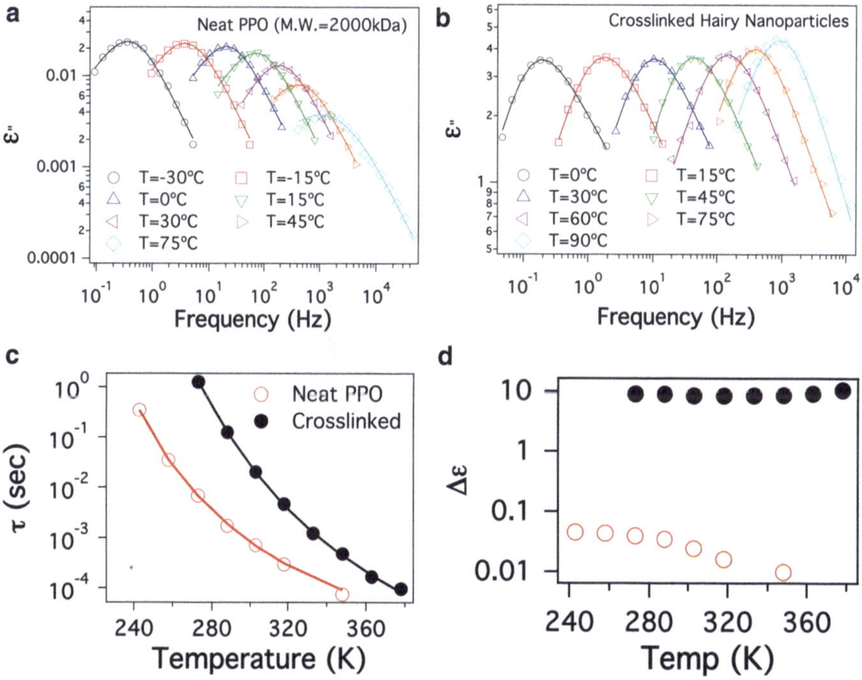

Supplementary Fig. 4.1 Dielectric relaxation of cross-linked hairy nanoparticles: (**a**, **b**) Dielectric permittivity at various temperatures fitted with the H-N model for Neat PPO and cross-linked hairy nanoparticles, respectively; (**c**) polymer relaxation time as a function of temperatures fitted with a VFT model; (**d**) comparison between the dielectric strength between the neat PPO and cross-linked PPO; the symbols are the same as in part c

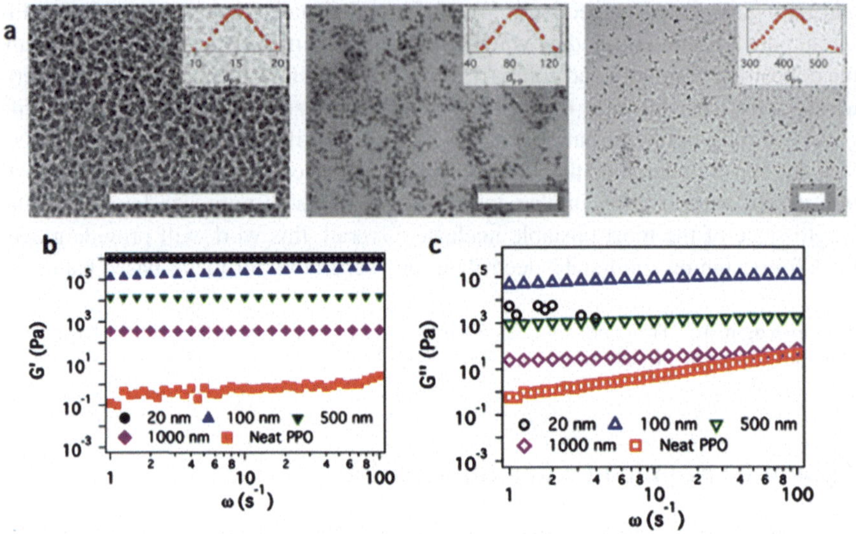

Supplementary Fig. 4.2 Characterization of pore architecture in the cross-linked structure: (**a**) TEM micrographs of cross-linked nanoparticles. The scale bar represents 200 nm. From left to right, the samples are r.c.p. pore sizes 20 nm, 100 nm, and 500 nm. (**b**) Storage modulus and (**c**) loss modulus obtained through frequency sweep measurements at strain of 5% for different cross-linked samples and neat PPO at 60 °C

Supplementary Fig. 4.3 Normal distribution of interparticle distances obtained by analysis of TEM images for cross-linked hairy nanoparticles with different random closed packing pore sizes

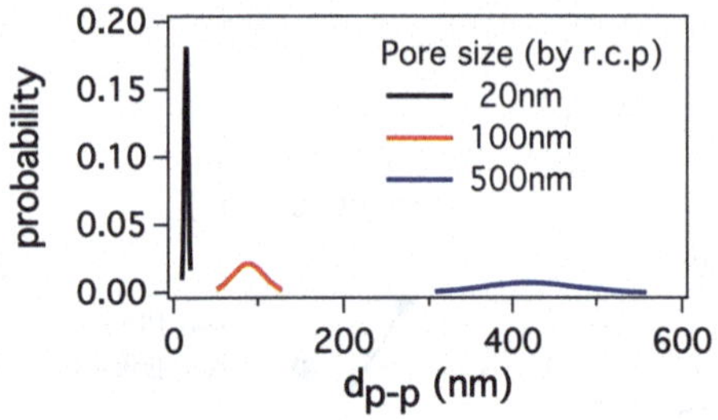

Supplementary Fig. 4.4 Polarization of a symmetric lithium cell using the cross-linked hairy nanoparticle electrolyte (r.c.p. = 20 nm) soaked with the electrolyte 1 M EC/DMC $LiPF_6$ at 20 mV. The inset shows the impedance results before and after polarization

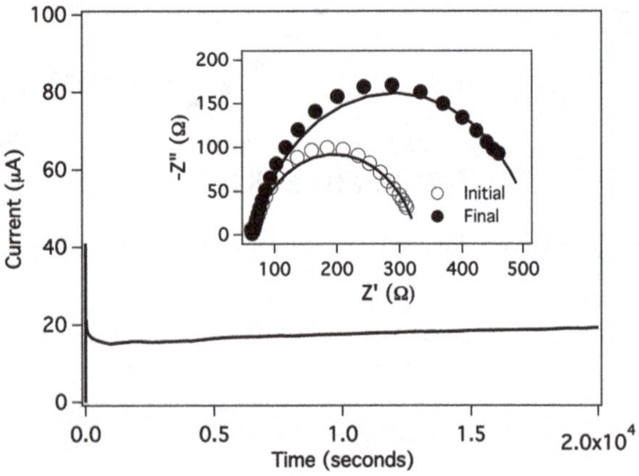

Supplementary Fig. 4.5 Room temperature conductivity and limiting current density variation with different pore-sized membranes. The arrows show the corresponding values for a neat electrolyte of 1 M EC/MC $LiPF_6$

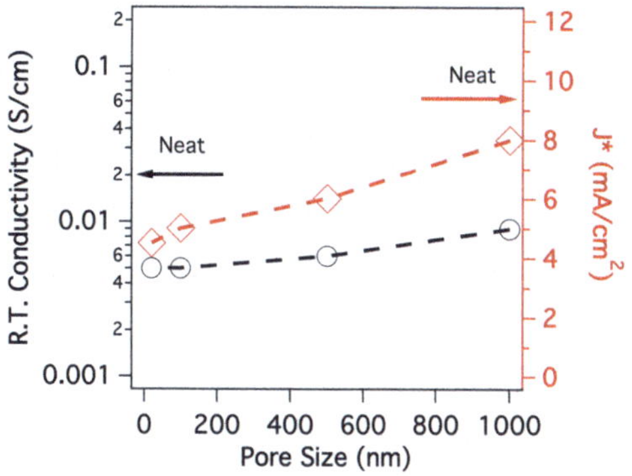

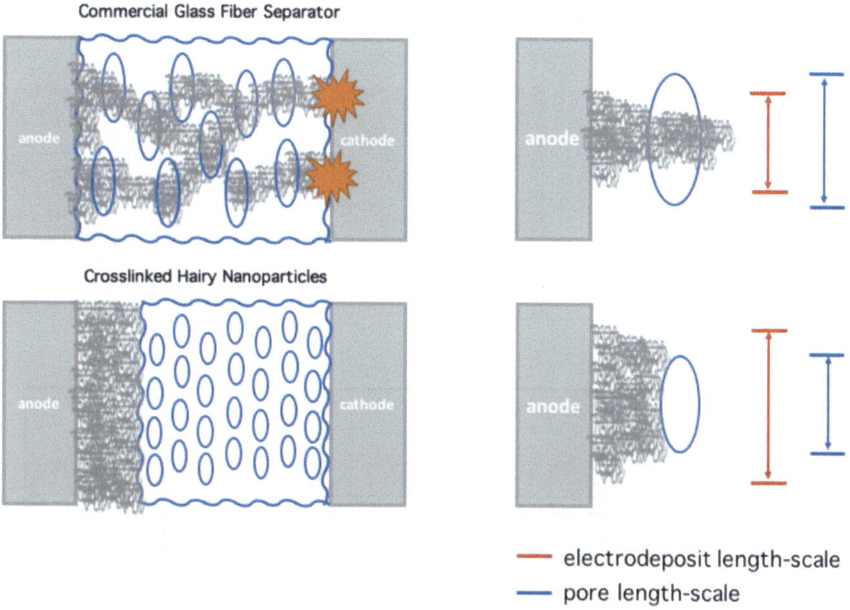

Supplementary Fig. 4.6 Schematic representing the idea that the pore size of the electrolyte/separator is important and related to the stability of electrodeposition. In this figure, the cross-linked nanoparticles have random closed packing pore size of 20 nm

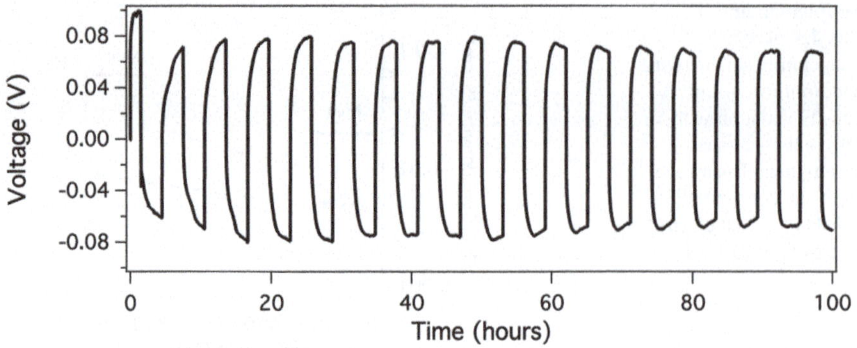

Supplementary Fig. 4.7 Charge and discharge cycles in a symmetric lithium coin cell using the cross-linked hairy nanoparticles electrolyte with pore size of 20 nm. The battery was operated at a current density of 0.1 mA/cm^2 with each half cycle being 3-h long

Supplementary Fig. 4.8 (**a**, **b**) SEM images of lithium electrode surface before and after cycling in a symmetric lithium cell for 100 h at 0.1 mA/cm^2

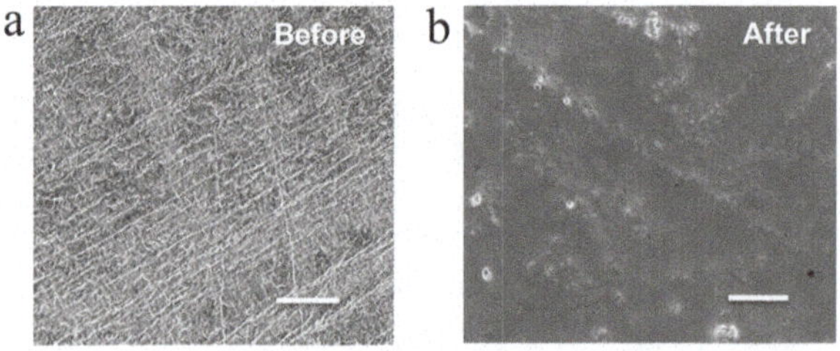

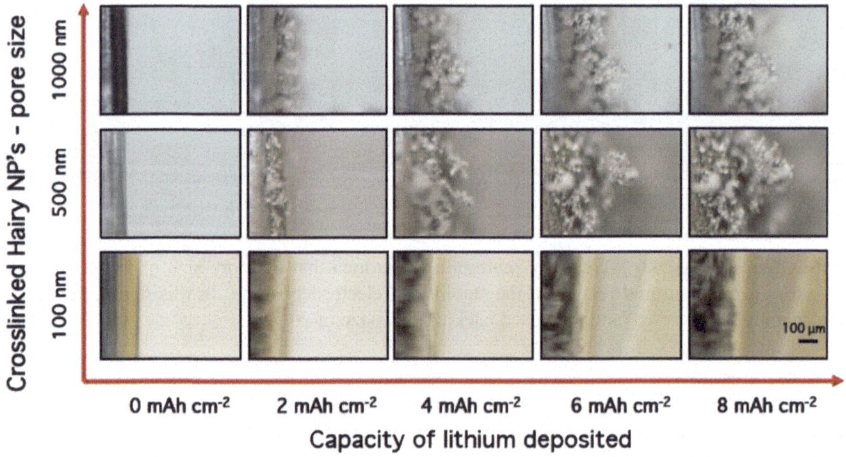

Supplementary Fig. 4.9 Electrodeposition with different pore sizes of the cross-linked nanoparticles: Snapshots of the electrode and cross-linked electrolyte with pore sizes of 1000, 500, and 100 nm in every 15 min during charging at the rate of 8 mA cm^{-2}

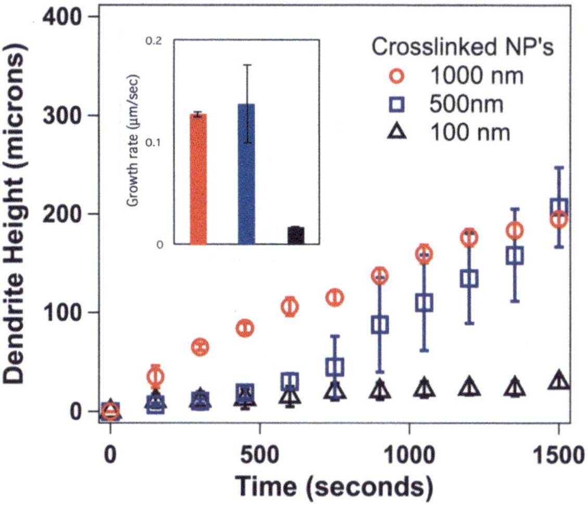

Supplementary Fig. 4.10 Height of dendrite at various points of the electrode for the initial 1500 s. The inset compares the growth rate by assuming a linear growth for the visualization experiment in Figure S9 at a current density of 8 mA cm^{-2}

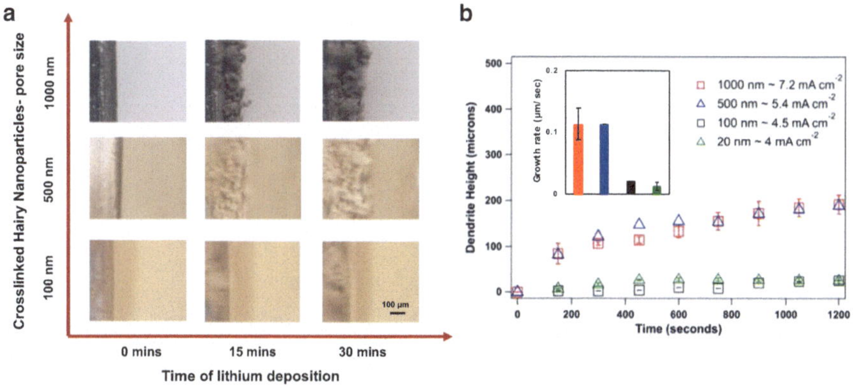

Supplementary Fig. 4.11 (**a**) Snapshots of electrodeposition with different cross-linked membrane pore sizes at variable current density such that the $J/J*$ is maintained at 0.9 for each case; (**b**) height of dendrite as a function of time for different samples. The absolute values of current densities are reported in the label that correspond to $J/J = 0.9$. The inset shows the comparison of the dendrite growth rates for the respective pore sizes reported in the main figure

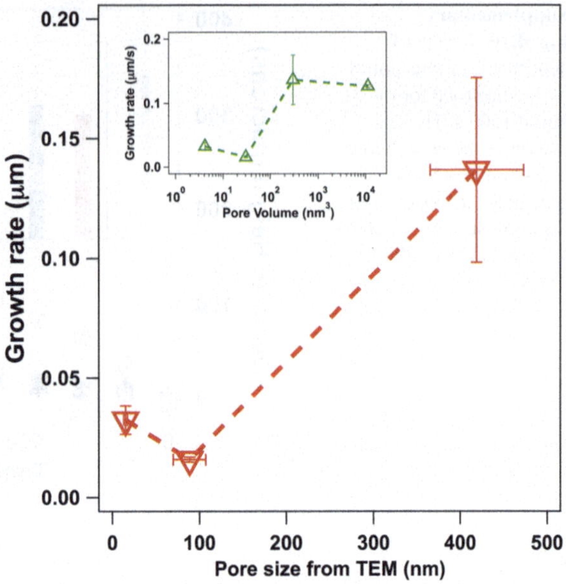

Supplementary Fig. 4.12 Pore size dependence of dendrite growth at a current density of 8 mA/cm^2, where the pore size is obtained from the TEM analysis. The inset shows the growth rate as a function of pore volume obtained from the plateau modulus using the equation (*kT/G'*)

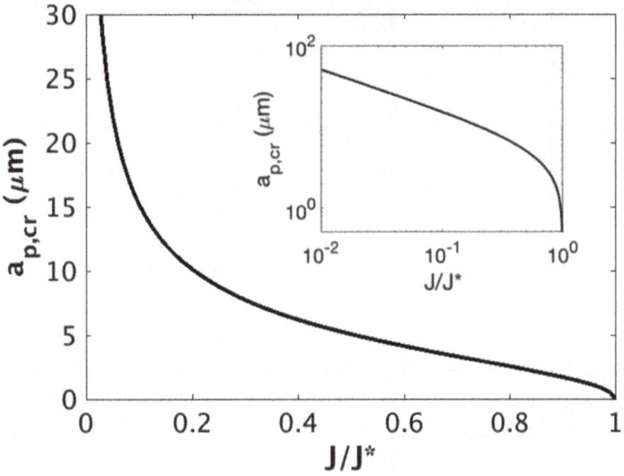

Supplementary Fig. 4.13 Critical pore size, representative of crossover from positive to zero growth rate, at various normalized current density. The inset shows the same graph in semilogarithmic scale

Supplementary Table 4.1 VFT parameters for fitting the dielectric relaxation times at different temperatures for the cross-linked and neat PPO

Sample	*A* (sec)	*B* (kj/mole)	T_0(K)
Crosslinked PPO	1.7×10^{-9}	20	155.4
Neat PPO	1.6×10^{-7}	9.8	162.0

Supplementary Table 4.2 Weight fraction of silica nanoparticles corresponding to the effective pore size obtained by the random packing fraction spherical particles

Effective Pore size [$d_{p\text{-}p}$] (nm)	Silica weight fraction (%)
20	6.3081
100	0.1333
500	0.0013

References

1. Lin, D., Liu, Y., Cui, Y.: Reviving the lithium metal anode for high-energy batteries. Nat. Nanotechnol. **12**, 194–206 (2017)
2. Cheng, X., et al.: A review of solid electrolyte interphases on lithium metal anode. Adv. Sci. **3**, 1–20 (2016)
3. Cheng, X.-B., Zhang, R., Zhao, C.-Z., Zhang, Q.: Toward safe lithium metal anode in rechargeable batteries: a review. Chem. Rev. **117**, 10403–10473 (2017)
4. Ma, L., Hendrickson, K.E., Wei, S., Archer, L.A.: Nanomaterials: science and applications in the lithium–sulfur battery. Nano Today. **10**, 315–338 (2015)
5. Zhang, W., et al.: Design principles of functional polymer separators for high-energy, metal-based batteries. Small. 1703001–n/a. https://doi.org/10.1002/smll.201703001
6. Wei, S., Choudhury, S., Tu, Z., Zhang, K., Archer, L.A.: Electrochemical interphases for high-energy storage using reactive metal anodes. Acc. Chem. Res. (2017). https://doi.org/10.1021/acs.accounts.7b00484
7. Tikekar, M.D., Archer, L.A., Koch, D.L.: Stabilizing electrodeposition in elastic solid electrolytes containing immobilized anions. Sci. Adv. **2**, e1600320 (2016)
8. Chazalviel, J.-N.: Electrochemical aspects of the generation of rampified metallic electrodeposits. Phys. Rev. A. **42**, 7355–7367 (1990)
9. Bai, P., Li, J., Brushett, F.R., Bazant, M.Z.: Transition of lithium growth mechanisms in liquid electrolytes. Energy Environ. Sci. **9**, 3221–3229 (2016)
10. Choudhury, S., et al.: Electroless formation of hybrid lithium anodes for fast interfacial ion transport. Angew. Chem. Int. Ed. **56**, 13070–13077 (2017)
11. Choudhury, S., Wei, S., Ozhabes, Y., Gunceler, D., Nath, P.: Designing solid-liquid interphases for sodium batteries. Nat. Commun. **8**, 898 (2017)
12. Wood, K.N., et al.: Dendrites and pits: untangling the complex behavior of lithium metal anodes through operando video microscopy. ACS Cent. Sci. **2**, 790–801 (2016)
13. Zheng, G., et al.: Interconnected hollow carbon nanospheres for stable lithium metal anodes. Nat. Nanotechnol. **9**, 618–623 (2014)
14. Monroe, C., Newman, J.: The effect of interfacial deformation on electrodeposition kinetics. J. Electrochem. Soc. **151**, A880 (2004)
15. Monroe, C., Newman, J.: The impact of elastic deformation on deposition kinetics at lithium/polymer interfaces. J. Electrochem. Soc. **152**, A396 (2005)
16. Stone, G.M., et al.: Resolution of the modulus versus adhesion dilemma in solid polymer electrolytes for rechargeable lithium metal batteries. J. Electrochem. Soc. **159**, A222–A227 (2012)
17. Agrawal, A., Choudhury, S., Archer, L.A.: A highly conductive, non-flammable polymer–nanoparticle hybrid electrolyte. RSC Adv. **5**, 20800–20809 (2015)
18. Choudhury, S., Mangal, R., Agrawal, A., Archer, L.A.: A highly reversible room-temperature lithium metal battery based on crosslinked hairy nanoparticles. Nat. Commun. **6**, 10101 (2015)
19. Lu, Y., Korf, K., Kambe, Y., Tu, Z., Archer, L.A.: Ionic-liquid-nanoparticle hybrid electrolytes: applications in lithium metal batteries. Angew. Chem. **126**, 498–502 (2014)

20. Zhang, J., Sun, B., Huang, X., Chen, S., Wang, G.: Honeycomb-like porous gel polymer electrolyte membrane for lithium ion batteries with enhanced safety. Sci. Rep. **4**, 6007 (2014)
21. Stephan, A.M.: Review on gel polymer electrolytes for lithium batteries. Eur. Polym. J. **42**, 21–42 (2006)
22. Wu, H., et al.: Stable Li-ion battery anodes by in-situ polymerization of conducting hydrogel to conformally coat silicon nanoparticles. Nat. Commun. **4**, 1943 (2013)
23. Khurana, R., Schaefer, J.L., Archer, L.A., Coates, G.W.: Suppression of lithium dendrite growth using cross-linked polyethylene/poly(ethylene oxide) electrolytes: a new approach for practical lithium-metal polymer batteries. J. Am. Chem. Soc. **136**, 7395–7402 (2014)
24. Porcarelli, L., Gerbaldi, C., Bella, F., Nair, J.R.: Super soft all-ethylene oxide polymer electrolyte for safe all-solid lithium batteries. Sci. Rep. **6**, 19892 (2016)
25. Long, L., Wang, S., Xiao, M., Meng, Y.: Polymer electrolytes for lithium polymer batteries. J. Mater. Chem. A. **4**, 10038–10069 (2016)
26. Gurevitch, I., et al.: Nanocomposites of titanium dioxide and polystyrene-poly(ethylene oxide) block copolymer as solid-state electrolytes for lithium metal batteries. J. Electrochem. Soc. **160**, A1611–A1617 (2013)
27. Zhao, C.-Z., et al.: An anion-immobilized composite electrolyte for dendrite-free lithium metal anodes. Proc. Natl. Acad. Sci. **114**, 11069 LP-11074 (2017)
28. Tu, Z., et al.: Designing artificial solid-electrolyte interphases for single-ion and high-efficiency transport in batteries. Joule. (2017). https://doi.org/10.1016/j.joule.2017.06.002
29. Choudhury, S., et al.: Designer interphases for the lithium-oxygen electrochemical cell. Sci. Adv. **3**, e1602809 (2017)
30. Ma, L., Nath, P., Tu, Z., Tikekar, M., Archer, L.A.: Highly conductive, sulfonated, UV-cross-linked separators for Li–S batteries. Chem. Mater. **28**, 5147–5154 (2016)
31. Lu, Y., et al.: Stable cycling of lithium metal batteries using high transference number electrolytes. Adv. Energy Mater. **5**, 1402073 (2015)
32. Oh, H., et al.: Poly(arylene ether)-based single-ion conductors for lithium-ion batteries. Chem. Mater. **28**, 188–196 (2016)
33. Bouchet, R., et al.: Single-ion BAB triblock copolymers as efficient electrolytes for lithium-metal batteries. Nat. Mater. **12**, 452–457 (2013)
34. Choudhury, S., Archer, L.A.: Lithium fluoride additives for stable cycling of lithium batteries at high current densities. Adv. Electron. Mater. 1–6 (2015). https://doi.org/10.1002/aelm.201500246
35. Lu, Y., Tu, Z., Archer, L.A.: Stable lithium electrodeposition in liquid and nanoporous solid electrolytes. Nat. Mater. **13**, 961–969 (2014)
36. Seh, Z.W., Sun, J., Sun, Y., Cui, Y.: A highly reversible room-temperature sodium metal anode. ACS Cent. Sci. **1**, 449–455 (2015)
37. Zhang, X., Cheng, X., Chen, X., Yan, C., Zhang, Q.: Fluoroethylene carbonate additives to render uniform Li deposits in lithium metal batteries. Adv. Funct. Mater. **27**, 1605989 (2017)
38. Tikekar, M.D., Archer, L.A., Koch, D.L.: Stability analysis of electrodeposition across a structured electrolyte with immobilized anions. J. Electrochem. Soc. **161**, A847–A855 (2014)
39. Choudhury, S., Agrawal, A., Kim, S.A., Archer, L.A.: Self-suspended suspensions of covalently grafted hairy nanoparticles. Langmuir. **31**, 3222–3231 (2015)
40. Choudhury, S., Agrawal, A., Wei, S., Jeng, E., Archer, L.A.: Hybrid hairy nanoparticle electrolytes stabilizing lithium metal batteries. Chem. Mater. **28**, 2147–2157 (2016)
41. Wei, S., et al.: Highly stable sodium batteries enabled by functional ionic polymer membranes. Adv. Mater. **29**, 1605512 (2017)
42. Stalin, S., Choudhury, S., Zhang, K., Archer, L.A.: Multifunctional cross-linked polymeric membranes for safe, high-performance lithium batteries. Chem. Mater. **30**, 2058–2066 (2018)
43. Agarwal, P., Kim, S.A., Archer, L.A.: Crowded, confined, and frustrated: dynamics of molecules tethered to nanoparticles. Phys. Rev. Lett. **109**, 258301 (2012)
44. Ding, Y., et al.: Dielectric spectroscopy investigation of relaxation in C 60 - polyisoprene nanocomposites. 3201–3206 (2009). https://doi.org/10.1021/ma8024333

45. Adachi, K., Kotaka, T.: Dielectric normal mode relaxation. Prog. Polym. Sci. **18**, 585–622 (1993)
46. Nyman, A., Behm, M., Lindbergh, G.: Electrochemical characterisation and modelling of the mass transport phenomena in LiPF6-EC-EMC electrolyte. Electrochim. Acta. **53**, 6356–6365 (2008)
47. Valoen, L.O., Reimers, J.N.: Transport properties of LiPF6-based Li-ion battery electrolytes. J. Electrochem. Soc. **152**, A882 (2005)
48. Tu, Z., Nath, P., Lu, Y., Tikekar, M.D., Archer, L.A.: Nanostructured electrolytes for stable lithium electrodeposition in secondary batteries. Acc. Chem. Res. **48**, 2947–2956 (2015)
49. Tu, Z., et al.: Nanoporous hybrid electrolytes for high-energy batteries based on reactive metal anodes. Adv. Energy Mater. **7**, 1602367 (2017)

Chapter 5
Lithium Fluoride Additives for Stable Cycling of Lithium Batteries at High Current Densities

5.1 Introduction

Rechargeable batteries are key components in a growing list of technologies where portability and reliability are required. The commercial success of one particular family of rechargeable batteries, the lithium-ion battery (LIB), has played a crucial role in the development of portable device technology that offers consistent performance, power, lifetime, as well as enhanced safety. In a rechargeable lithium metal battery (LMB), one replaces the inert carbon anode with metallic lithium and in so doing creates an energy storage platform with significantly improved energy density, portability, and power [1–10]. Two specific examples of such batteries, the Li-sulfur and Li-air batteries, are currently under active investigation worldwide as potential platforms for increasing range, performance, and lifetime costs in next-generation-electrified vehicle technologies, including electric cars [11]. It is understood, however, that even pairing a metallic lithium anode with any of the currently used LIB cathode materials (e.g., Li/$LiFePO_4$, Li/$LiCoO_2$, Li/LiNiCoO, etc.) provides more straightforward opportunities for engineering LMBs with energy storage and performance characteristics that are superior to today's workhorse LIB technologies.

A key barrier to progress in development of rechargeable batteries in any of these configurations is the complex electrochemistry of deposition at metallic surfaces in a liquid electrolyte. In course of successive charge-discharge processes in LMBs, uneven plating of the anode can be caused by electroconvection and other deposition instabilities leading to premature cell failure [4, 8]. In the most extreme cases, the uneven electrodeposition on the anode results in the formation and growth of dendritic structures that ultimately bridge the interelectrode space and short-circuit the cell [4, 12, 13]. However, the most common mode of cell failure occurs by loss of the active electrode material by various interrelated electrochemical and interfacial processes. For example, the ohmic energy in and fragility of a growing dendrite

S. Choudhury, *Rational Design of Nanostructured Polymer Electrolytes and Solid–Liquid Interphases for Lithium Batteries*, Springer Theses,
https://doi.org/10.1007/978-3-030-28943-0_5

can cause it to break before it spans the interelectrode space. This produces regions of electrically disconnected or dead lithium that is no longer able to exchange electrons with the electrode mass and contributes to a low coulombic efficiency and shortened lifetime of a battery. An even more challenging failure mode results from the loss of lithium metal and electrolyte due to side reactions at the roughened electrode-electrolyte interface. Indeed while such reactions always occur when liquids are in conformal contact with reactive metals, such as Li, during each cycle of deposition, the roughened metal surface exposes new Li metal to the electrolyte that leads to continuous formation of new so-called SEI layer that ultimately depletes the electrolyte and causes formation of "mossy" lithium [13]. All of these situations are exacerbated by the common use of flammable organic liquids as electrolyte solvents in rechargeable batteries to improve ionic conductivity of the electrolyte, which now add the threat of fire or even explosion to the list of potential failure modes [4, 14, 15].

Over the years, several efforts have been made to eliminate the possibility of dendrite-induced short circuits in batteries by designing high modulus electrolytes through which a growing metal dendrite cannot penetrate [16–18]. These efforts have largely met with, at most, limited success because of the fundamental difficulty in designing materials that are simultaneously mechanically strong enough to stop dendrites, but in which fast ionic transport needed to sustain battery performance can be achieved at moderate temperatures. A notable exception is the work of Tu et al. [19], which shows that a Al_2O_3 ceramic separator with uniform, nanometer-sized pores that host a liquid electrolyte in its pores is able to perform both functions. However, as none of these approaches address the root cause of the electrodeposition instabilities that trigger dendrite formation, more elegant solutions in the form of SEI additives have been sought to stop dendrites at the nucleation stage. There are generally two approaches that have been investigated in the previous literature:

1. Direct addition, wherein specific chemical agents are used as electrolyte additives to promote stable SEI formation but do not take part in the bulk ion transfer. Hydrofluoric acid [20], vinylene carbonate [21–23], lithium bis(oxalato) borate [24, 25], lithium nitrate [26], and organic sultones [27, 28]have all been reported to function in this way. While each of these additives has been shown to improve the cell stability to an extent, wider use of all is challenged by the associated decrease in ionic conductivity and gradual loss of efficacy due to decomposition [26]
2. Indirect formation of stabilizing layers. This approach involves the formation of a stable layer by internal reactions of two or more components added to the electrolyte. A recent example by Miao et al. [29] showed that a binary mixture of LiTFSI and LiFSI can significantly improve the cyclability of a LMB due to the formation of a LiF layer by degradation of these salts on the surface of the lithium metal anode. Also, recently Qian et al. [30] showed similar improvement in performance by use of LiFSI in excess concentration, which may also be rationalized as producing LiF at the electrolyte/electrode interface as seen using XPS, and predicted previously by ab initio calculations that show the tendency of LiFSI to reduce to LiF [31]

5.2 Experimental Section

5.2.1 Materials

Lithium foil was bought from Alfa Aesar. 1 M EC:DMC-$LiPF_6$, high-purity lithium fluoride were obtained from and opened as received in an argon-filled glove box. All coin cell parts and $LiFePO_4$ sheets (~2 mAh/cm^2) were bought from MTI Corporation. Polypropylene-based separator with commercial name Celgard 3501 was obtained from Celgard LLC.

5.2.2 Methods

For the preparation of modified electrolyte system, 30 ml of 1 M EC:DMC-$LiPF_6$ and 0.5% (by wt.) of LiF were added to it. This particular additive approximately equals to the 30% molar of LiF previously used as reinforced electrolyte system used by Lu et al. [12] This mixture of electrolyte additive was continuously stirred overnight to ensure proper mixing owing to the fact that LiF is partially insoluble in this electrolyte. Coin cells of different types were prepared in 2032 type configurations. For symmetric cells, both electrodes comprised of lithium foil, whereas for coulombic efficiency test, thoroughly cleaned stainless steel plate was taken as one of the electrodes. For making half-cells, $LiFePO_4$ sheets were punched and kept in vacuum for 12 h, and electrodes were weighted individually, before using as cathode. For all kinds of configurations (in exception of those made for impedance spectroscopy measurements), the batteries were allowed to rest for a period of 7 days before any electrochemical analysis.

5.2.3 Electrochemical Characterization

Impedance spectroscopy measurements were done using a Solatron frequency analyzer. The measurements were done in a symmetric cell configuration at a frequency range of 10^{-3} Hz to 10^7 Hz. All galvanostatic measurements were done using a Neware CT-3008 battery tester. Coulombic efficiency tests were done in a Li-stainless steel configuration, where the batteries were initially charged and discharged between 0.5 V and 0 V for 10 cycles at 0.01 mA/cm^2 to enable a stable SEI formation. Next, they were discharged at different current densities with a time control and then charged until the voltage was 0.5 V. The coulomic efficiency was calculated from ratio of charge to discharge capacity. Strip-plate measurements were done symmetric cells with time-controlled charge and discharge cycles. The half-cells comprising of lithium anode and $LiFePO_4$ cathode were cycled in the voltage window of 3.8–2.5 V. The SEM analysis of the electrode surface was done using a LEO155FESEM instrument after the disassembling of lithium symmetric cells, which were subjected to polarization at different current densities.

5.3 Results

Ion transfer through SEI layer has been an extensive topic of research, and recent findings from "joint density-functional theory (JDFT) studies by the Arias and co-workers underscore the importance of SEI additives such as LiF in controlling surface diffusion of metal ions during electrodeposition [32, 33]. Figure 5.1 is a schematic of the mechanism through which an enhanced Li-ion surface mobility would enable stable electrodeposition. The JDFT analysis [33], in fact, shows that the barrier energy for surface diffusion of Li ions over a surface of LiF is lower by 0.09 eV (= 3.5 kT), compared to that of lithium carbonate, which is the most abundant component in a SEI layer. This means that the rate of transport of Li ions on a LiF surface is more than 30 times larger than a $LiCO_3$ substrate. It is proposed that this enhanced surface diffusion is likely to promote smooth lithium plating in contrast to the more typical rough, dendritic electroplating. Additionally, the JDFT calculations by Ozhabes et al. [33] show that the surface energy of a LiF surface is more than three times that of $LiCO_3$ and close to eight times of LiOH. This means that a lithium metal surface with a coating of LiF is much more resistant to roughening than one with a similar coating of $LiCO_3$ or LiOH.

In a departure from the previous "sacrificial" methods and motivated by the JDFT calculations, Lu et al. [12] reported a potentially simpler approach based on partially reinforcing the liquid electrolyte with halide salt additives that can deposit on the anode surface. This method was reported to produce significant improvements in the short circuit time of LMBs cycled at moderate to low current densities. However, the study did not investigate the effect of these additives at current densities typical for electrochemical transport in batteries and also did not evaluate the effect of LiF on cell failure modes associated with interfacial reactions and degradation of the electrolyte at the lithium metal anode. Additionally, the study by Lu et al. considered electrolytes primarily based on propylene carbonate (PC) as solvent. This choice is a limitation because PC is a poor solvent for LiF and is typically not used as the electrolyte solvent in lithium batteries because of its reactivity towards carbon. Finally, in an effort to demonstrate the effectiveness of the LiF salt additives, the study by Lu et al. interrogated cells using a so-called membraneless

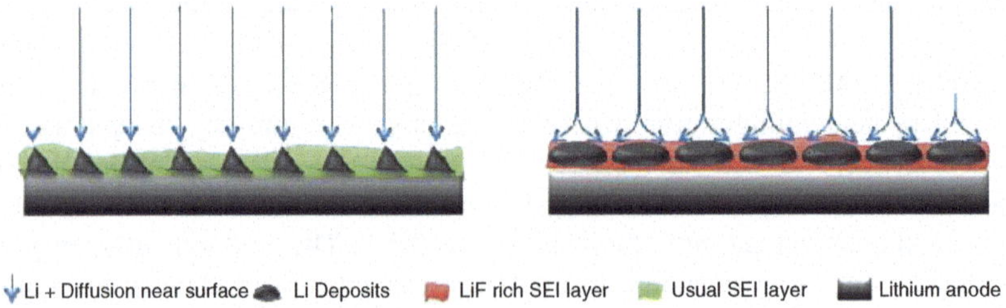

Fig. 5.1 Pictorial representation of proposed mechanism: Lithium diffusion near the surface of electrodes represented by blue arrows. LiF rich SEI is shown in red, while usual SEI is indicated by green color. Due to lower lateral diffusion barrier on LiF crystals, the Li + ions form smooth electrodeposits, while in usual SEI needlelike lithium plating is expected

configuration in which O-ring-shaped separators are employed, which is again not a widely practiced approach.

Herein we report on the deposition of metallic lithium at low and high current densities in LMBs containing conventional carbonate electrolytes, EC:DMC-$LiPF_6$, reinforced with halide additives. In an attempt to advance fundamental understanding of how and why LiF salt additives work to impede dendrite proliferation and electrochemical degradation under typical cell running conditions, and to evaluate the broader relevance of salt additives to LMBs, the present work systematically removes all of the aforementioned constraints in the study by Lu et al. In so doing, we find that electrolytes containing as little as 0.5% by wt. LiF in a one molar electrolyte solution simultaneously offer remarkable dendrite suppression abilities and enhanced electrolyte stability during cell cycling. The results reported in the present work therefore expand significantly upon current knowledge and define a path towards LMBs with lifetime and safety characteristics compatible with requirements for commercially important cells.

Various recent works have considered the role of electrolyte chemistry on coulombic efficiency of Li-metal electrodes. The electrolyte blend DOL:DME-LiTSFI has received significant attention because it is known to exhibit good stability in the presence of metallic lithium and high coulombic efficiency due to its ability to form a stable SEI layer, particularly in the presence of $LiNO_3$ [13]. In a departure from this approach, we focus instead on the carbonate electrolyte EC:DMC-$LiPF_6$, which is more commonly used in a high-voltage lithium-ion battery, owing to its high vaporization temperature, cost, and compatibility with high-voltage cathode materials. In order to monitor the stability of the SEI layer in a LMB, symmetric cells were constructed using neat and LiF reinforced electrolytes, and impedance spectroscopy measurements were performed at different time intervals to characterize the interfacial resistance in the cells. Figure 5.2a shows the Nyquist plots for the corresponding impedance spectra. The interfacial resistance obtained by fitting the plots to the equivalent circuit depicted in Supplementary Fig. 5.1 is plotted against time in Fig. 5.2b. At initial times, the interfacial resistance for batteries containing the neat (no LiF) electrolytes is over 350 Ωcm^2, consistent with the formation of a thick carbonate-rich SEI. In contrast, electrolytes containing LiF-based electrolyte demonstrated a much lower values, close to 200 Ωcm^2. Furthermore, the interfacial resistance is shown to increase with time for the neat electrolyte system, which is a clear trait of an unstable SEI formation that leads to side reaction between the bare lithium and electrolyte creating an insulating interface. On comparison, cells having LiF-based electrolytes show slight increase in the interfacial resistance in initial time; thereafter it becomes constant. LiF salt being partially insoluble in the used electrolytes forms a thin coating over the surface of lithium, thus stabilizing the SEI layer and preventing any side reaction.

The coulombic efficiency for batteries consisting of electrolytes with/without LiF additive was examined in a two-electrode setup consisting of metallic lithium and stainless steel (SS) electrodes. Since a symmetric cell has virtually infinite lithium source, the lithium reserve in the electrode can compensate for any Li loss in forming the SEI or in side reactions with the electrolyte during cell operation. Thus, it is not

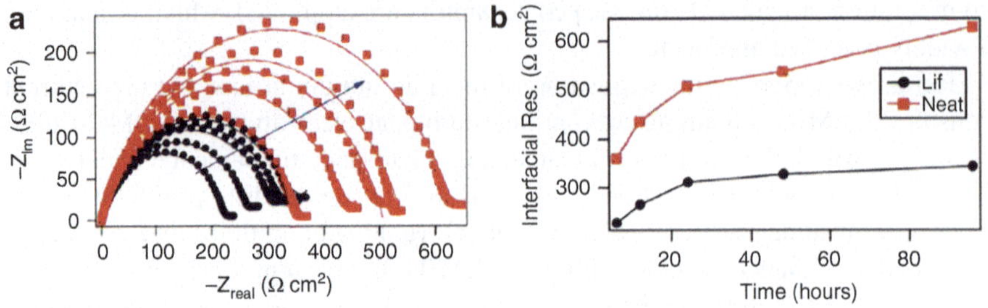

Fig. 5.2 Impedance spectroscopy as a function of time: (**a**) Nyquist plots obtained from impedance spectrocopy for symmetric cells having neat electrolyte (shown in red) and modified electrolyte with LiF additive (shown in black). The blue arrow shows the direction of increasing time from 6 to 96 h. (**b**) Interfacial resistance for neat and LiF-based batteries shown as a function of time. It is obtained by fitting to an equivalent circuit model, and approximately it corresponds to the width of the semicircle in the Nyquist diagram

possible to quantify the effects of many of the failure modes outlined in the introduction in a symmetric cell. However, in the asymmetric Li-SS cells used in the present study, the lithium loss in SEI formation or by side reactions can be quantified by the difference between the amount of lithium plated and successively stripped. Recent reports by Cui and co-workers [13] have shown ~99% efficiency in a composite electrolyte of DOL:DME-LiTFSI+$Li2S_6$ + $LiNO_3$; however with carbonate-based electrolyte such as PC, the coulombic efficiency from similar tests [34] has been reported to rarely exceed ~77%, which is attributed to the relatively poor SEI forming characteristics of the carbonate solvents in comparison to DOL:DME. Figure 5.3a, b reports the coulombic efficiencies of cells with and without LiF additive at rates ranging from 0.25 mA/cm^2 and 0.50 mA/cm^2, respectively, with 1.00 mAh/cm^2 of lithium cycled in each run. It is seen that at 0.25 mA/cm^2, the coulombic efficiency for the neat electrolyte is close to 80% in agreement with reported values in literature [34], while for cells with LiF additive, the coulombic efficiency exceeds 90%. In addition, it is observed that for at least 120 cycles, the values remain essentially constant for cells containing the LIF salt-reinforced electrolyte. In contrast for the control batteries that do not include LiF in the electrolyte, the coulombic efficiency is observed to begin fluctuating beyond 50 cycles. Also, at current density of 0.50 mA/cm^2, the LiF-reinforced electrolytes are seen to exhibit coulombic efficiency close to 90%, while their unreinforced counterpart shows poor performance. The respective voltage profile as a function of areal capacity is shown in Supplementary Fig. 5.2. It is clear that the LiF additive is enabling a stable SEI formation and flat electrodeposition, hence improving the coulombic efficiency by at least 10% compared to convention battery. It is remarkable that addition of just 0.5% by weight of LiF additive can significantly improve the lifetime and performance of a battery without the need of any other modification of the conventional electrolyte.

In order to more carefully investigate the aforementioned characteristics of lithium deposition onto the electrode, scanning electron microscopy was performed on the lithium metal surface after Li deposition as shown in Fig. 5.4. For these experiments, symmetric cells were constructed, and a fixed amount of lithium (4 mAh/cm^2) was

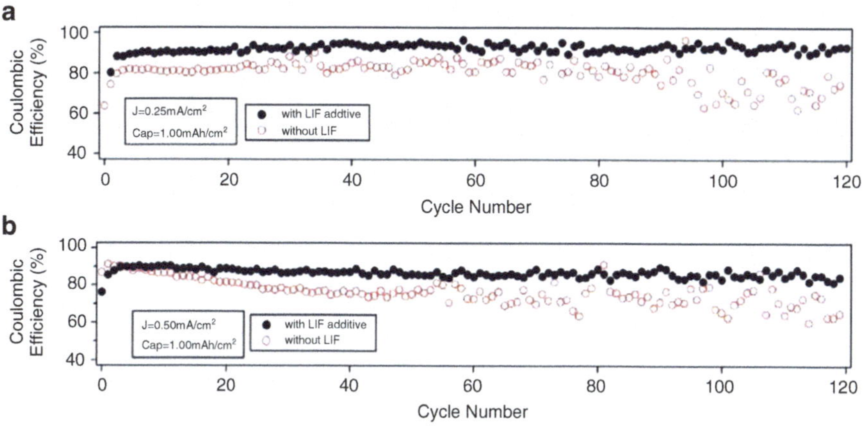

Fig. 5.3 Coulombic efficiency test of lithium cells: Batteries with neat electrolyte (shown in red) and with 0.5% LiF additive (shown in black) were cycled at different current densities (**a**) at current density of 0.25 mA cm^{-2} with capacity of 1 mAh cm^{-2}; (**b**) at current density of 0.25 mA cm^{-2} with capacity of 1 mAh cm^{-2}

passed from one electrode to the other at current densities of 2 mA/cm^2 and 4 mA/cm^2. It is seen that the lithium deposits obtained using the neat electrolyte are uneven and needlelike, while for the LiF-reinforced electrolytes, the Li deposition is significantly flatter. The SEM images of the lithium anodes after deposition at 2 mA/cm^2 were analyzed to determine the size of the deposits as given in the Supplementary Fig. 5.4. It is seen that the electrode with neat electrolyte had dendrites in mostly conical or cylindrical shape with mean length of ~20 μm, while for the electrolyte with LiF additive, the deposits were mostly spherulites with a mean diameter of ~11.5 μm. A mechanism that explains these large effects in terms of enhanced lateral mobility of Li ions on a LiF surface is provided in Fig. 5.1. The surface morphology of the lithium electrode in the presence of LiF electrolytes is qualitatively consistent with expectations based on this concept.

Further, to evaluate the hypothesis that LiF-reinforced electrolytes yield LMBs with higher resistance to dendrite formation and thus more resistant to failure by dendrite-induced short circuits, "strip-plate" measurements were carried out in symmetric lithium cells. In these experiments the Li/Li cells are subjected to alternating periods of charge and discharge at a range of current densities. The experiments were deliberately performed under relatively harsh conditions, such that 4 mAh/cm^2 of lithium was deposited in each charge and discharge cycles, which is at least twice as high as the areal capacity of commercially available cathode sheets. Figure 5.5 shows the voltage profiles as a function of time for batteries with and without LiF additive at current densities of 1, 2, and 4 mA/cm^2. It is seen that at the highest current density of 4 mA/cm^2, the cells that do not contain LiF additives in the electrolyte short-circuit within 80 h of the start of the test. In contrast, those in which LiF is used show lifetimes in excess of 140 h. The cell lifetime is also seen to double at a current density of 2 mA/cm^2 upon addition of 0.5 wt% LiF to the electrolyte; at a current density of 1 mA/cm^2, the neat electrolyte-based battery fails within 900 h, while its counterpart

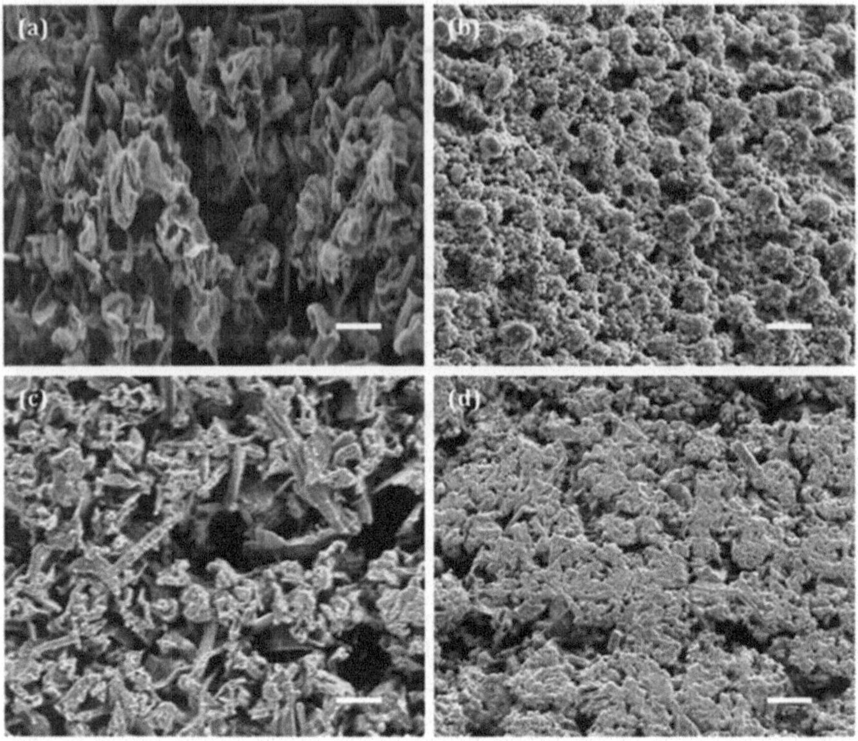

Fig. 5.4 Surface morphology of lithium anode: Lithium symmetric cells were polarized at different conditions before disassembling them for SEM analysis of the surface features (**a**) without LiF at 2 mA cm^{-2} for 2 h; (**b**) with 0.5% LiF at 2 mA cm^{-2} for 2 h; (**c**) without LiF at 4 mA cm^{-2} for 1 h; (**d**) with 0.5% LiF at 4 mA cm^{-2} for 1 h. All scale bars: 20 μm

continues to cycle to more than 1700 h without any sign of short circuit. Thus, it is evident that LiF additive can not only prolong short circuit time at high current density; it can also prevent short circuit at low current densities, making lithium metal batteries viable for practical applications. Another important conclusion that can be drawn from the Fig. 5.5 is the fact that the voltage profiles for neat electrolytes show gradual increment with time, which is an indication of electrolyte degradation by side reactions. This is attributed to the unstable SEI formation in usual lithium batteries as mentioned earlier, while LiF-based batteries show significant improvement in the stability of voltage profiles.

The practical viability of LiF as an electrolyte for LMBs was analyzed in the simplest LMBs in which a metallic lithium anode is paired with a $LiFePO_4$ cathode as shown in Fig. 5.6. Figure 5.6a, b reports the voltage profiles obtained from galvanostatic measurements for different cycle numbers for Li|EC:DMC-$LiPF_6$|$LiFePO_4$ and Li|EC:DMC-$LiPF_6$–0.5%LiF|$LiFePO_4$, respectively, while Fig. 5.6c shows the cyclability of these cells. The active material loading of the cathode used in this experiment of ~2 mAh/cm^2 is higher than most previous studies, and the current density used in experiment is ~0.50 mA/cm^2, corresponding to a C-rate of C/4.

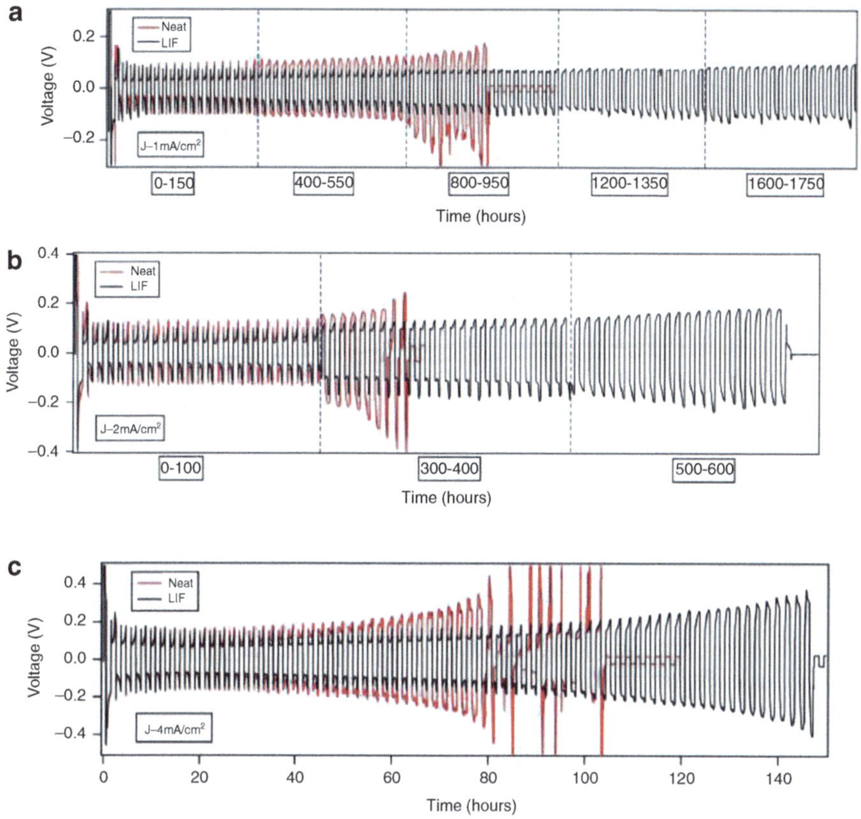

Fig. 5.5 Strip-plate tests: Symmetric lithium cells were charged and discharged cycles with neat electrolyte (shown in red) and with 0.5% LiF additive (shown in black) at different conditions: (**a**) at current density of 1 mA cm^{-2} with capacity of 4 mAh cm^{-2}; (**b**) at current density of 2 mA cm^{-2} with capacity of 4 mAh cm^{-2}; (**c**) at current density of 4 mA cm^{-2} with capacity of 4 mAh cm^{-2}

Both types of cells show coulombic efficiency of ~99%, which is to an extent intuitive; there is a reservoir of virtually infinite lithium at the anode. In the initial cycle, both the batteries are observed to have a discharge capacity of 120 mAh/gm. In successive cycles, this value increases to about 130 mAh/gm as the battery reaches steady state. However, the discharge capacity for the neat electrolyte systems shows a gradual decrement as a result of side reactions in the electrolyte and uneven electrodeposition. Beyond 100 cycles, it is seen that the neat electrolyte system becomes significantly unstable and ultimately fails. The LiF-based battery shows good performance for at least 150 cycles with a near constant discharge capacity, which is certainly a remarkable result considering that the battery operates at room temperature and at a relatively high current density. Results for batteries operating at even higher current density of 1 mA/cm^2 are shown in Fig. 5.3 of supplementary information, where a similar behavior is observed for the two types of batteries.

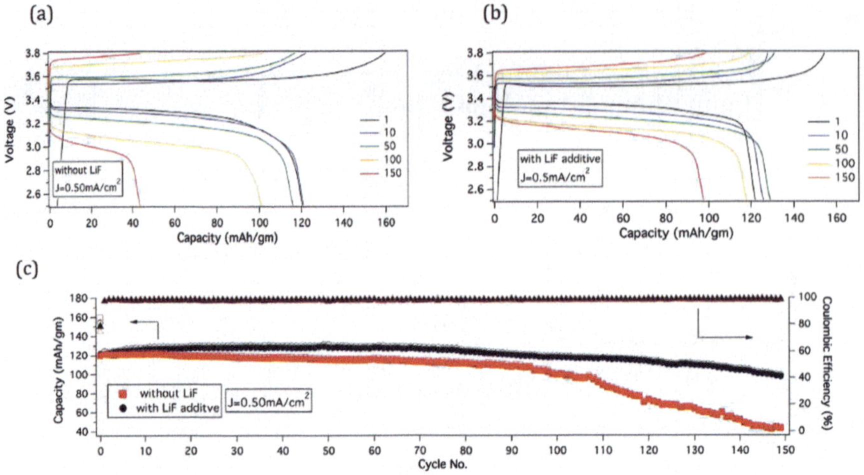

Fig. 5.6 Half-cell tests: Lithium metal batteries comprising of lithium anode and LiFePO 4 were charged and discharged with neat electrolyte and with 0.5% LiF additive; (**a**) voltage profile at current density of 0.5 mA cm^{-2} for neat electrolyte; (**b**) voltage profile at current density of 0.5 mA cm^{-2} with LiF additive; (**c**) cycling performance of these batteries up to 150 cycles

5.4 Conclusion

In summary, we have demonstrated that addition of 0.5% by weight of LiF to a conventional electrolyte can significantly improve the stability and reversibility of a battery. The rationality behind this observation is attributed to the recent JDFT calculations [33] that predicted high surface diffusivity of Li ions over a layer of LiF crystals as well as its higher surface stability over other SEI components. The batteries with LiF additive show low interfacial resistance and more than 90% coulumbic efficiency owing to better protection of lithium metal compared to usual electrolytes. In addition, it enables suppression of dendrite growth by facilitating smooth electrodeposition, thus increasing the lifetime of cells to hundreds of hours. The LiF additive is also successful in improving the discharge capacity and cyclability of a $LiFePO_4$ half-cell for at least 150 cycles, while a usual battery fails within 100 cycles. All these features of LiF additive, in addition to the convenience of usage, make it a viable option for practical applications.

Acknowledgments The material reported in this paper is based on work supported as part of the Energy Materials Center at Cornell, an Energy Frontier Research Center funded by the US Department of Energy, Office of Science, Office of Basic Energy Sciences under Award Number DESC0001086. Financial support from the National Science Foundation, Partnerships for Innovation (Grant#IIP-1237622), is also gratefully acknowledged. Electron microscopy, X-ray diffractometry, and X-ray spectroscopy facilities available through the Cornell Center for Materials Research (CCMR) were used for this work (NSF Grant DMR-1120296).

Appendix: Supplementary Information

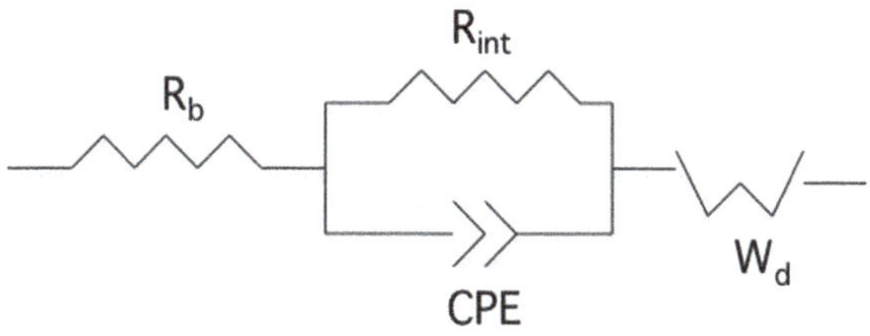

Supplementary Fig. 5.1 Equivalent circuit: The impedance spectroscopy results are fitted with equivalent circuit model comprising of a bulk impedance, interfacial resistance, and a solid-state diffusion element

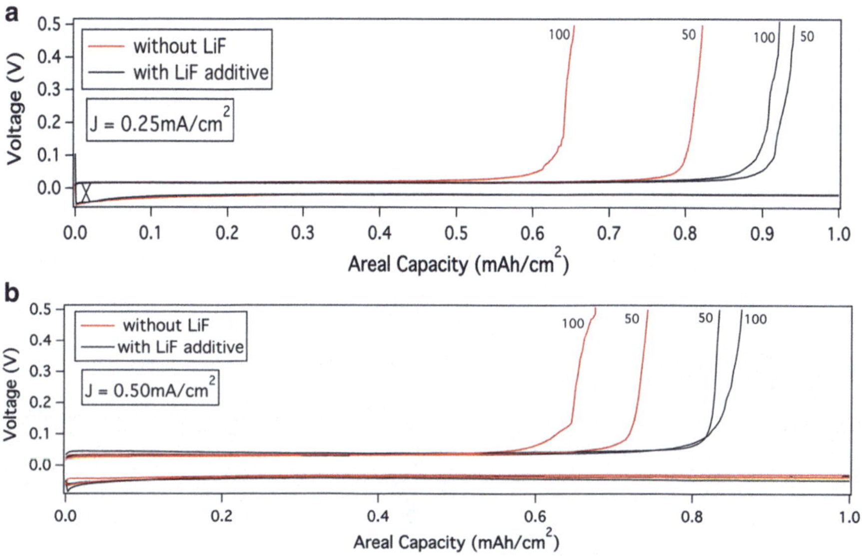

Supplementary Fig. 5.2 Coulombic Efficiency Test: Voltage profiles for batteries with Li-Stainless Steel configuration comprising of neat electrolyte (shown in red) and with 0.5% LiF additive (shown in black) at different conditions: (**a**) at current density of 0.25 mA/cm^2 for 1 mAh/cm^2 capacity; (**b**) at current density of 0.50 mA/cm^2 for 1 mAh/cm^2 capacity

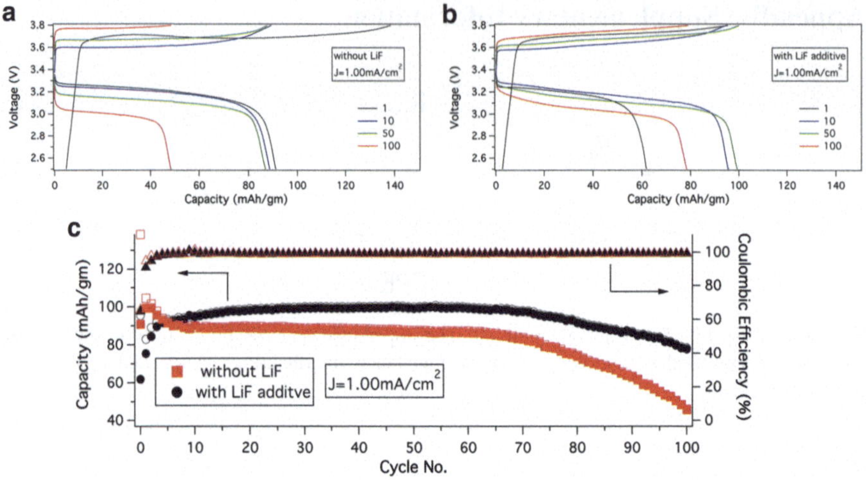

Supplementary Fig. 5.3 Half-cell test: Performance for batteries with Li-$LiFePO_4$ configuration comprising of neat electrolyte and 0.5% LiF additive; (**a**) voltage profiles of batteries with neat electrolytes at current density of 1.00 mA/cm² for 1 mAh/cm² capacity; (**b**) voltage profiles of batteries with LiF additive at current density of 1.00 mA/cm²; (**c**) cyclability of these batteries for 100 cycles

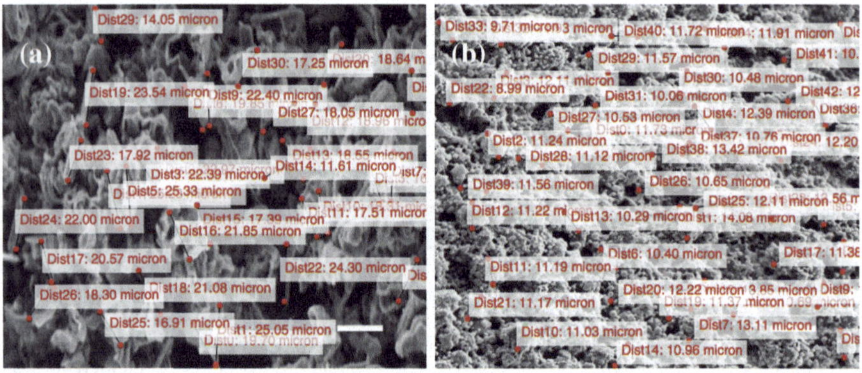

Supplementary Fig. 5.4 Postmortem analysis of electrodeposited anode: Anode surfaces electrodeposited at 2 mA/cm², (**a**) without and (**b**) with LiF additive are analyzed to obtain the morphology of dendrites. (**a**) With neat electrolytes, the deposits are mostly conical or cylindrical with a mean length of ~20 μm; (**b**) with LiF additive, the deposits are mostly spherulites with a mean diameter of ~11.5 μm

References

1. Dunn, B., Kamath, H., Tarascon, J.-M.: Electrical energy storage for the grid: a battery of choices. Science. **334**, 928–935 (2011)
2. Dresselhaus, M.S., Thomas, I.L.: Alternative energy technologies. Nature. **414**, 332–337 (2001)
3. Armand, M., Tarascon, J.-M.: Building better batteries. Nature. **451**, 652–657 (2008)
4. Tarascon, J.M., Armand, M.: Issues and challenges facing rechargeable lithium batteries. Nature. **414**, 359–367 (2001)
5. Aricò, A.S., Bruce, P., Scrosati, B., Tarascon, J., Schalkwijk, W.V.A.N., De Picardie, U., Verne, J., Umr-, C.: Nanostructured materials for advanced energy conversion and storage devices. Nat. Mater. **4**, 366 (2005)
6. Yang, P., Tarascon, J.-M.: Towards systems materials engineering. Nat. Mater. **11**, 560–563 (2012)
7. Kang, B., Ceder, G.: Battery materials for ultrafast charging and discharging. Nature. **458**, 190–193 (2009)
8. Goodenough, J.B., Kim, Y.: Challenges for rechargeable Li batteries †. Chem. Mater. **22**, 587–603 (2010)
9. Ji, L., Lin, Z., Alcoutlabi, M., Zhang, X.: Recent developments in nanostructured anode materials for rechargeable lithium-ion batteries. Energy Environ. Sci. **4**, 2682 (2011)
10. Whittingham, M.S.: Lithium batteries and cathode materials. Chem. Rev. **104**, 4271–4301 (2004)
11. Bruce, P.G., Freunberger, S.a., Hardwick, L.J., Tarascon, J.-M.: Li-O2 and Li-S batteries with high energy storage. Nat. Mater. **11**, 19–29 (2012)
12. Lu, Y., Tu, Z., Archer, L.A.: Stable lithium electrodeposition in liquid and nanoporous solid electrolytes. Nat. Mater. **13**, 961–969 (2014)
13. Zheng, G., Lee, S.W., Liang, Z., Lee, H.-W., Yan, K., Yao, H., Wang, H., Li, W., Chu, S., Cui, Y.: Interconnected hollow carbon nanospheres for stable lithium metal anodes. Nat. Nanotechnol. **9**, 618–623 (2014)
14. Agrawal, A., Choudhury, S., Archer, L.A.: A highly conductive, non-flammable polymer–nanoparticle hybrid electrolyte. RSC Adv. **5**, 20800–20809 (2015)
15. Kashiwagi, T., Du, F., Douglas, J.F., Winey, K.I., Harris, R.H., Shields, J.R.: Nanoparticle networks reduce the flammability of polymer nanocomposites. Nat. Mater. **4**, 928–933 (2005)
16. Monroe, C., Newman, J.: The effect of interfacial deformation on electrodeposition kinetics. J. Electrochem. Soc. **151**, A880 (2004)
17. Monroe, C., Newman, J.: The impact of elastic deformation on deposition kinetics at lithium/polymer interfaces. J. Electrochem. Soc. **152**, A396 (2005)
18. Hallinan, D.T., Mullin, S.a., Stone, G.M., Balsara, N.P.: Lithium metal stability in batteries with block copolymer electrolytes. J. Electrochem. Soc. **160**, A464–A470 (2013)
19. Tu, Z., Kambe, Y., Lu, Y., Archer, L.A.: Nanoporous polymer-ceramic composite electrolytes for lithium metal batteries. Adv. Energy Mater. **4**, n/a–n/a (2014)
20. Kanamura, K.: Electrochemical deposition of very smooth lithium using nonaqueous electrolytes containing HF. J. Electrochem. Soc. **143**, 2187–2197 (1996)
21. Aurbach, D., Gamolsky, K., Markovsky, B., Gofer, Y., Schmidt, M., Heider, U.: On the use of vinylene carbonate (VC) as an additive to electrolyte solutions for Li-ion batteries. Electrochim. Acta. **47**, 1423–1439 (2002)
22. Chen, L., Wang, K., Xie, X., Xie, J.: Effect of vinylene carbonate (VC) as electrolyte additive on electrochemical performance of Si film anode for lithium ion batteries. J. Power Sources. **174**, 538–543 (2007)
23. Guo, J., Wen, Z., Wu, M., Jin, J., Liu, Y.: Vinylene carbonate–LiNO3: a hybrid additive in carbonic ester electrolytes for SEI modification on Li metal anode. Electrochem. Commun. **51**, 59–63 (2015)

24. Xu, K., Zhang, S., Jow, T.R.: LiBOB as additive in LiPF[sub 6]-based lithium ion electrolytes. Electrochem. Solid-State Lett. **8**, A365 (2005)
25. Pieczonka, N.P.W., Yang, L., Balogh, M.P., Powell, B.R., Chemelewski, K., Manthiram, A., Krachkovskiy, S.a., Goward, G.R., Liu, M., Kim, J.H.: Impact of lithium Bis(oxalate) borate electrolyte additive on the performance of high-voltage spinel/graphite Li-ion batteries. J. Phys. Chem. C. **117**, 22603–22612 (2013)
26. Zhang, S.S.: A review on electrolyte additives for lithium-ion batteries. J. Power Sources. **162**, 1379–1394 (2006)
27. Pires, J., Timperman, L., Castets, A., Santos-pena, J., Dumont, E., Levasseur, S., Dedryvere, R., Tessier, C., Anouti, M.: Role of propane sultone as additive to improve the performance of lithium-rich cathode material at high potential. RSC Adv. **5**, 42088 (2015)
28. Li, B., Xu, M., Li, T., Li, W., Hu, S.: Prop-1-Ene-1,3-Sultone as SEI formation additive in propylene carbonate-based electrolyte for lithium ion batteries. Electrochem. Commun. **17**, 92–95 (2012)
29. Miao, R., Yang, J., Feng, X., Jia, H., Wang, J., Nuli, Y.: Novel dual-salts electrolyte solution for dendrite-free lithium-metal based rechargeable batteries with high cycle reversibility. J. Power Sources. **271**, 291–297 (2014)
30. Qian, J., Henderson, W.a., Xu, W., Bhattacharya, P., Engelhard, M., Borodin, O., Zhang, J.-G.: High rate and stable cycling of lithium metal anode. Nat. Commun. **6**, 6362 (2015)
31. Philippe, B., Gorgoi, M., Edstro, K.: Improved performances of nanosilicon electrodes using the salt LiFSI : a photoelectron spectroscopy study. J. Am. Chem. Soc. **135**, 9829–9842 (2013)
32. Gunceler, D., Letchworth-Weaver, K., Sundararaman, R., Schwarz, K.a., Arias, T.a.: The importance of nonlinear fluid response in joint density-functional theory studies of battery systems. Model. Simul. Mater. Sci. Eng. **21**, 074005 (2013)
33. Ozhabes, Y., Gunceler, D., Arias, T.A.: Stability and surface diffusion at lithium-electrolyte interphases with connections to dendrite suppression. arXiv. **1504.05799**, 1–7 (2015)
34. Ding, F., Xu, W., Graff, G.L., Zhang, J., Sushko, M.L., Chen, X., Shao, Y., Engelhard, M.H., Nie, Z., Xiao, J., et al.: Dendrite-free lithium deposition via self-healing electrostatic shield mechanism. J. Am. Chem. Soc. **135**, 4450–4456 (2013)

Chapter 6
Designing Solid-Liquid Interphases for Sodium Batteries

6.1 Introduction

Rechargeable batteries based on lithium and sodium metal anodes are of interest for high-energy storage solutions in portable and stationary applications [1, 2]. Although sodium-based batteries predate those based on lithium [3], Li has received more recent attention for a variety of reasons, including its greater electronegativity, higher specific energy, low atomic radius [4, 5], and the commercial success of related Li-ion battery technology. The greater natural abundance of sodium and its availability in regions all over the world provide significant cost advantages over Li that have within the last decade helped re-ignite interest in Na-based batteries [6–8]. Metallic sodium has other attractive features as a battery anode, including its relatively high electronegativity and low atomic weight, which combine to give the Na anode a specific capacity (1166 mAh g^{-1}) that is competitive with Li (3860 mAh g^{-1}) in many applications [6]. Additionally, recent studies have shown that rechargeable batteries that pair a Na anode with highly energetic O_2-based cathodes are intrinsically more stable during discharge than their Li analogs because the species generated electrochemically in the cathode, the metal superoxide, is more stable when the anode is Na, as opposed to Li [9, 10].

As with rechargeable batteries comprising Li metal anodes, the Achilles heel of the rechargeable sodium battery is the anode's susceptibility to failure during the charging process. Specifically, during battery recharge Na ions deposit in rough, low-density, and uneven patches on the negative electrode, even at current densities below the limiting current where classical instabilities such as electroconvection that drive rough, dendritic deposition are expected to be unimportant [11, 12]. Instead, dendrites on Na (and Li) arise from inhomogeneities in the resistance of the solid electrolyte interphase (SEI), formed spontaneously on the anode surface when in contact with an electrolyte. The resultant concentration of electric field lines on faster growing regions of the interface drives the morphological instability

S. Choudhury, *Rational Design of Nanostructured Polymer Electrolytes and Solid–Liquid Interphases for Lithium Batteries*, Springer Theses,
https://doi.org/10.1007/978-3-030-28943-0_6

loosely termed dendritites [12, 13]. At later stages, uncontrolled dendritic deposition leads to metallic structures able to bridge the interelectrode space, ultimately short-circuiting the cell. Short circuits lead to two catastrophic failure mechanisms: (i) thermal runaway which drives chemical reactions in the electrolyte, ending the cell life by fire, explosion, or both [12, 14–16], and (ii) melting and breakage of the dendrites, which electrically disconnects the material from the electrode mass [4, 17], causing rapid or gradual reduction in the storage capacity of the anode. Unlike Li, where dendrite-induced short circuits are considered the dominant failure mode, chemical reaction between the electrolyte and metal anode are regarded as the most important mechanism of cell failure for batteries based on a Na anode. Na also has a lower melting point than Li, which makes batteries based on Na more prone than their Li counterparts to failure by thermal runaway and/or dendrite breakage [6, 18, 19].

Few studies have addressed the challenges associated with stabilizing a Na anode [18]. In contrast, several approaches have been reported for preventing/retarding Li dendrite proliferation in Li metal batteries [11, 12]. Some of the approaches include using high-modulus electrolyte or nanoporous/tortuous separator [14, 20–22], modifying the ion transport in electrolytes by using single ion conductors and ionic liquids [23–27], or forming a stable electrode-electrolyte interface to suppress the nucleation of dendrites [4, 13, 28–30]. In addition to preventing dendrite-induced short circuits, the last approach may impede unwanted parasitic reactions between the electrode and electrolyte that lead to formation of insulating products and loss of electrochemically active material, causing decay in the battery capacity with increasing charge-discharge cycles [12]. A common approach for the formation of artificial SEI on the metal involves use of special electrolyte additives such as vinylene carbonate [31, 32], fluoroethylene carbonate [33], dioxane [34], sultones [30, 35], or functional ionic liquids [7] which can electropolymerize on the surface of electrode to form an elastic coating that protects the metal surface and accommodate volume changes in the electrode during charge and discharge. There are also recent reports of protecting the electrode interface by direct formation of a barrier layer by deliberate reaction between electrodes and reactive species in electrolytes [36–38]. Various indirect methods have also been reported for stabilizing a Li anode during battery recharge. These include use of a functional nanoparticles [6, 11, 23, 24], and mixtures of salts (e.g., LiTFSI-LiFSI) [39], use of concentrated electrolytes (e.g., 5 M LiFSI in DME) [40], or use of polysulfides and $LiNO_3$ [29] as electrolyte additives. A common feature of these methods is that they produce lithium fluoride (LiF) in the SEI. In recent studies [13, 22], direct incorporation of LiF as an additive in liquid electrolytes was reported to yield dramatic enhancements to battery lifetime in Li metal cells at both high and low current densities. Cui and co-workers [18] were among the first to show that application of this concept to Na metal batteries, through electrolyte additives that generate sodium fluoride (NaF), leads to markedly higher coulombic efficiencies (as high as 99%) in Na||Cu cells.

With the specific aim of developing rational strategies for stabilizing the anode of Na batteries during cell recharge, we herein investigate how the chemistry of the SEI alters ion transport at Na- and Li-electrolyte interfaces by means of joint density-functional theory (JDFT) calculations and experiment. Our focus on interfacial transport derives from the observation that magnesium metal anodes, which do not form uneven deposits under charging at currents below the limiting current [41], present the lowest barriers for interfacial metal ion transport [42]. Remarkably, we find that a Na metal anode protected by NaBr presents a barrier of only ~0.02 eV/atom (i.e., comparable to Mg metal) for interfacial ion transport. By means of direct visualization studies and electrochemical analysis, we investigate the stability imparted to the Na electrode by a NaBr protective coating.

6.2 Methods

6.2.1 Materials

Sodium cubes, bromopropane, chloropropane, iodopropane, propylene carbonate, ethylene carbonate (EC), diglyme, dimethoxyethane, sodium hexafluorophosphate, and magnesium(II) bis(trifluoromethanesulfonyl)imide were all purchased from Sigma-Aldrich. Celgard 3501 separator was obtained from Celgard Inc. Glass fiber separator was bought from Whatman Inc. All the chemicals were used as received in after rigorous drying in a ~0 ppm water level and <5 ppm oxygen glove box, in order to make sure the sodium metal is not oxidized.

6.2.2 Sodium Bromide and Other Halide Coating Formation

Sodium-cube pieces were taken out of mineral oil and cleaned with Kimwipes. Then with a sharp knife, thin slices of sodium pieces were cut before punching with ¼ inch punch. For the coating of solid electrolyte interface, 15 μl of bromopropane was added to the sodium electrode before drying in vacuum antechamber for 5 min. It is known that the reaction is instantaneous due to high reactivity of sodium metal. Further the by-product obtained by reaction – hexane – is believed to vaporize rapidly in vacuum owing to its low boiling point characteristics. Coating with NaCl and NaI was done in exact same procedure, however, with chloropropane and iodopropane, respectively.

6.2.3 Physical Characterization

XPS was conducted using Surface Science Instruments SSX-100 with operating pressure of ~2×10^{-9} torr. Monochromatic Al K-α X-rays (1486.6 eV) with beam diameter of 1 mm were used. Photoelectrons were collected at an emission angle of

55°. A hemispherical analyzer determined electron kinetic energy, using pass energy of 150 V for wide survey scans and 50 V for high-resolution scans. Samples were ion-etched using 4 kV Ar ions, which were rastered over an area of 2.25 × 4 mm with total ion beam current of 2 mA, to remove adventitious carbon. Depth profile was obtained by ion etching at 2 kV, 22 μA over 2 * 3 mm, which yielded an etch rate of approximately 5 nm min^{-1}. The etching was done for 395 min. Spectra were referenced to adventitious C 1s at 284.5 eV. CasaXPS software was used for XPS data analysis with Shelby backgrounds. Br 3d was assigned to double peaks ($3d_{5/2}$ and $3d_{3/2}$) for each bond with 1.05 eV separation. Residual SD was maintained close to 1.0 for the calculated fits. Samples were exposed to air only during the short transfer time to the XPS chamber (less than 5 s).

XRD was carried out on a Scintag Theta-Theta X-ray diffractometer using Cu K_α radiation at $\lambda = 1.5406$ Å. All samples were covered with Kapton tape to ensure that the sodium metal is not oxidized in air. For the sodium metal samples after cycling, the symmetric cell of NaBr-coated sodium was charged and discharged five times at a current density of 0.5 mA cm^{-2}.

6.2.4 Electrochemical Characterization

The impedance spectroscopy measurement ionic conductivity of electrolytes was measured as a function of temperature using a Novocontrol N40 Broadband Dielectric instrument. Symmetric cells were prepared using two sodium pristine metal pieces using the electrolyte 1 M $NaPF_6$ EC/PC with a glass fiber separator. For understanding the impedance of halide-based interfacial layer, both sodium pieces were coated with respective halides using the same method (as described for NaBr coating section) before performing the experiment. For the symmetric cell with Mg electrodes, the electrolyte 0.3 M $Mg(TFSI)_2$ in DME/diglyme was used with a glass fiber separator. The measurements were done in a frequency range from 10^{-3} to 10^7 Hz.

6.2.5 Scanning Electron Microscopy

Postmortem characterization of the sodium metal electrodes was done to understand the morphology of sodium metal deposition and also failure mechanisms involved in short circuits. For this reason, the cells with symmetric sodium cells with or without NaBr layer were charged at a current density of 1 mA cm^{-2} for 2 h before disassembling the cell inside the glove box. The charged sodium metal pieces were washed with the electrolyte solvent and were transferred carefully to the microscopy facility minimizing the exposure of sodium metal to atmosphere. The SEM analysis was done using the LEO155FESEM instrument.

6.2.6 Focused Ion Beam/Scanning Electron Microscopy

An FEI Strata 400 Focused Ion Beam (FIB) was used for the FIB/SEM experiments. The FIB is fitted with a Quorum PP3010T Cryo-FIB/SEM Preparation System that enables cryogenic experiments. For room-temperature experiments, the sample was removed from an inert environment and loaded onto an aluminum stub. The stub was subsequently attached to a shuttle and placed in a loadlock that pumps down to vacuum before inserting into the FIB. For cryogenic experiments, the sample was removed from an inert environment, attached to a stub, and immediately plunged into slush nitrogen. This shortened exposure time and minimized any reactions with air or moisture. The sample was then transferred into the FIB in a transfer device at liquid nitrogen temperature and subsequently maintained at −165 °C in the cryo-FIB.

6.2.7 In Situ Visualization Studies

The visualization experiment was carried out for understanding the in-operando observation of electrodeposition in sodium metal batteries. In all experiments the electrolyte – 1 M $NaPF_6$-EC/PC – was used. For control experiments, pristine sodium was used as both electrodes, while for understanding the role of stable SEI, NaBr-coated sodium pieces were used. The sodium metal pieces were attached to current collectors and fixed in an airtight cuvette chamber as shown in Supplementary Fig. 6.3. In these experiments, it is made sure that the electrodes are fully facing each other and then one electrode is continuously charged with a current density of 1 mA/cm^2. The electrode, being charged, was monitored over time, and images of sodium deposition at different intervals were captured from an optical microscope.

6.2.8 Cell Lifetime and Failure Studies

Symmetric 2032 type Na|Na coin cells with and without NaBr coating containing liquid electrolyte of 1 M $NaPF_6$ EC/PC (1:1 v/v) inside an argon-filled glove box. The cells were evaluated using galvanostatic (strip-plate) cycling using a Neware CT-3008 battery tester. In the "strip-plate" experiments, the batteries were repeatedly charged and discharged with each half-cycle 0.5 h long. Failure was deduced from irregularities in the voltage profile as well as excessive increment in the over-potential indicating excessive formation of electrolyte by-products completely insulating the electrodes.

6.2.9 Sulfur-PAN Cathode Cycling

Galvanostatic measurements with the sulfur-PAN composite cathode were done in a 2032-type coin cell comprising a sodium metal anode with and without the NaBr coating. Celgard 3501 polypropylene membranes were used as the separator. 40 μl 1 M $NaClO_4$ in a mixture of ethylene carbonate (EC) and propylene carbonate (PC) (v:v = 1:1) was used as the electrolyte. The cathode consisted of 70 wt% of the active material, 15 wt% of carbon black (Super-P Li from TIMCAL) as a conductivity aid, and 15 wt% of polymer binder (PVDF, polyvinylidene fluoride, Aldrich). A carbon-coated aluminum foil used for current collector. The mass loading of the cathode was ~0.85 mg SPAN cm^{-2}. Detailed synthesis of the sulfur-PAN composite can be found in recent paper by Wei et al. [43]. Cell assembly was carried out in an argon-filled glove box. The measurements were done in Neware CT-3008 battery tester.

6.3 Results

6.3.1 Joint Density-Functional Theory (JDFT) Study of SEI

It has been argued that the concentration of electric field lines at protrusions on the electrode surface leads to nonuniform ion distribution and deposition rate, which serves to seed dendrites [44]. We have previously proposed through density-functional simulations that enhanced surface diffusion at the electrode-electrolyte interface could serve as a counter mechanism by smoothing protrusions on the surface and thus prevent formation of dendrites [45–47].

To understand these effects in the context of the sodium anode, we simulated sodium adatoms on the surface of different passivated sodium electrodes using a similar methodology as described in detail in our previous work focused on density-functional calculations of transport barriers for halogenated SEI salt layers in lithium metal batteries [45]. The surface diffusion barrier is affected by the presence of the liquid electrolyte at the interface, and its calculation is thus nontrivial. Regular density-functional theory can provide the total energy of a given configuration (or snapshot) at fixed atomic positions, but to accurately compute the free energy of a solid-liquid interface, one must also sample the configuration space of the liquid. While this can be done by molecular dynamics methods (e.g., QM/MM), such calculations are computationally demanding and to date have not been reported for the systems of interest. Joint density-functional theory [46], which works with thermodynamic averages of the fluid variables, provides an economical alternative to molecular dynamics and provides direct access to free energies without the need for sampling.

We performed all electronic structure calculations with JDFTx [47], an open-source implementation of joint density-functional theory. To account for the effect of the electrolyte, we used the nonlinear polarizable continuum model [46] generalized to nonaqueous solvents [48], which is taken to be acetonitrile in the present work. All calculations employ a plane wave basis with a cutoff of 20 hartrees, and we use ultrasoft pseudopotentials from GBRV library [49].

Results for these calculations are reported in Fig. 6.1. The binding energy of a Na adatom depends on where it binds onto the Na-halide surface (as shown in Fig. 6.1a). For the smallest halide, F, the minimum energy position for the sodium adatom is directly on top of the fluoride ion; we refer to this site as the "anion site." Again, for F the saddle point on the diffusion path is in the middle of two neighboring anion sites; we call this middle point the "in-between site." Traversing down the periodic table, it is observed that with increasing anion size, the binding energy of the in-between site becomes relatively closer to the anion site, and eventually the saddle point becomes the minimum. This transition happens with NaBr and results in the lowest diffusion barrier for interfacial ion transport.

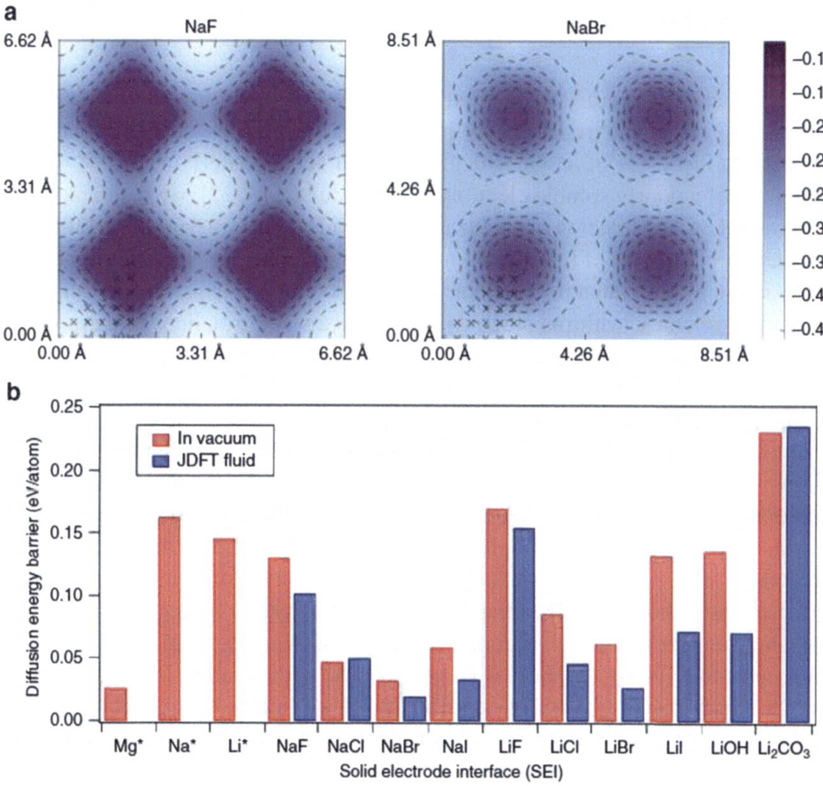

Fig. 6.1 Surface diffusion barriers calculated using joint density-functional theory. (**a**) Surface-binding energy vs. binding site for NaF (left) and NaBr (right) obtained from JDFT analysis of adatom diffusion. The entire contour plot is generated by symmetry using the data points indicated by cross symbols. (**b**) Diffusion energy barriers computed for Mg, Na, and Li adatoms on surfaces with the chemistries noted. The red bars denote surface in contact with vacuum and blue bars indicate the same in presence of acetonitrile. The $*$ symbol marks the data points obtained from Ref. [42]

Comparing the surface diffusion barriers of NaBr with other sodium halide salts and lithium salts as well as pure elements, lithium, sodium, and magnesium (Fig. 6.1b) places our finding in perspective with recent theory-supported strategies for suppressing dendritic deposition at metal electrodes. It is notable that the diffusion barrier for NaBr adatoms is substantially lower than for NaF and even in a liquid electrolyte is comparable to those computed for Mg in vacuum. Based on earlier reports that LiF coatings on Li metal dramatically stabilize electrodeposition of Li [13, 22] and that NaF coatings on Na has a similar large stabilizing effect on Na deposition [18], we hypothesize that a Na anode protected by a coating of NaBr would be particularly attractive for room-temperature sodium batteries employing liquid electrolytes.

6.3.2 Formulation and Stability of a NaBr-Based SEI Layer on Sodium Metal

To evaluate the JDFT prediction, we first developed a method for uniformly coating NaBr on a Na metal electrode. Unlike previous experiments, where the source of halide salts in SEI layer of anode is degradation of active materials [18, 39, 40] or precipitation of a poorly soluble electrolyte salt additive [13, 22], we here employ a well-known chemical reaction to create a layer of NaBr at the interface. Specifically, we carried out a reaction of the sodium metal anode with bromopropane to undergo Wurtz reaction as illustrated in Fig. 6.2a. This reaction is widely used for production of symmetric alkanes, with the side product being a sodium halide. In the present case, along with NaBr, hexane is formed, which is removed by evaporation. X-ray diffraction (XRD) analysis (Fig. 6.2b) for pristine and treated sodium metal confirms that crystalline NaBr is formed on the surface of the Na electrode surface. The morphology of the NaBr layer was interrogated using survey scanning electron microscopy (SEM) for different exposure times of sodium metal piece with 1-bromopropane. This exposure time corresponds to the time the sodium metal piece was dipped into the 1-bromopropane liquid (which we define that as the nominal reaction time, though the actual time of reaction is difficult to quantify). Figure 6.2c and f show the SEM images of the sample surface for exposure times of 5 min and 1 min, respectively. Figure 6.2c clearly shows large regions of a dense, smooth deposit of NaBr on Na that is interspersed with smaller, less well-coated regions as confirmed by EDX. In contrast, fewer well-coated regions are seen in Fig. 6.2f. More in-depth information about the Na electrode coatings was obtained using cryo-focused ion beam-scanning electron microscopy (cryo-FIB-SEM), and the results are shown in Fig. 6.2d and e after 5 and 1 min of treatment of the electrodes with 1-bromopropane, respectively. The thicknesses and depth-dependent composition of the coating layers were determined by SEM imaging and energy dispersive X-ray (EDX) mapping (Supplementary Fig. 6.1a) of cross sections produced by FIB milling. The EDX element mapping confirms that the top layer is

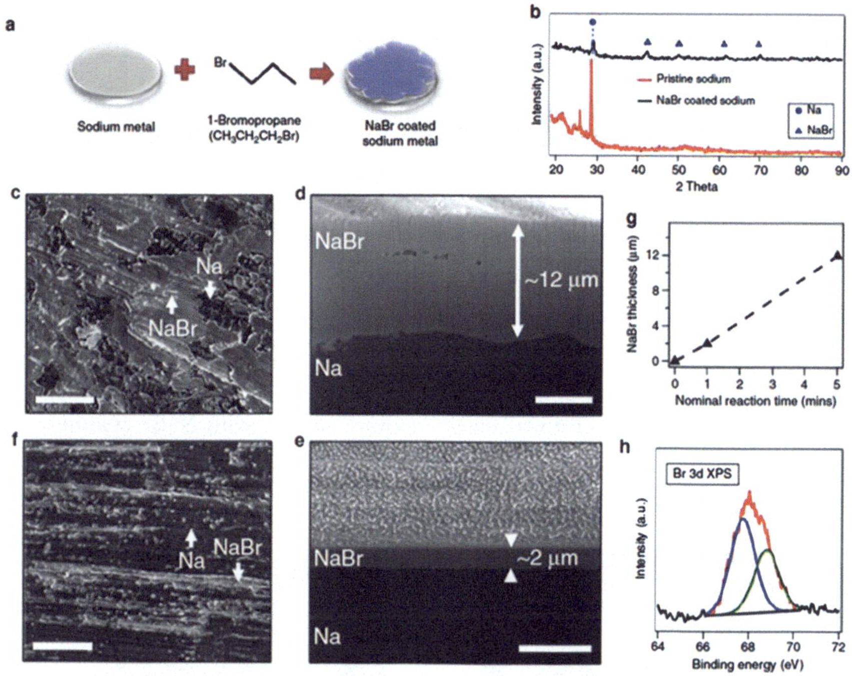

Fig. 6.2 Formulation and characterization of sodium bromide layer. (**a**) Schematic showing the procedure used to coat Na with NaBr. (**b**) XRD analysis of pristine and NaBr-coated sodium showing that NaBr exists in crystalline form in the coatings on Na. (**c**) Cryo-SEM image of ~12 µm thick NaBr coating on sodium metal surface. Light regions are NaBr coating and dark regions are sodium metal, as confirmed by EDX; scale bar, 200 µm. (**d**) Cryo-SEM image of a cross section through the NaBr coating obtained by focused ion beam milling under cryogenic conditions. Cross-sectional SEM imaging was used for layer thickness analysis, and the layer composition was confirmed by EDX mapping; scale bar, 5 µm. (**e, f**) Complementary images and scale bars as (**d, c**), respectively, but for a Na substrate exposed to 1-bromopropane for 1 min. A thinner (2 µm) NaBr coating is formed in this case. (**g**) Thickness of the NaBr-based SEI layer on sodium metal at various nominal reaction times. (**h**) X-ray photoelectron spectrum centered on the Br 3d bands confirms the existence of metallic bromide bond

predominantly Br, while the bottom comprises of essentially pure Na metal. The thickness of the NaBr layer is plotted as a function of nominal reaction time in Fig. 6.2g. For further studies, NaBr-coated sodium samples with nominal reaction times between 1 and 5 min were utilized. X-ray photoelectron spectroscopy (XPS) of the Br 3d peaks was used to more carefully analyze the bromine containing compounds in contact with the sodium metal surface. A high-resolution scan was performed after 45 s sputtering to remove any oxide layer that may form when transferring samples to the XPS chamber. As shown in Fig. 6.2h, two deconvoluted peaks for Br 3d 5/2 and 3/2 at 68.8 eV and 67.7 eV, respectively, are observed, which correspond to the predominant presence of metallic bromide bonds on the sodium surface [50–52].

The effectiveness of the NaBr coating in protecting sodium can be most easily evaluated by comparing SEM images of the pristine (Fig. 6.3a) and NaBr-protected (Fig. 6.3b) Na electrodes following brief air exposure during transfer to FIB/SEM chamber at room temperature. The former is seen to be covered with a porous oxide layer, which is entirely absent from the NaBr-coated Na. The oxide layer on the pristine sodium surface was ~5 μm thick, as seen from the cross-sectional image in Supplementary Fig. 6.1b. The stability of the NaBr interphase layer during electrochemical cycling was characterized by postmortem analysis of the Na anode after five cycles of charge and discharge in a symmetric cell at a fixed current density of 0.5 mA cm^{-2}. Figure 6.3c shows that the NaBr crystal structure is retained, confirming that the anode protection mechanism is sustained. Results from SEM analysis of the cycled anodes are reported in Fig. 6.3d. The surface morphology is seen to remain relatively flat and compact. XPS analysis for the Br 3d peaks (inset of Fig. 6.3e) was further performed on the sodium sample with NaBr coating after

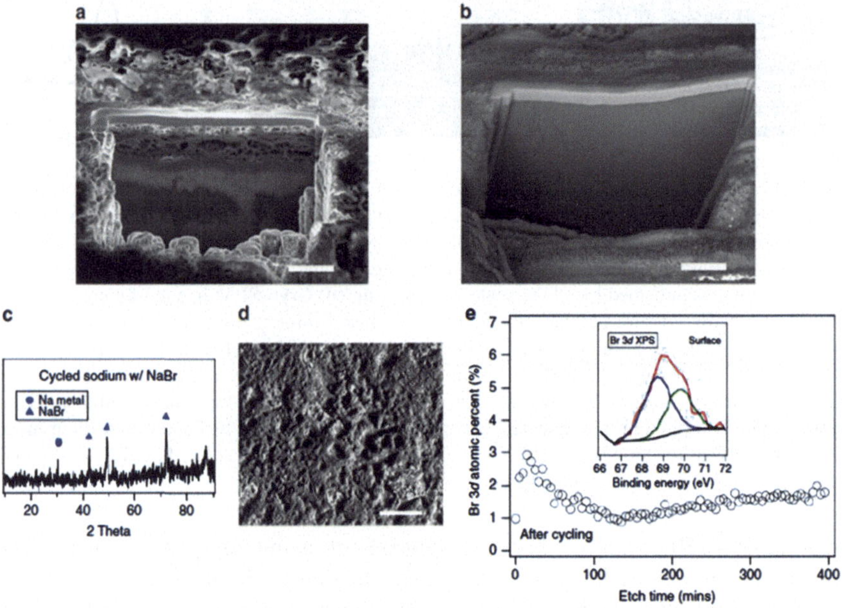

Fig. 6.3 Air sensitivity and stability of sodium bromide interphase. (**a**) Milled region on pristine sodium metal obtained using focused ion beam milling at room temperature. The porous layer indicates severe oxidation of sodium during sample transfer; scale bar, 5 μm. (**b**) Same as (**a**) but with NaBr coating. In both (**a, b**) the thin top solid layer is a platinum protective coating, deposited inside the FIB prior to milling. (**c**) XRD showing intensity of Na metal and NaBr peaks for sodium anode with NaBr coating and after cycling at 0.5 mA cm^{-2} for five times in a symmetric sodium cell. (**d**) SEM image of NaBr-coated sodium metal anode after cycling; scale bar, 200 μm. (**e**) Depth profiling of cycled anode obtained by ion etching and XPS measurements. The atomic content of Br 3d is shown as a function of etch time. The etching rate is ~5 nm/min. The inset shows the Br 3d XPS result for the cycled sodium anode surface (before etching)

cycling. The depth profile was obtained by etching the surface at 2 kV, 2 μA over an area of 2 mm × 3 mm, at a rate of 5 nm min^{-1} for 395 min. It is seen from Fig. 6.3e that the Br 3d atomic content decreases for the first 167 min, followed by a steady state Br 3d atomic content; this indicates that at least 2 μm thick NaBr layer is retained even after cycling. The surface atomic composition of the cycled sample is deduced from both EDX mapping and XPS analysis (prior to etching) as seen in Supplementary Fig. 6.2. In both cases, the Br element is seen to coexist with other elements (phosphorus, fluorine, carbon, oxygen), typical for degradation of the EC/PC $NaPF_6$ electrolyte. It is also seen that with the exception of carbon, the atomic compositions deduced from the two techniques are comparable.

Quantitative assessment of interfacial transport of Na ions in sodium halide coatings was made using impedance spectroscopy. These experiments were performed using symmetric sodium cells with/without halide salt coatings on Na and, for comparison, symmetric magnesium cells. Figure 6.4a and b reports Nyquist plots at different temperatures for pristine sodium and NaBr-coated sodium metal symmetric cells, respectively. By fitting the Nyquist plots to an equivalent circuit model (Supplementary Fig. 6.3), it is possible to deduce from the data the bulk resistance (R_b), representing ion transport in the electrolyte, and two interfacial resistances (R_{int1} and R_{int2}) representing ion transport through the passivating layer on Na as well as electronic transport. The temperature dependence of the interfacial ion conductivity can be used to extract information about how the halide coating alters the energy barrier for transport. The reciprocal of bulk impedance and net interfacial resistance are plotted with temperature in Arrhenius form as in Fig. 6.4c. The temperature-dependent analogs of these plots for NaCl- and NaI-coated sodium as well as that of Mg are provided in Supplementary Fig. 6.4. The lines through the data in Fig. 6.4c are fits obtained using the Vogel-Fulcher-Tammann (VFT) formula [1], $\sigma = A \exp(-E_a/R(T-T_0))$, commonly used for modeling ion transport in liquid electrolytes. Here A is the prefactor, E_a is the apparent activation energy for ion transport, R is the universal gas constant, and T_0 is the reference temperature. The respective VFT coefficients for all materials used in the study are tabulated in Supplementary Table 6.1. It is seen that the bulk impedance for sodium cells utilizing pristine and halide-coated sodium are similar, indicating that such coatings have at best a minimal effect on ion transport in the liquid electrolyte. In contrast, the interfacial conductivity and its temperature dependence are seen to be very sensitive to the chemistry of the SEI. Figure 6.4c, for example, shows that, whereas $1/R_{int1}$ for pristine Na is higher than for the NaBr-coated material, it decreases more rapidly with temperature. This latter behavior can be captured in terms of the apparent activation energy E_a for interfacial ion transport, which is reported in Fig. 6.4d for various pristine and halide-coated sodium electrodes, as well as for Mg. It is observed that E_a for the pristine sodium (~0.175 eV $atom^{-1}$) is higher by a factor of around 3 than the corresponding NaBr- or NaCl-coated metal. This means that at any temperature transport of Na ions is 20 times or faster in a SEI composed of NaBr or NaCl, in comparison to the SEI formed spontaneously at the pristine Na electrode. These experiments also reveal that the apparent interfacial activation energy for the Mg-symmetric cell (~0.02 eV $atom^{-1}$) is around 10 times lower than for pristine Na,

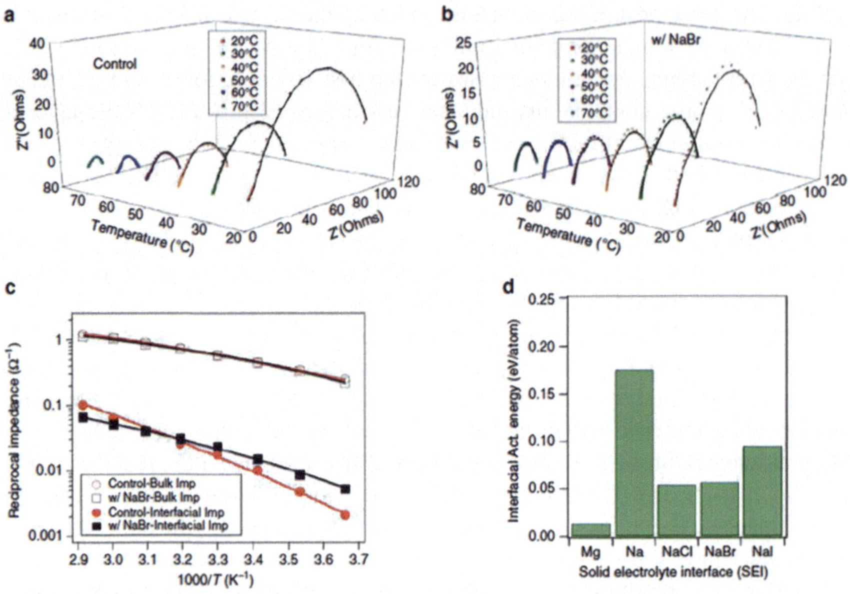

Fig. 6.4 Electrochemical impedance spectroscopy analysis temperature-dependent Nyquist plots for a symmetric sodium cell with (**a**) pristine sodium and (**b**) NaBr-coated sodium. (**c**) Reciprocal bulk and interface impedance as a function of reciprocal temperature; the lines are VFT model fits. (**d**) The apparent interfacial activation energy obtained from VFT fits of the temperature-dependent reciprocal resistance for various interphase chemistries. Here, Na and Mg symbolize results from symmetric cell studies with pristine sodium and magnesium electrodes, while NaCl, NaBr, and NaI are for the corresponding salt-coated Na electrodes

although the value of interfacial resistance for Mg is two orders of magnitude higher. As illustrated in Supplementary Fig. 6.5, these conclusions are broadly insensitive to the model (VFT or Arrhenius) used to extract E_a from the temperature-dependent electrochemical impedance measurements.

A lower E_a for a halide-rich SEI on Na means that at any current density deposition of the ions at the Na interface is less restricted. This result is consistent with the earlier prediction based on our JDFT analysis and is interpreted here to mean that a low diffusion energy barrier for halide adatom transport contributes to the low interfacial activation energy measured here. Additionally, the fact that we experimentally capture both the trends with Na-halide salts and the dramatically lower E_a values for Mg-electrolyte interphases predicted by the JDFT analysis implies that the diffusion barrier to adatom transport dominates the overall energy barrier to ion transport at the solid electrolyte interphase. Comparison of the E_a values in bulk electrolyte and at interphases composed of halide salts (see Supplementary Table 6.1) indicates that there is only a modest change in the barrier for ion movement as ions leave the bulk electrolyte and cross the electrolyte-electrode interface during deposition, which would be expected to favor more stable deposition. This inference is confirmed by the fact that the difference in energy barrier is lowest for Mg cells, which do not form dendrites.

6.3.3 Electrodeposition of Sodium Metal with NaBr-Coated Anode

It is known that sodium metal is more reactive than lithium [6], thus in contact with a liquid electrolyte it is expected to fail more easily by the first of the three mechanisms discussed in the introduction. Protecting the Na surface with a coating like NaBr that does not increase the barrier for Na ion transport at the interface should therefore inhibit failure by this mechanism. Additionally, since rough deposition is triggered by formation of an inhomogeneous SEI, cell failure by dendrite-induced short circuits should also be lower. To evaluate these statements, we directly visualize the surface of the Na anode with/without NaBr coatings during electrodeposition and quantify the growth of surface roughness as a function of time. A symmetric cuvette-type optical cell (see Supplementary Fig. 6.6) was used for this component of the study. In a typical experiment, the cell is polarized with a fixed current density of 1 mA cm^{-2} and the morphology of the electrode surface viewed in an upright Olympus optical microscope outfitted with 6.10 mm Extra Long Working Distance (EWLD) 10× objectives. Videos of the visualization experiment are provided (with a 300× speed). Figure 6.5a reports images obtained from these measurements at discrete time points separated by a fixed 10-min interval. It is observed that a pristine Na electrode is prone to form moss-like dendritic structures. In comparison, the NaBr-coated sodium metal is seen to electrodeposit without the formation any dendritic structure; however, it is seen that the overall volume (thickness) of the electrode is increasing over time due to the deposition under the NaBr coating. Figure 6.5b plots the tip length of the dendrites over time. It is seen that for the pristine Na electrode, dendrites begin to grow immediately upon imposition of the current and grow throughout the electrode surface. The number density and reactivity of dendrites can be estimated from the brightness of the respective images. Reaction of Na with electrolyte is known to cause the metal to lose its shiny appearance. This, along with the much greater number density of dendritic structures, is the source of the darker appearance of the images obtained using the pristine Na electrode. Figure 6.5c reports the comparative lightness ($L*$) of different spots on the sodium metal for both cases. Here, lightness ($L*$) is defined as the relative brightness of a spot, such that a white spot would correspond $L*$ of 100 and that of a black spot would be zero. $L*$ of pristine sodium drops down close to zero within 1 min of electrodeposition, while that of sodium with NaBr maintains $L* > 90$ for entire time of measurement. It is observed that the reduction in brightness (or synonymously formation of undesired by-products) is synchronous to the growth of dendrite-like structure. It can be hypothesized that the distribution of the insulating products is heterogeneous, which leads to a nonuniformity of local current density on the electrode surface that ultimately causes the formation of rough and needlelike structures. These dendrites further increment the local current densities due to their sharp edges causing a cascade of instabilities.

Figure 6.6 reports the electrochemical performance of sodium metal cells in different configurations. To simulate the performance of a working Na metal battery, a

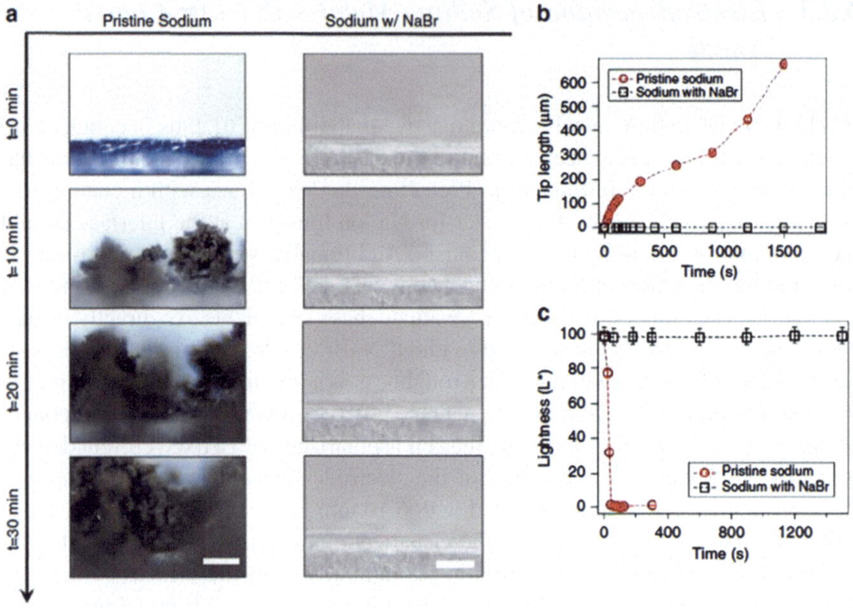

Fig. 6.5 Visualization of sodium electrodeposition. (**a**) Snapshots from video microscopy of the pristine Na- (left column) and NaBr-coated Na- (right column) electrolyte interface at 0, 10, 20, and 30 min (top to bottom) after onset of Na deposition at $J = 1$ mA cm^{-2}; scale bar, 100 μm. (**b**) Na dendrite tip length as a function of time. (**c**) Relative brightness ($L*$) of electrolyte near the electrode-electrolyte interface as a function of time. This variable captures the obvious darkening of the Na deposits in the pristine case and the complete absence of this effect for the NaBr-coated electrodes. In parts (**b**, **c**), the error bars represent the standard deviation of measurements taken at different points in the same image

symmetric Na coin cell was cycled galvanostatically at various fixed current densities of 0.25, 0.5, and 1.0 mA cm^{-2} and the voltage profile is reported as a function of time in Fig. 6.6a, b, and c, respectively. For all the different current density values, the voltage profiles are flat. The voltage hysteresis, of the sodium metal plating and stripping, represented by the mid-voltage value of charge and discharge, is plotted in Fig. 6.6d. On comparing results for pristine and NaBr-coated Na electrodes (in Fig. 6.6b, d), it is seen that for cell with pristine Na, the voltage diverges to >0.5 V after 30 h of cycling, meaning that the effective interfacial resistance is >15 kΩ. In contrast, the cell comprising NaBr-coated Na is stable for at least 250 h with minimal rise in cell voltage and hence effective resistance. The divergence in cell resistance is most likely a result of formation of a thick and insulating SEI layer; such an observation with pristine sodium anodes complements the result obtained from visualization experiments explained in Fig. 6.5. However, it is remarkable that with the NaBr coating the overpotential is nearly constant for 250 h, indicating that the NaBr coating produces nearly a tenfold improvement in the cell lifetime.

Postmortem analysis of polarized sodium anodes was done using SEM after continuously electrodepositing at the rate of 1 mA cm^{-2} for 2 h. Figure 6.6e shows the

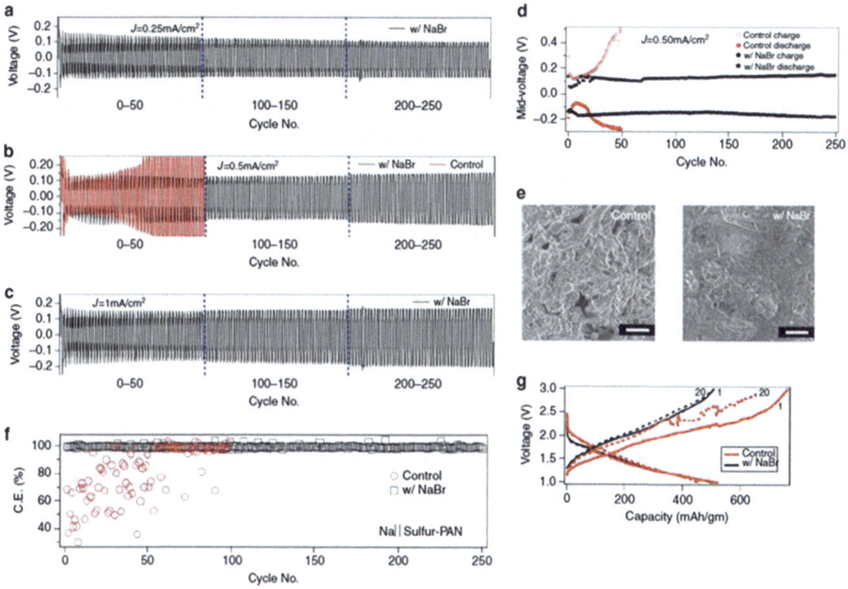

Fig. 6.6 Galvanostatic cycling performance of Na anodes. Voltage profile obtained by consecutively charging and discharging a symmetric sodium cell at current densities of (**a**) 0.25 mA cm^{-2}, (**b**) 0.5 mA cm^{-2}, and (**c**) 1 mA cm^{-2}. In each experiment, one complete cycle was 1 h long, with each charge and discharge time is half an hour. The profiles in black represent sodium metals with NaBr coating and red stands for pristine sodium (control). (**d**) The voltage hysteresis represented by the mid-voltage values of charge and discharge is plotted as a function of cycle no. for NaBr-coated and control sodium cells corresponding to (**b**). (**e**) Morphology of sodium metal electrode obtained from postmortem SEM analysis. The cells were charged at a rate of 1 mA cm^{-2} for 2 h before being taken apart for the postmortem analysis. The left image are results for pristine Na, while the image to the right is for NaBr-coated Na; for both images: scale bar, 5 μm. (**f**) Coulombic efficiency of a Na||Sulfur-PAN composite half-cell as a function of cycle number. (**g**) Voltage profile of Na||Sulfur-PAN during charging and discharging at the 1st and 20th cycle numbers. The red lines/symbols represent control, while black shows the result for NaBr-coated sodium

morphology of the sodium metal with and without NaBr coating. It is seen that the pristine Na electrode develops a nonuniform surface with few protruding sharp structures compared to a relatively smooth surface for the NaBr-coated electrode. It is important to note that in contrast to visualization experiments the general electrodeposition is more uniform, which can be attributed to the stabilizing effect of compression forces exerted by the separator on Na electrodes in the coin cells.

Finally, the effectiveness of the NaBr-coated Na metal was evaluated in a full Na||S electrochemical cell, comprising a sulfur-polyacrylonitrile composite (SPAN) cathode. In this cathode, molecular sulfur is covalently trapped in a PAN framework, which has been reported to completely eliminate polysulfide dissolution and shuttling effects with carbonate based solvents in lithium-sulfur cells [43]. However, unlike their Li counterparts, the anodes in Na||SPAN cells have been reported to develop "black mossy dendrites" within a few cycles, which results in an unstable voltage profile during the cell recharge and low coulombic efficiency [53]. This effect is apparent in the inset to Fig. 6.6g for the Na||SPAN cells based on pristine Na anodes. The lower coulombic efficiency measured in Na||SPAN cells employing pristine Na as anode is also observed from Fig. 6.6f. Because the reactivity of sodium with electrolytes increases with voltage, we conclude that Na||SPAN cell configuration provides a more rigorous assessment of electrode reactivity and stability, in comparison to the Na||Cu type cells [18] used in previous work. Here we observe that coating Na with NaBr protects the metal and in a liquid electrolyte, and in the presence of a conversion cathode, yields coulombic efficiency >99% for at least 250 cycles with minimal fade in discharge capacity as plotted in Supplementary Fig. 6.7a. Also, the SEM image of the Na anode (with NaBr protection) is shown in Supplementary Fig. 6.7b along with the EDX mapping of elements. No dendritic features are seen even after the long-term cycling, and in addition, the sulfur species are absent, indicating immunity of the anode from side reactions. This observation can be attributed to the near complete protection of the metal from parasitic reactions, without compromising ion transport across the SEI. Thus, we conclude that consistent with the JDFT calculations a NaBr-coated Na anode opens the possibility of stable, room-temperature rechargeable sodium metal batteries able to operate using liquid electrolytes.

6.4 Discussion

We used density-functional theory calculations to analyze the surface diffusion barriers for Na and Li adatom transport on various salts. It was observed that the usually formed SEI components on Li, including LiOH and Li_2CO_3, have very high activation energy barriers, which is thought to increase the propensity of the metal to form needlelike, dendrite nucleates. In contrast, energy barriers for adatom diffusion on metal halide salts, including LiF, NaF, and NaBr, are low, with the energy barrier for diffusion on NaBr as low as that of magnesium, which is known to form spherical nucleates on charging. We evaluate these predictions using Na electrodes on which an artificial SEI composed of pure NaBr is used to protect the metal. By means of XRD, EDAX, XPS, and cryo-FIB-SEM measurements, we confirm that NaBr coatings with thicknesses ranging from 2 to 12 μm are achieved on Na. Further, impedance spectroscopy measurements at different temperatures show that NaBr-coated sodium anodes exhibit at least three times lesser interfacial ion transport activation energy compared to pristine sodium. In situ visualization was performed to contrast the electrodeposition stability with and without NaBr layer on sodium anodes. It showed that the NaBr coating not only restricts dendrite formation but also prevents unwanted side reactions between the electrode and electrolyte. This observation is in line with the charge-discharge measurements in symmetric cells as well as in coulombic measurement tests in Na||SPAN type half-cells. Thus, we think that this rational analysis of SEI layers in reactive metal batteries, as well as the methodology of incorporating the desired component in the SEI, can provide a new outlook towards low-cost and long-lasting secondary batteries.

Acknowledgments This work was supported by the Department of Energy, Advanced Research Projects Agency – Energy (ARPA-E) through award #DE-AR0000750. The work made use of electrochemical characterization facilities in the KAUST-CU Center for Energy and Sustainability, supported by the King Abdullah University of Science and Technology (KAUST) through Award # KUS-C1-018-02. Electron microscopy facilities at the Cornell Center for Materials Research (CCMR), an NSF-supported MRSEC through Grant DMR-1120296, were also used for the study. M.J.Z. and L.F.K. acknowledge support by the NSF (DMR-1654596).

Appendix: Supplementary Information

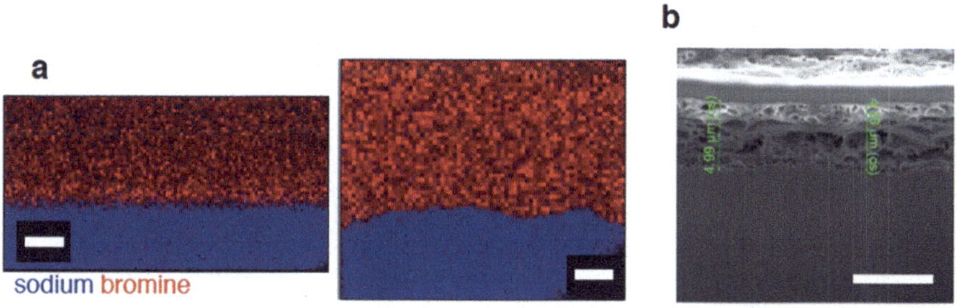

Supplementary Fig. 6.1 Cross-sectional images of NaBr-coated sodium and pristine sodium. (**a**) EDX mapping of cross sections of sodium anodes with NaBr surface layers after reacting for 1 min (left) and 5 min (right). The cross sections were obtained by cryo-focused ion beam milling; scale bar, 3 μm. (**b**) Room-temperature FIB-SEM cross-sectional imaging of pristine sodium shows a 5 μm thick oxidation layer on the surface; scale bar, 5 μm

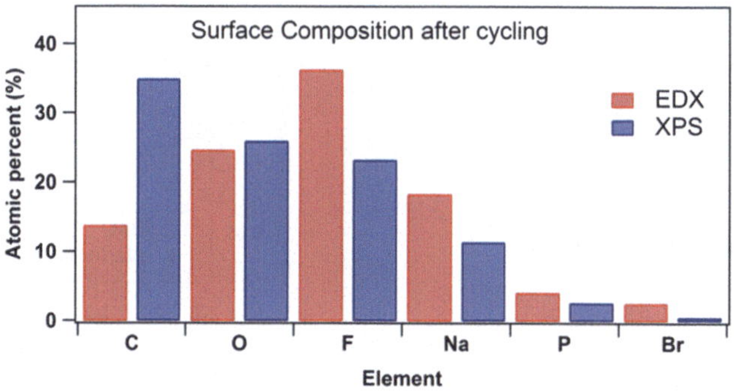

Supplementary Fig. 6.2 Surface composition of cycled sodium metal with NaBr coating obtained by two separate methods of EDX and XPS

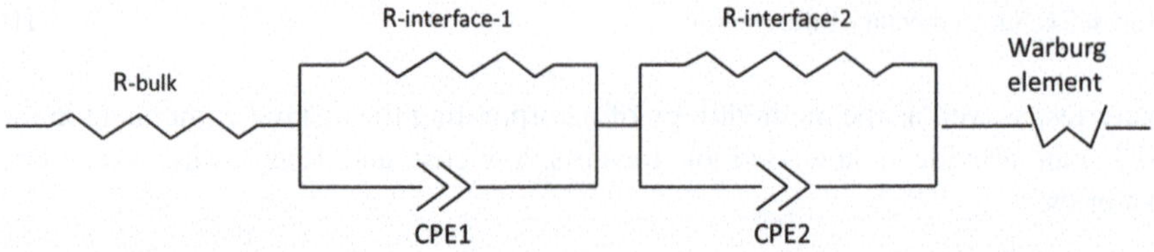

Supplementary Fig. 6.3 Equivalent circuit model for fitting Nyquist plots of impedance measurements. Here, *R*-bulk represents the ion transport in bulk electrolyte. *R*-interface$_1$ represents interfacial resistance associated with the passivation layer between electrode and electrolyte. *R*-interface$_2$ denotes the electronic transport in the interface. CPE1 and CPE2 represent the constant phase elements. Warburg element stands for solid-state diffusion contribution

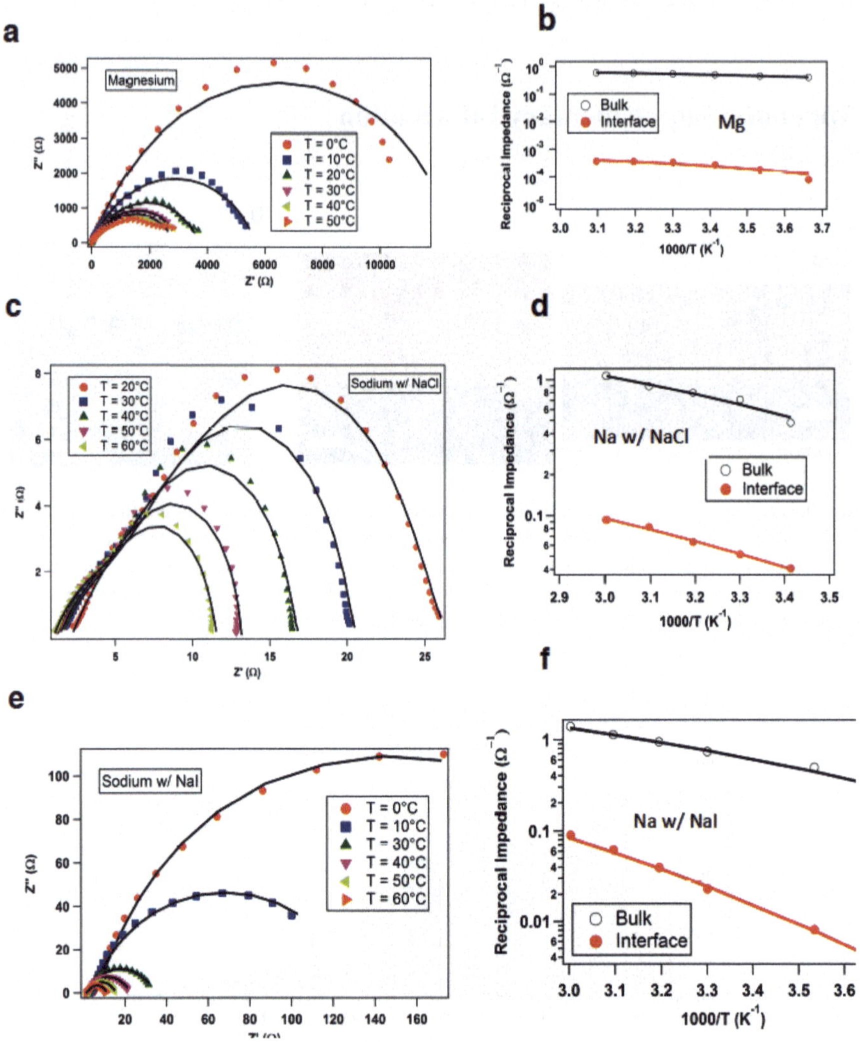

Supplementary Fig. 6.4 Impedance spectroscopy for different anode Nyquist plots at various temperatures for (**a**) Mg, (**c**) Na with NaCl, and (**e**) Na with NaI. Figures (**b**), (**d**), and (**f**) represent the temperature dependence of the reciprocal impedances (both bulk and interface) which is plotted as a function of Arrhenius temperature; the lines represent corresponding VFT fits. The labels represent the type of electrode interface used for the experiment

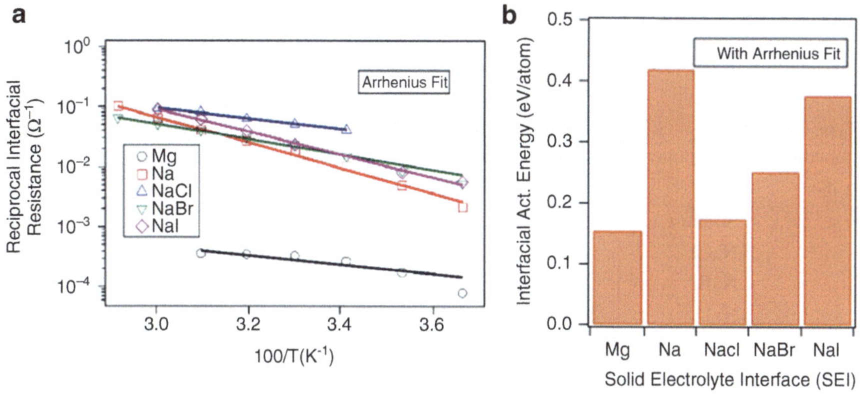

Supplementary Fig. 6.5 Activation energy obtained by Arrhenius analysis. (**a**) Reciprocal of interfacial resistance plotted as a function of inverse temperature. The fits represent prediction from Arrhenius equation; (**b**) the activation energy is plotted for different interface or metal electrodes, obtained by Arrhenius fitting. Mg and Na represent cells with magnesium and sodium electrodes, respectively, without any modification, while NaCl, NaBr, and NaI represent data for respective halide-coated sodium metal

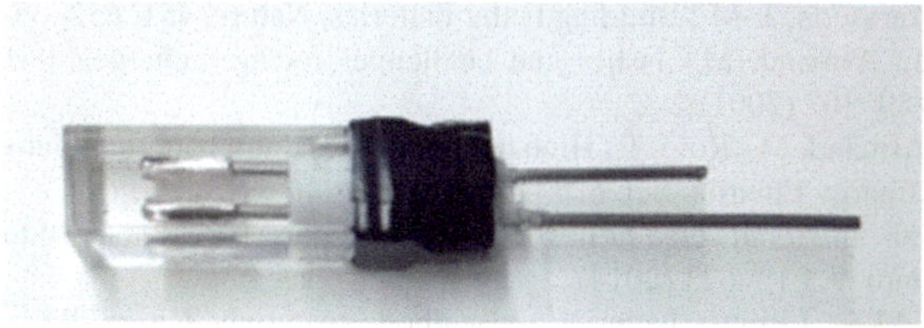

Supplementary Fig. 6.6 Electrochemical setup for in situ visualization of sodium electrodeposition consisting of an airtight cuvette and two rods serving as current collectors for attaching the sodium electrodes. The cap is well sealed with a black tape to ensure that there is no leakage of electrolyte or air contamination

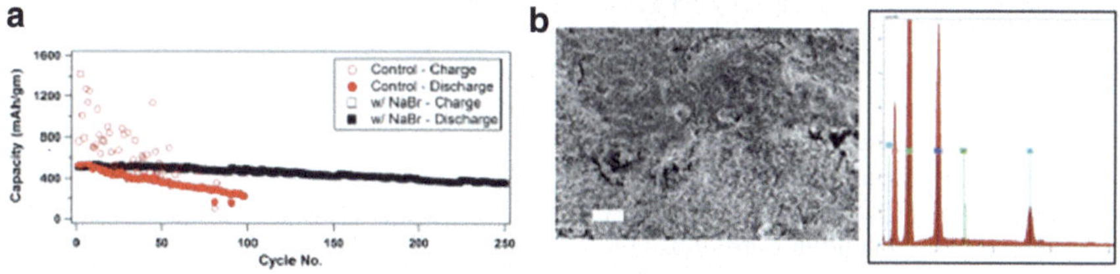

Supplementary Fig. 6.7 Performance and characterization of Na||SPAN cell. (**a**) Charge and discharge capacity as a function of cycle no. for pristine sodium-based half-cell and NaBr-coated sodium-based half-cell comprising of the SPAN cathode. (**b**) SEM image of sodium metal after cycling in Na||SPAN cell with NaBr coating anode; scale bar, 100 μm. The adjacent image shows the EDX mapping of the elements of Na anode

Supplementary Table 6.1 Parameters of VFT analysis bulk impedance represents the ion transport in the electrolyte media, while interfacial impedance indicates the ion transport across electrode-electrolyte interface. The temperature-dependent data is fitted to VFT model

Type of impedance	Component	Prefactor (A) ($'\Omega^{-1}$)	Activation energy (E_a) (kJ mole^{-1})	Activation energy (E_a) (eV atom^{-1})	Reference temp. (T_o) (K)	Goodness of fit (χ^2)
Bulk	Mg	1.38	1.095	0.011	164.88	6.9×10^{-5}
	Sodium	14.86	3.61	0.037	167.25	1.2×10^{-3}
	NaCl	7.64	2.51	0.026	180	6.4×10^{-3}
	NaBr	6.73	2.14	0.022	197.82	1.9×10^{-3}
	NaI	20.30	3.91	0.041	160	2.1×10^{-2}
Interface	Mg	0.0016	1.32	0.014	211	6.1×10^{-9}
	Sodium	755.81	16.94	0.18	112.84	1.4×10^{-4}
	NaCl	2.44	5.20	0.054	140.43	1.4×10^{-5}
	NaBr	2.70	5.48	0.057	166.40	5.4×10^{-6}
	NaI	37.42	9.14	0.095	152.99	2.0×10^{-5}

References

1. Armand, M., Tarascon, J.-M.: Building better batteries. Nature. **451**, 652–657 (2008)
2. Tarascon, J.M., Armand, M.: Issues and challenges facing rechargeable lithium batteries. Nature. **414**, 359–367 (2001)
3. Hueso, K.B., Armand, M., Rojo, T.: High temperature sodium batteries: status, challenges and future trends. Energy Environ. Sci. **6**, 734–749 (2013)
4. Zheng, G., et al.: Interconnected hollow carbon nanospheres for stable lithium metal anodes. Nat. Nanotechnol. **9**, 618–623 (2014)
5. Whittingham, M.S.: Lithium batteries and cathode materials. Chem. Rev. **104**, 4271–4301 (2004)
6. Wei, S., et al.: A stable room-temperature sodium–sulfur battery. Nat. Commun. **7**, 11722 (2016)
7. Wei, S., et al.: Highly stable sodium batteries enabled by functional ionic polymer membranes. Adv. Mater. **29**, 1605512 (2017)
8. Cohn, A.P., Muralidharan, N., Carter, R., Share, K., Pint, C.L.: Anode-free sodium battery through in situ plating of sodium metal. Nano Lett. **17**, 1296–1301 (2017)
9. Choudhury, S., et al.: Designer interphases for lithium-oxygen electrochemical cell. Sci. Adv. **3**, e1602809 (2017)
10. Yadegari, H., Sun, Q., Sun, X.: Sodium-oxygen batteries: a comparative review from chemical and electrochemical fundamentals to future perspective. Adv. Mater. **28**, 7065–7093 (2016)
11. Tu, Z., Nath, P., Lu, Y., Tikekar, M.D., Archer, L.A.: Nanostructured electrolytes for stable lithium electrodeposition in secondary batteries. Acc. Chem. Res. **48**, 2947–2956 (2015)
12. Tikekar, M.D., Choudhury, S., Tu, Z., Archer, L.A.: Design principles for electrolytes and interfaces for stable lithium-metal batteries. Nat. Energy. **1**, 16114 (2016)
13. Choudhury, S., Archer, L.A.: Lithium fluoride additives for stable cycling of lithium batteries at high current densities. Adv. Electron. Mater. **2**, 1500246 (2015)
14. Choudhury, S., Mangal, R., Agrawal, A., Archer, L.A.: A highly reversible room-temperature lithium metal battery based on crosslinked hairy nanoparticles. Nat. Commun. **6**, 10101 (2015)
15. Agrawal, A., Choudhury, S., Archer, L.A.: A highly conductive, non-flammable polymer–nanoparticle hybrid electrolyte. RSC Adv. **5**, 20800–20809 (2015)

16. Wong, D.H.C., et al.: Nonflammable perfluoropolyether-based electrolytes for lithium batteries. Proc. Natl. Acad. Sci. U. S. A. **111**, 3327–3331 (2014)
17. Wu, H., et al.: Stable Li-ion battery anodes by in-situ polymerization of conducting hydrogel to conformally coat silicon nanoparticles. Nat. Commun. **4**, 1943 (2013)
18. Seh, Z.W., Sun, J., Sun, Y., Cui, Y.: A highly reversible room-temperature sodium metal anode. ACS Cent. Sci. **1**, 449–455 (2015)
19. Pan, H., Hu, Y.-S., Chen, L.: Room-temperature stationary sodium-ion batteries for large-scale electric energy storage. Energy Environ. Sci. **6**, 2338 (2013)
20. Tu, Z., et al.: Nanoporous hybrid electrolytes for high energy batteries based on reactive metal anodes. Adv. Energy Mater. **7**, 1602367 (2017)
21. Hallinan, D.T., Mullin, S.A., Stone, G.M., Balsara, N.P.: Lithium metal stability in batteries with block copolymer electrolytes. J. Electrochem. Soc. **160**, A464–A470 (2013)
22. Lu, Y., Tu, Z., Archer, L.A.: Stable lithium electrodeposition in liquid and nanoporous solid electrolytes. Nat. Mater. **13**, 961–969 (2014)
23. Lu, Y., Das, S.K., Moganty, S.S., Archer, L.A.: Ionic liquid-nanoparticle hybrid electrolytes and their application in secondary lithium-metal batteries. Adv. Mater. **24**, 4430–4435 (2012)
24. Schaefer, J.L., Yanga, D.A., Archer, L.A.: High lithium transference number electrolytes via creation of 3-dimensional, charged, nanoporous networks from dense functionalized nanoparticle composites. Chem. Mater. **25**, 834–839 (2013)
25. Bouchet, R., et al.: efficient electrolytes for lithium-metal batteries. Nat. Mater. **12**, 452–457 (2013)
26. Feng, S., et al.: Single lithium-ion conducting polymer electrolytes based on poly[(4-styrenesulfonyl)(trifluoromethanesulfonyl)imide] anions. Electrochim. Acta. **93**, 254–263 (2013)
27. Bertasi, F., et al.: Single-ion-conducting nanocomposite polymer electrolytes for lithium batteries based on lithiated-fluorinated-iron oxide and poly(ethylene glycol) 400. Electrochim. Acta. **175**, 113–123 (2015)
28. Guo, J., Wen, Z., Wu, M., Jin, J., Liu, Y.: Vinylene carbonate–$LiNO_3$: a hybrid additive in carbonic ester electrolytes for SEI modification on Li metal anode. Electrochem. Commun. **51**, 59–63 (2015)
29. Li, W., et al.: The synergetic effect of lithium polysulfide and lithium nitrate to prevent lithium dendrite growth. Nat. Commun. **6**, 7436 (2015)
30. Pires, J., et al.: Role of propane sultone as additive to improve the performance of lithium-rich cathode material at high potential. RSC Adv. **5**, 42088–42094 (2015)
31. Aurbach, D., et al.: On the use of vinylene carbonate (VC) as an additive to electrolyte solutions for Li-ion batteries. Electrochim. Acta. **47**, 1423–1439 (2002)
32. Chen, L., Wang, K., Xie, X., Xie, J.: Effect of vinylene carbonate (VC) as electrolyte additive on electrochemical performance of Si film anode for lithium ion batteries. J. Power Sources. **174**, 538–543 (2007)
33. Etacheri, V., et al.: Effect of fluoroethylene carbonate (FEC) on the performance and surface chemistry of Si-nanowire li-ion battery anodes. Langmuir. **28**, 965–976 (2012)
34. Miao, R., et al.: A new ether-based electrolyte for dendrite-free lithium-metal based rechargeable batteries. Sci. Rep. **6**, 21771 (2016)
35. Li, B., Xu, M., Li, T., Li, W., Hu, S.: Prop-1-ene-1,3-sultone as SEI formation additive in propylene carbonate-based electrolyte for lithium ion batteries. Electrochem. Commun. **17**, 92–95 (2012)
36. Choudhury, S., et al.: Designer interphases for the lithium-oxygen electrochemical cell. Sci. Adv. **3**, 1–12 (2017)
37. Li, N., Yin, Y., Yang, C., Guo, Y.: An artificial solid electrolyte interphase layer for stable lithium metal anodes. Adv. Mater. **28**, 1853–1858 (2016)
38. Ye, H., et al.: Synergism of Al-containing solid electrolyte interphase layer and Al-based colloidal particles for stable lithium anode. Nano Energy. **36**, 411–417 (2017)
39. Miao, R., et al.: Novel dual-salts electrolyte solution for dendrite-free lithium-metal based rechargeable batteries with high cycle reversibility. J. Power Sources. **271**, 291–297 (2014)

40. Qian, J., et al.: High rate and stable cycling of lithium metal anode. Nat. Commun. **6**, 6362 (2015)
41. Ha, S., et al.: Magnesium (II) bis(trifluoromethane sulfonyl) imide-based electrolytes with wide electrochemical windows for rechargeable magnesium batteries. ACS Appl. Mater. Interfaces. **6**, 4063–4073 (2014)
42. Jäckle, M., Groß, A.: Microscopic properties of lithium, sodium, and magnesium battery anode materials related to possible dendrite growth. J. Chem. Phys. **141**, 174710 (2014)
43. Wei, S., et al.: Metal-sulfur battery cathodes based on PAN-sulfur composites. J. Am. Chem. Soc. **137**, 12143–12152 (2015)
44. Chazalviel, J.-N.: Electrochemical aspects of the generation of rampified metallic electrodeposits. Phys. Rev. A. **42**, 7355–7367 (1990)
45. Ozhabes, Y., Gunceler, D., Arias, T.A.: Stability and surface diffusion at lithium-electrolyte interphases with connections to dendrite suppression. arXiv. **1504.05799**, 1–7 (2015)
46. Gunceler, D., Letchworth-Weaver, K., Sundararaman, R., Schwarz, K.A., Arias, T.A.: The importance of nonlinear fluid response in joint density-functional theory studies of battery systems. Model. Simul. Mater. Sci. Eng. **21**, 74005 (2013)
47. Gunceler, D., et al.: Nonlinear solvation models: Dendrite suppression on lithium surfaces. 16th International Workshop on Computation Physics and Materials Science: Total Energy and Force Methods, 10 (2013)
48. Gunceler, D., Arias, T.A.: Universal iso-density polarizable continuum model for molecular solvents. arXiv. **1403.6465**, 1–11 (2014)
49. Garrity, K.F., Bennett, J.W., Rabe, K.M., Vanderbilt, D.: Pseudopotentials for high-throughput DFT calculations. Comput. Mater. Sci. **81**, 446–452 (2014)
50. Zangmeister, D.Z., Turner, A.T., Pemberton, E.P.: Segregation of NaBr/NaCl crystals grown from aqueous solutions: Implications for sea salt surface chemistry. Geophys. Res. Lett. **28**, 995–998 (2001)
51. Wang, X., et al.: A chelation strategy for in-situ constructing surface oxygen vacancy on {001} facets exposed BiOBr nanosheets. Sci. Rep. **6**, 24918 (2016)
52. Naumkin, A.V., Kraut-Vass, A., Powell, C.J.: NIST X-ray Photoelectron Spectroscopy Database, Version 4.1. National Institute of Standards Technology, Gaithersburg (2012)
53. Wang, J., et al.: Room temperature Na/S batteries with sulfur composite cathode materials. Electrochem. Commun. **9**, 31–34 (2007)

Chapter 7
Electroless Formation of Hybrid Lithium Anodes for High Interfacial Ion Transport

7.1 Introduction

Secondary batteries comprising of metallic anodes (e.g., Li, Na, Al) have drawn significant recent attention due to their promise for enabling large (as high as tenfold) increase in the anodic capacity, in comparison to state-of-the-art lithium-ion anode chemistry [1, 2]. A major barrier to the successful implementation of such anodes is the uneven electrodeposition of the metal in the charging process, which leads to formation of rough, so-called densdrite structures. These structures are diffusion-limited and as such under extended battery operation can grow to fill the interelectrode space, short-circuiting the battery (often accompanied with fire and/or cell explosion) [3–5]. Previously, using continuum modeling, Tikekar et al. [6, 7] proposed that dendrite-induced short circuits can be prevented using electrolytes with moderate mechanical modulus (10 MPa) if a fraction of the anions are tethered to prevent cell polarization at high current densities and to maintain electrolyte conductivity in the region near the anode. There is also a large body of experimental work showing that dendrite growth can be regulated using solid-state electrolytes [8], nanoporous polymers, or ceramics that host liquids in their pores [9–11] and cross-linked polymers [12, 13] as well as with high transference number gel-like electrolytes [14–17]. Although researchers are now able to address the safety concerns by variety of means, this progress has difficult challenges associated with uncontrolled parasitic reactions between liquid electrolytes and a reactive metal electrode [18, 19].

In rechargeable batteries that employ metallic anodes, particularly those based on Li and Na, the electrolyte would ideally react with the metal surface to form a stable and self-limiting passivation layer termed the solid electrolyte interphase (SEI). Ion and mass transport through this poorly understood interfacial layer are crucial for the long-term stability of any rechargeable battery that utilizes a reactive metal anode. Continuous expansion and contraction of the electrode during cycles of battery charge and discharge unfortunately degrade the SEI, exposing the native metal to

S. Choudhury, *Rational Design of Nanostructured Polymer Electrolytes and Solid–Liquid Interphases for Lithium Batteries*, Springer Theses,
https://doi.org/10.1007/978-3-030-28943-0_7

the electrolyte, thus forming further reaction products [3, 4]. The parasitic reactions can be checked by minimizing the exposed surface to the electrolyte, which can be attained by either implementing an artificial protection film on the electrode [17, 20] or through physical means that enable smooth (flat) electrodeposition using appropriate salt chemistries [21, 22]. Zheng et al. [23], for example, utilized interconnected hollow carbon nanospheres for fabricating a protective film able to shield a Li metal anode from the electrolyte media resulting in high coulombic efficiencies. Similar methodologies based on ex situ coating processes such as atomic layer deposition of metal oxides, most notably alumina [24] on Li and electrospun polymer fibers [25], have also been reported to protect Li from reacting with liquid electrolytes. Further, there are several recent studies based on in situ generation of specialized interface materials using electrolyte additives including liquids, such as fluoroethylene carbonate (FEC) [26] and salts like LiFSI [27], $LiPF_6$ [28], LiF [29], and CsF [21].

The successes of these methods in simultaneously suppressing dendritic deposition and parasitic side reactions all appear to hinge upon formation of a SEI enriched with species such as LiF that facilitate fast Li-ion diffusion at the electrolyte-metal interface. It also underscores the importance of fundamentally based strategies able to control the dendrite nucleation processes at reactive metal/liquid electrolyte interfaces. Previous work has shown that low energy barriers for surface diffusion, i.e., high predicted diffusion rates, correlate well with long short circuit times in experiment, i.e., low dendrite formation rate, for lithium [20, 30] as well as sodium batteries [31]. Low diffusion barriers also correlate well with low surface energies [30, 32, 33], presumably because this results in weaker binding of electrodeposited atoms overall, with small energy differences between the weakly and tightly bound spots, and therefore a low barrier for diffusion. These approaches therefore all appear to take advantage of processes that compete with growth of high-curvature regions on the electrode surface.

Here, we report a new approach to diffusion barrier minimization that exploits the effects of the solvent (and electrolyte) at the interface. The key idea is to utilize strong interactions of the solvent with the electrodeposited atom to weaken its binding to the electrode surface and flatten the energy landscape for atom motion in the plane. Aprotic solvents used in battery electrolytes will interact most strongly with charged species. Stable charging of the surface atom should be possible under these conditions by employing a difference in electropositivity between the deposited atom and the electrode. We illustrate these ideas using indium metal coatings on lithium metal anodes by an in situ electroless plating technique. The high electropositivity of lithium relative to indium is expected to result in (partially) positively charged lithium atoms on the In surfaces.

7.2 Results

In order to evaluate the energetic landscapes at the electrode-electrolyte interphase, we perform density functional theory calculations of lithium atom diffusion on the surface of indium, both in vacuum and in an electrolyte environment described using a previously established continuum solvation methodology [33, 34]. The two most stable surfaces of indium, (011) and (001), are very close in surface energy, so we examine lithium ion diffusion on both these surfaces (Supplementary Table 7.1 and Fig. 7.1). In vacuum, we find a large difference between the diffusion barriers on the two surfaces. The highest-density In(011) surface has a small diffusion barrier of 0.04 eV (less than 2 k_BT) between adjacent hollow sites, whereas the In(001) surface has a much larger barrier of 0.30 eV (~11 k_BT), also in a path connecting adjacent hollow sites over a bridge site. Supplementary Table 7.1 compares the predicted surface energies of the low-index surfaces of indium in vacuum as well as solution, with previous calculations for other metal electrodes. Importantly, the surface energies of indium are comparable to the other metals (larger than Na but smaller than Li and Mg) and are not significantly affected by the solvent.

The solvent dramatically alters the energy landscape for Li diffusion on both these surfaces. For the (011) surface, the minimum energy path is qualitatively similar (connecting adjacent hollow sites), but the barrier reduces to 0.015 eV ($<k_BT$). For the (001) surface, the change is much more dramatic: the stable binding sites switch from the hollow to the atop sites because the latter are more accessible by the solvent and hence stabilized further. The barrier drops to 0.013 eV ($<k_BT$) for this surface as well. Lowdin charge analysis suggests a charge ranging from +0.3 to +0.5 for the Li atom on the solvated In surfaces; while the exact value of atomic charges is not particularly meaningful, this does agree with the qualitative picture of solvent stabilization in this case. In summary, our JDFT calculations reveal a dramatic reduction in the diffusion barriers for Li on solvated *In* surfaces due to solvent stabilization. For the two most stable surfaces of In, the barriers are less than $k_BT \sim 0.026$ eV at room temperature and smaller than those reported for other electrode or coating materials [32]. Significantly, as seen from the bar chart in Fig. 7.1e, the diffusion barrier is much lower than the commonly generated SEI component lithium carbonate but is comparable to that of magnesium, which is known to electrodeposit at moderate currents in the diffusion-controlled regime, without dendrite formation.

We choose to focus on indium for this first study because in addition to its low surface diffusion barrier of indium metal, an additional advantage is its property of alloying with lithium. Commonly used intercalating materials like graphite, tin and silicon are known to form carbonates and oxides by side reactions. Here, we utilize indium metal coating, which is relatively inert and nonreactive with most electrolyte components. Figure 7.2a shows the schematic of the proposed lithium metal protection technique using indium metal. In this method, $In(TFSI)_3$ salt is used in the electrolyte as additive to form the electroless coating by reduction reaction given by: $3Li + In(TFSI)_3 \rightarrow 3LiTFSI + In$. The predicted field lines of Li ions indicate the enhanced surface diffusion, followed by alloying in the indium buffer layer before electrodeposition. To demonstrate the electroless plating method, lithium metal anodes were dipped in a solution of 12 mM $In(TFSI)_3$ in EC/DMC (1:1) and allowed to incubate for 6 h, followed by washing and drying under vacuum. Figure 7.2b shows the treated lithium surface using scanning electron microscopy (SEM). The chemical composition was mapped using energy-dispersive X-ray spectroscopy (EDX) shown in Fig. 7.2c. It is seen that indium is evenly plated on the lithium surface. Further, X-ray diffraction (XRD) was utilized to analyze the surface of

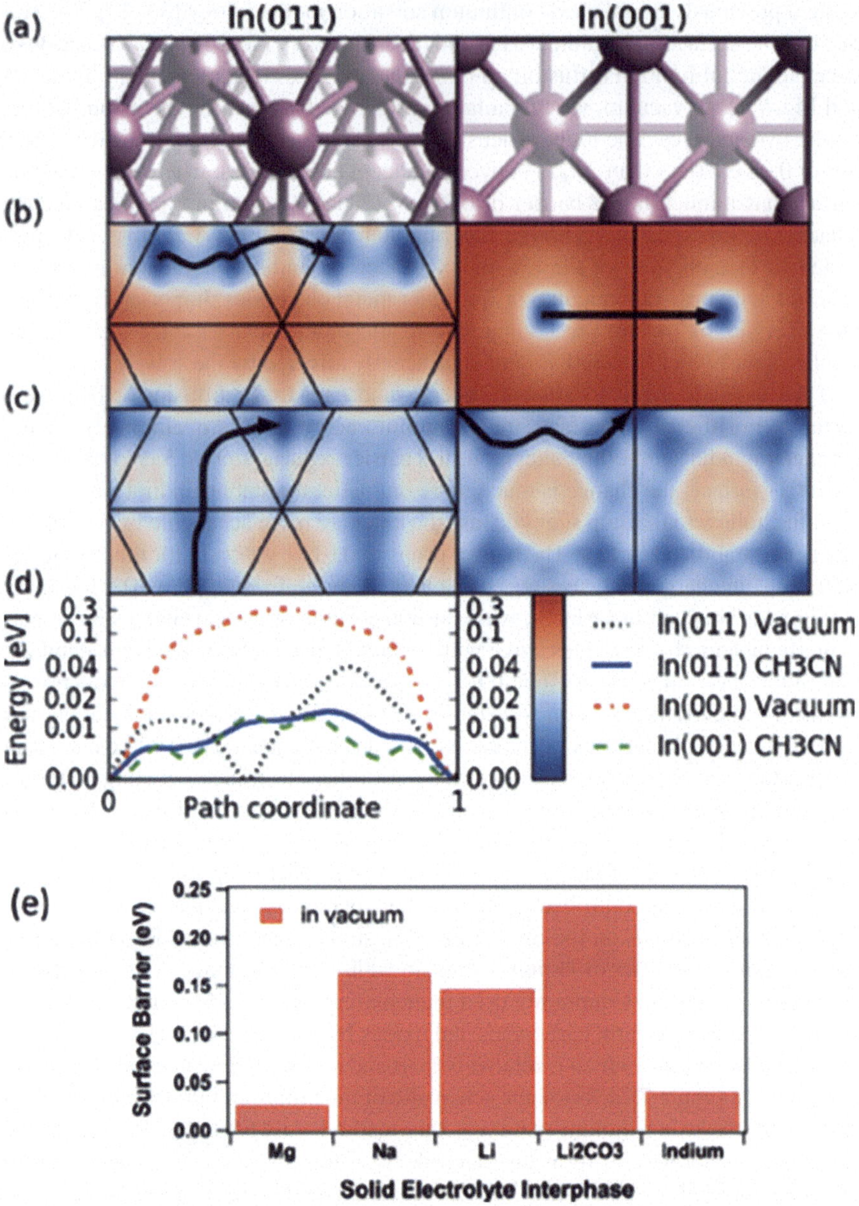

Fig. 7.1 Joint density-functional theory for ion transport on indium surface: (**a**) surface structures and energy landscapes of Li diffusion in (**b**) vacuum and (**c**) acetonitrile solvent on the In(011) surface (left panels) and the In(001) surface (right panels). The arrows annotate the minimum energy paths and (**d**) compares the energy along the path for all four cases from (**b**) and (**c**). Note that the energy axis and color scale are mapped nonlinearly (as $y = E/(E + k_BT)$) to show features on the k_BT scale while capturing the larger energy range of the data. Solvation substantially reduces the energy barriers for Li diffusion on In(011) as well as In(001); (**e**) comparison of surface diffusion barrier of indium with other interphases reported before, including Mg, Na, Li, and Li_2CO_3

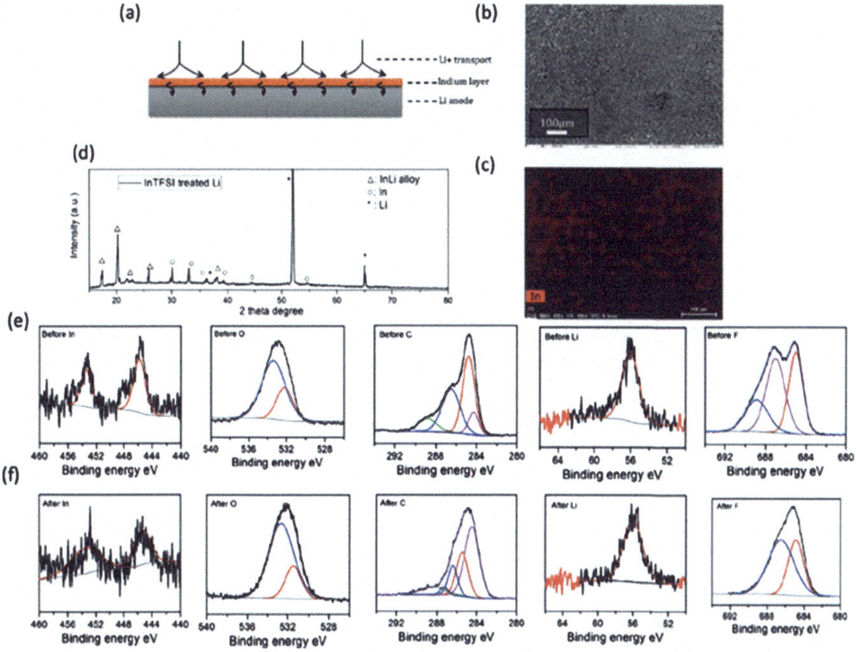

Fig. 7.2 Surface characterization of indium metal-coated lithium anode: (**a**) schematic of predicted Li^+ ion transport near the interface and through indium layer on lithium. (**b**) SEM image of lithium metal after treatment with *In(TFSI)₃* solution. (**c**) EDX mapping of indium atom of lithium after surface treatment. (**d**) XRD showing presence of indium metal and In-Li alloy phases on the surface of lithium. Here ∗ indicated Li, ○ is Indium, and Δ stands for In-Li alloy. (**e**) Energy spectra obtained by XPS measurement of lithium electrode surface after cycling at low current density of 0.1 mA cm^{-2}. (**f**) Energy spectra of lithium anode after cycling at high current density of 1 mA cm^{-2} for ten cycles

indium-coated lithium metal as shown in Fig. 7.2d. The presence of In and In-Li alloy peaks in addition to lithium XRD peak confirms the indium metal formation by the reaction of $In(TFSI)_3$ with lithium. We later show that it is possible to create a self-healing version of these coatings by adding 12 mM $In(TFSI)_3$ as a co-salt in a standard battery electrolyte.

Figure 7.2e shows the X-ray photon spectroscopy (XPS) results for the lithium electrode after conditioning at a low current density of 0.1 mA cm^{-2} for ten cycles in a symmetric cell with the electrolyte 1 M $LiPF_6$ EC/DMC + 12 mM $In(TFSI)_3$. Figure 7.2f reports the analogous results when the electrodes are cycled for ten times at 1 mA cm^{-2}. The In 3d spectra shows two fully split peaks of $3d_{3/2}$ and $3d_{5/2}$ at 352 eV and 344 eV, respectively, for both before and after cycling at high current densities [35]. The presence of metallic indium peaks even after cycling indicates that the coating does not break down during the expansion and contraction of lithium electrode during charge and discharge, respectively. Typical peaks of 55 eV in Li 1s representing organometallic compounds are further seen. The C1s spectra peaks (284.8, 285.5, 286, and 288.5 eV) confirm the presence of $ROCO_2Li$ and polycarbonates before and after cycling [35–37]. An additional peak of 299.9 eV is seen in the C 1s spectra, after the high current density cycling of the cell; this represents additional formation of carbonates [37]. The peaks of 533.5 and 531 eV in case of O 1s spectra also confirm the presence of carbonates and $ROCO_2Li$ species [38]. In case of F1 spectra, 685 eV peak indicates LiF, while that at 688–689 eV represents organo-fluorides (C–F bond), with a shift in the energy for CF_2 and CF_3 as seen in the spectra before cycling. The LiF here is predicted to be generated by the interaction between the anion species of $TFSI^{-1}$ and PF_6^{-1}. As LiF generation is often associated with degradation of PF_6^{-1}, its absence confirms the ability of indium layer in preventing side reactions. Overall, it can be concluded that indium metal forms a conformal coating on the lithium anode along with other carbonate species. Further it is confirmed that the chemical composition of this coating is unperturbed even after long-term cycling. The approach therefore provides a facile method for creating In-Li electrodes that take advantage of the electrochemical attributes of both metals.

We performed impedance spectroscopy measurements for control electrolyte comprising of 1 M $LiPF_6$ EC/DMC and electrolyte reinforced with 12 mM $In(TFSI)_3$. Figure 7.3a shows the Nyquist diagram obtained at 30 °C; the lines are fits obtained using the equivalent circuit diagram shown in Supplementary Fig. 7.1, while Supplementary Fig. 7.2 shows the Nyquist plots at various temperatures. It is seen that the interfacial resistance of the cell using the control electrolyte is essentially unchanged relative to the one where which $In(TFSI)_3$ was introduced. However, after five cycles at a low current density (0.1 mA cm^{-2}), the interfacial resistance of the latter cells is seen to reduce significantly, by about ten times. The high interfacial resistance initially observed is consistent with the blockage of direct electrolyte access to the Li electrode. The fact that the cell without initial conditioning has the same interfacial resistance as control cell is also unsurprising because it is known that lithium metals form a native coating of oxides and carbonates even with slightest exposure of organic solvent and oxygen (even inside glove box). The low current density conditioning thus appears to have the effect of cleaning off the surface for in situ reaction fresh Li with indium salts. Further the interfacial conductance or reciprocal of interfacial resistance is plotted in Fig. 7.3b for all three cases and fitted by Arrhenius rule, $R_{int}^{-1} = A \exp(-B/T)$, and the values of activation energy B, which are tabulated in Supplementary Table 7.2. The results show that the interfacial activation energy for a Li anode in the presence of indium salts without conditioning (0.39 eV/atom) is similar to that of pristine lithium (0.43 eV/atom) in the same baseline electrolyte. After generation of the indium coating on anode, however, the value drops to 0.29 eV/atom, which is a substantial decrease in comparison with the thermal energy kT. As previously reported [31], the lower interfacial activation energy reflects the reduced diffusion barrier for surface transport of Li ad atoms on the In surface.

Figure 7.3c reports results of the cyclic voltammetry analysis in a two-electrode cell comprised of Li foil as reference as well as counter electrode and indium metal (100 μm) as the working electrode using 1 M $LiPF_6$ EC/DMC electrolyte. On backward scanning, a positive current peak of -4 mA/cm^2 appears at ~0.5 V (vs. Li/Li^+),

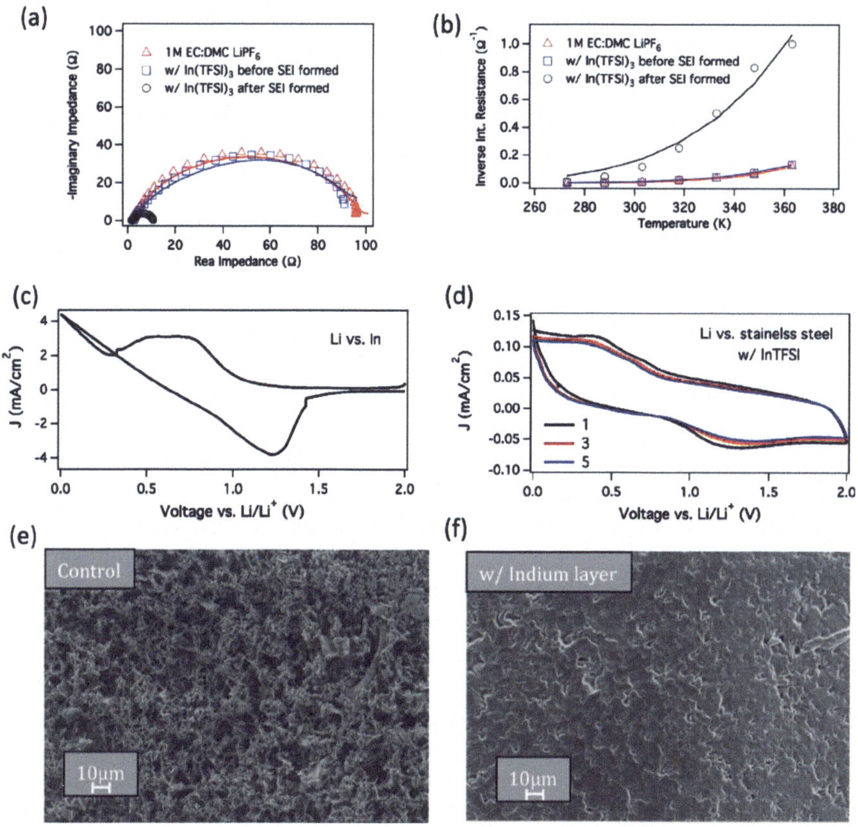

Fig. 7.3 Electrochemical and morphological analysis of interfacial ion transport: (**a**) Nyquist plots for control electrolyte as well as *In(TFSI)*$_3$-added electrolytes before and after condition at low current density obtained by impedance spectroscopy measurements. (**b**) Reciprocal of interfacial impedance obtained by circuit model fitting of impedance curves. The solid lines represent Arrhenius plot for interfacial impedances. (**c**) Cyclic voltammetry for Li- vs. indium-type cell performed at the rate of 1 mV/s. (**d**) Cyclic voltammetry of Li vs. stainless steel battery with *In(TFSI)*$_3$–added electrolyte. The scan rate is 1 mV/s. Morphology of lithium deposition on stainless steel at the rate of 0.5 mA/cm^2 for 6 h with (**e**) control electrolyte and (**f**) electrolyte with addition of *In(TFSI)*$_3$. Images obtained using scanning electron microscopy

indicative of lithium alloying in indium metal, while on reverse scanning, a deintercalation peak appears at ~1.2 V (vs. Li/Li$^+$). A similar experiment was conducted in a lithium versus stainless steel configuration using 12 mM In(TFSI)$_3$ salt additive, and the results are plotted in Fig. 7.3d. Current peaks at 0.5 and 1.2 V (vs. Li/Li$^+$) are also observed, however, with lower magnitude. These peaks are associated with the lithium alloying and de-alloying from the in situ formed indium layer on stainless steel. During battery operation, this layer is hypothesized to act as buffer passage for consequent plating on underlying Li layer. Further, it is seen that the peak heights do not change significantly over five cycles, indicating good stability of indium layer during the charge-discharge processes. In comparison, on cycling the thick indium electrode, there is significant shift in the current peaks, and it becomes noisy (see Supplementary Fig. 7.3). A simplistic calculation was done to calculate the thickness of the in situ formed In layer based on the measured capacity of the indium foil (100 μm), and it was found to be 1.25 μm thick. This layer is evidently much thicker than spontaneously formed SEI coatings on reactive metals such as Li, which leads us to describe the resultant anodes as In-Li *hybrid anodes* because of their ability to utilize both reversible alloying and electroplating of Li ions for energy storage.

The morphology of Li deposits, with and without indium coating, was obtained using scanning electron microscopy (SEM) after plating stainless steel electrode at the rate 0.5 mA/cm^2 for 6 h. It is seen that for the control electrolyte, lithium deposition is rough and needlelike, while with indium protection, it is smooth and uniform. Since the lithium deposits with control electrolyte are more perforated, there is a higher degree of contact between the electrolyte and electrode resulting in high side reaction sites; also the lithium spikes can cause rupture in the SEI layer. Also, it is important to note that the interconnected and flat deposits achieved with In protection testify to the low energy barrier for Li-ion diffusion on indium surface, similar to previously observed magnesium electrodeposition [39].

It is hypothesized that the macroscopic morphological evolution during electrodeposition is dependent on the interfacial ion mobility and initial nucleation process. We performed direct visualization experiments under optical microscope to monitor the electrodeposition on longer time scales. Specifically, an in-house built visualization setup was utilized (as shown in Supplementary Fig. 7.4) comprised of a lithium rod as one electrode and a stainless disc as the counter electrode, with electrolyte filled in the center tube. Using a current density of 8 mA cm^{-2}, morphological changes of the electrode were recorded at equal time intervals up to 15 min (2 mAh cm^{-2}), as shown in Fig. 7.4a. It can be seen that for the control electrolyte (1 M LiTFSI-PC), deposition morphology is uneven and rough compared to the smooth and compact deposits with the indium coating. Further, in comparison to the electrodeposition with indium salt additives, the lithium deposits with the PC-based neat electrolyte are seen to be darker in color; this is an indication of instantaneous side reactions because of an unstable SEI [3]. The quantitative comparison of the morphological changes was done by plotting the height of deposit at different times as shown in Fig. 7.4b. Using a linear fit, the growth rate of pristine lithium deposits were found to be ~90 nm/s, while that of In-Li electrode was ~12 nm/s. Thus, it is evident that the indium layer results in improved long-term electrodeposition stability.

Figure 7.4c shows the voltage profile of a galvanostatic charge-discharge experiment for a symmetric lithium cell, where the cell was charged and discharged for 1 h at a current density of 1 mA/cm^2. The control electrolyte is 1 M LiTFSI in PC, while in the other case, 12 mM *In(TFSI)*$_3$ salt was added. It can be seen that within ~30 cycles the overpotential of the control cell increases up to over twice its initial value. This can be attributed to the formation of an insulating SEI layer by continuous side reactions between Li and the electrolyte, which hinder ion transport at the

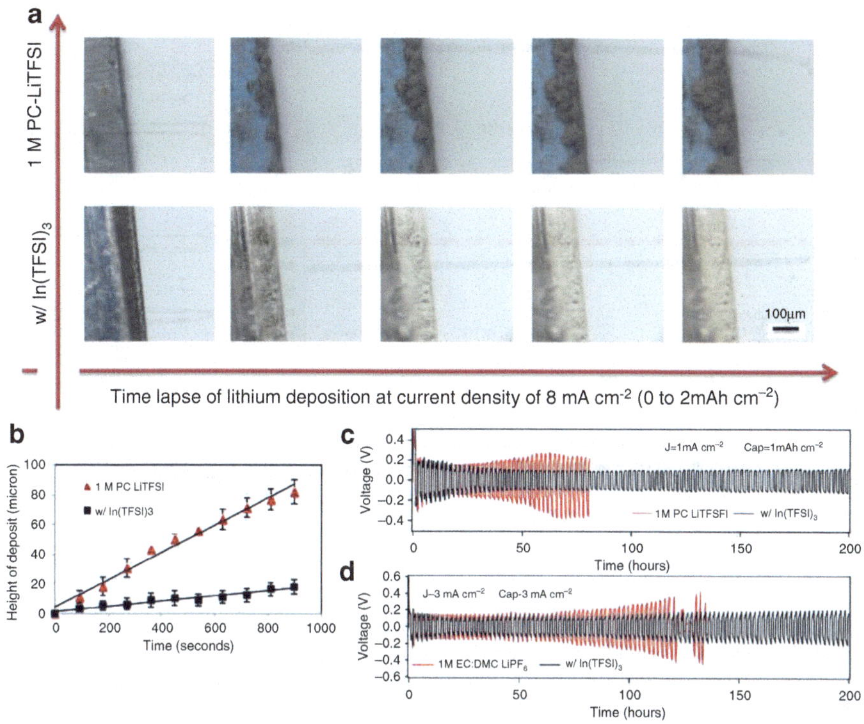

Fig. 7.4 Inhibition of dendrite growth during electrodeposition process. (**a**) Snapshots of stainless steel electrode during electrodeposition of lithium, obtained by direct visualization under optical microscope. The first row represents results with the electrolyte 1 M LiTFSi in PC, while in the second row, 12 mM *In(TFSI)$_3$* was added. The electrochemical cells were operated at a current density of 8 mA cm^{-2}. The scale bar is for all the images. (**b**) Height of individual dendrite as a function of time for control and *In(TFSI)$_3$*-added electrolyte, obtained by visualization experiments. (**c**) Voltage profile for symmetric lithium cell at current density of 1 mA/cm^2 with each half cycle is 1 h. The red curve represents control cell with the electrolyte 1 M LiTFSI in PC, while the black is with *12 mM In(TFSI)$_3$* added in the same electrolyte. (**d**) Voltage profile of symmetric cells cycled at 3 mA/cm^2 with each half cycle is 1 h. Here the control electrolyte, shown in red, is 1 M $LiPF_6$ EC/DMC, and the black lines represent cell with the same electrolyte and *12 mM In(TFSI)$_3$* additive

interface. In contrast, cells based on the In-Li anodes continue to cycle with lower overpotentials for over 200 h. A similar experiment was done using the base electrolyte comprised of commercial-grade 1 M LiPF6 in EC/DMC (1:1 by vol.), and the cells were cycled at a current density of 1 mA/cm^2 for the same charge and discharge time of 1 h (Fig. 7.4d). It is seen that after 120 h of cycling, there is a sudden drop in the voltage, following rise in overpotential, which is indicative of internal short circuit; in contrast long-term stability is attained with indium coating.

Figure 7.5a reports the coulombic efficiency (CE) measurements in cells comprising of Li and stainless steel electrodes. In this experiment, lithium is deposited

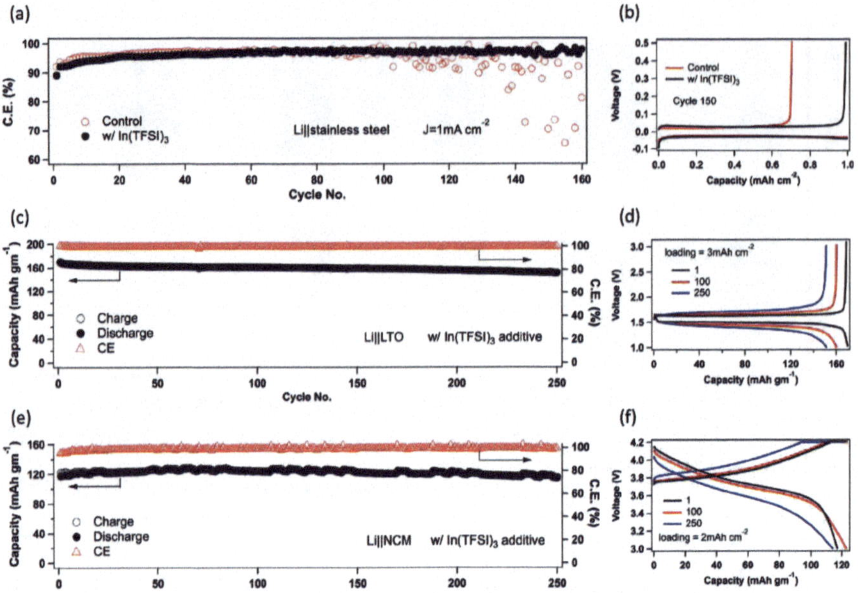

Fig. 7.5 Long-term galvanostatic cycling performance: (**a**) coulombic efficiency measurement in Li||stainless steel configuration at a current density of 1 mA/cm^2 and capacity of 1 mAh/cm^2. The red marker is control electrolyte of 1 M $LiPF_6$ in EC/DMC, and red is with 12 mM *In(TFSI)*$_3$ additive. In both the types of electrolyte 10% (by vol.) of fluoroethylene carbonate (FEC) have been added. (**b**) Voltage profile of the Li||stainless steel cells for the 150th cycle. (**c**) Half-cell cycling with lithium titanate cathode using the electrolyte of 1 M EC/DMC $LiPF_6$ with the additives 12 mM *In(TFSI)*$_3$ and 10% (by vol.) FEC. The areal capacity is 3 mAh/cm^2 and the C-rate is 1 C. (**d**) Voltage profiles for the Li||LTO cell for cycle numbers 1, 100, and 250. (**e**) Cycling with lithium nickel cobalt manganese oxide cathode using the same electrolyte. The areal capacity is 2 mAh/cm^2. The charging cycling rate is C/2 and the discharging rate is 1 C. (**f**) Voltage profiles for the Li||NCM for 1st, 100th, and 250th

onto the stainless steel at a current density of 1 mA/cm^2 for 1 h and then stripped back at the same current density until the voltage rises to 0.5 V and the CE is defined as ratio of stripping and plating time. The base electrolyte (control) used here comprises of 1 M $LiPF_6$ in EC/DMC with addition of 10% (vol.) fluoroethylene carbonate (FEC), which is known to stabilize the electrode interphase by generating lithium fluoride [26], while in the other cell, the base electrolyte was reinforced with 12 mM *In(TFSI)*$_3$. It can be seen in Fig. 7.5a that the initial CE for both electrolytes is >95%; however for the control cell, the CE begins to fluctuate and reverts to <70% at higher cycle numbers (see Fig. 7.5b); in contrast cells based on the In-Li electrodes maintain high CE values (>95%) for over 150 cycles. The fluctuations in the CE can be shown to arise from sporadic electrical connections of broken pieces of lithium from previous cycles (so-called orphaned lithium) in the cell. The addition of *In(TFSI)*$_3$ salt is hypothesized to result in formation of an indium layer not only on the Li anode but also on the bare stainless steel electrode, thus preventing side reactions when fresh Li is deposited on either electrode. On this basis it is argued that the protection technique should be applicable for stabilizing lithium metal anodes in presence of reactive solvents like dimethylacetamide or dimethyl sulfoxide, which are commonly utilized in lithium-oxygen batteries [20, 40].

The effectiveness of protection mechanism was finally evaluated in full cells comprising of Li||LTO with a very high mass loading cathode (24 mg/cm^2 or 3 mAh/cm^2) at a rate of 1 C. Figure 7.5c and d report the capacity, coulombic efficiency and voltage profiles of these cells. It is seen that the cell shows over 85% capacity retention and high coulombic efficiency for at least 250 cycles. Further, we utilized the indium protection method with high-voltage nickel manganese cobalt oxide (NCM) cathode using the same electrolyte (1 M $LiPF_6$ EC/DMC (10% FEC, 12 mM *In(TFSI)*$_3$)) as shown in Fig. 7.5e and f. The cell was charged at a fixed rate of 0.5 C and discharged at a 1 C. Since, $TFSI^-$ anion is unstable at high voltages, a very long charge capacity is observed for the initial cycle, shown in Supplementary Fig. 7.5. Similar to cells based on LTO cathodes, the coulombic efficiency of the cell remains high (>98%) for at least 250 cycles, while there is close to 95% capacity retention. We believe that the long cycle life of the hybrid In-Li anode (paired with a high-voltage cathode) is an important development in comparison with recently reported protected anodes [41, 42], and it can serve as a suitable, high-energy replacement for the graphitic anode in currently used Li-ion batteries.

7.3 Conclusion

In conclusion, using JDT calculations, we show that Li ions are very loosely bound to the surface, thus enabling faster relaxation and migration to form more uniform electrodeposits. Experimentally, we demonstrated this idea by an electroless plating method to form a protective indium layer on lithium metal anode. Using characterization tools like electron microscopy and x-ray spectroscopy, it was shown that the layer is uniform and is stable even after battery cycling. Further, it was found the interfacial resistance is significantly reduced due to the presence of the coating, which also acts as a buffer layer, where lithium ions alloy and diffuse before depositing onto the underneath lithium electrode. As a result of the enhanced interfacial ion transport mechanism, electrodeposition at long time scales was visually observed to be compact and uniform. As a result of the stable electrodeposition using the electroless indium coating, lithium metal batteries with high loading cathodes of LTO and NCM could be stably cycled for over 250 cycles with close to 90% capacity retention.

Acknowledgments We are grateful to the Advanced Research Projects Agency- Energy (ARPA-E) through award number 1002-2265, DEFOA-001002 for supporting this study. The study also made use of the electrochemical characterization facilities of the KAUST-CU Center for Energy and Sustainability, which is supported by the King Abdullah University of Science and Technology (KAUST) through award number KUS-C1-018-02. Electron microscopy facilities at the Cornell Center for Materials Research (CCMR), an NSF-supported MRSEC through Grant DMR-1120296, were also used for the study.

Appendix: Supplementary Information

Supplementary Table 7.1 Comparison of predicted surface energies and Li diffusion barriers for various low-index surfaces of indium in vacuum and acetonitrile solvent (this work) against those for the most stable surfaces of lithium, sodium, and magnesium [32]

		Surface energy (J/m^2)		Li diffusion barrier (eV)	
	Surface	Vacuum	CH_3CN	Vacuum	CH_3CN
This work	In(011)	0.25	0.25	0.04	0.015
	In(001)	0.26	0.26	0.30	0.013
	In(110)	0.29	0.29	–	–
	In(100)	0.31	0.30	–	–
Ref. [1]	Li(001)	0.46	–	0.14	–
	Na(001)	0.22	–	0.16	–
	Mg(0001)	0.52	–	0.02	–

While surface energies and vacuum diffusion barriers are comparable to previous cases, dramatic reduction in diffusion barriers for Li on indium surfaces in solution suggests reduced dendrite formation on indium electrodes

Supplementary Table 7.2 Values of Arrhenius fitting parameters for inverse interfacial resistance

Type of cell	Prefactor (A) (Ω^{-1})	Activation energy (E_a) (eV/atom)
Control	123660.9	0.43
Within salt (before conditioning)	36997.2	0.39
Within salt (after conditioning)	9979.9	0.29

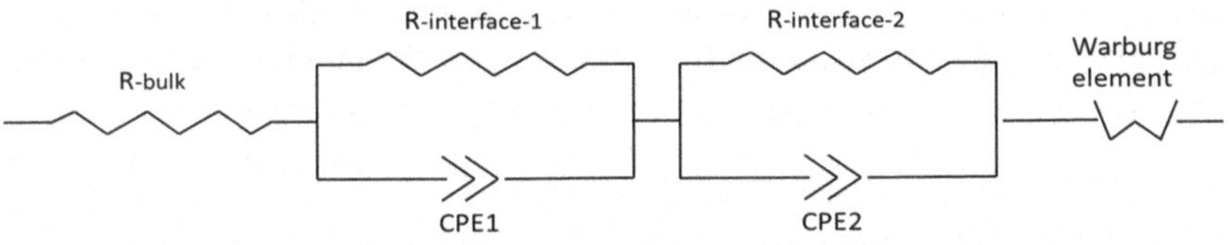

Supplementary Fig. 7.1 Equivalent circuit model for fitting Nyquist plots of impedance measurements. R-bulk represent bulk electrolyte. R-interface$_1$ represents interfacial resistance associated with the passivation layer, and R-interface$_2$ denotes the charge transfer resistance. CPE1 and CPE2 represent the constant phase elements. Warburg element is the solid-state diffusion contribution

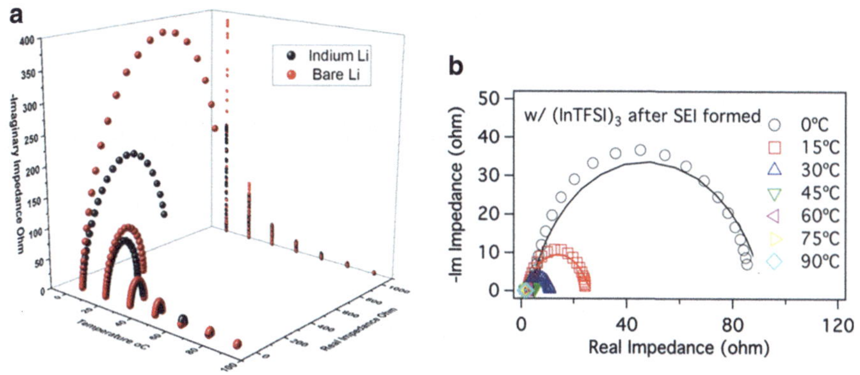

Supplementary Fig. 7.2 (**a**) Nyquist plots as a function of temperature for control electrolyte and *In(TFSI)*$_3$-added electrolyte before conditioning obtained by impedance spectroscopy measurements; (**b**) Nyquist diagram for *In(TFSI)*$_3$-added electrolyte, after cycling at low current densities

Supplementary Fig. 7.3 Cyclic voltammetry result of indium electrode (100 μm) vs. lithium electrode for five cycles. It is seen that the cycling result is "noisy" and there is significant shift in the current peaks

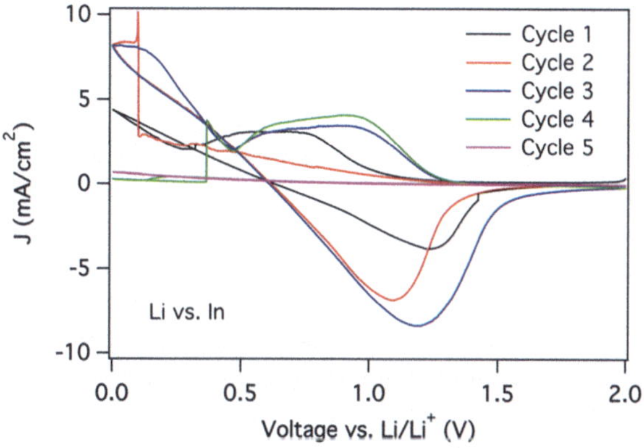

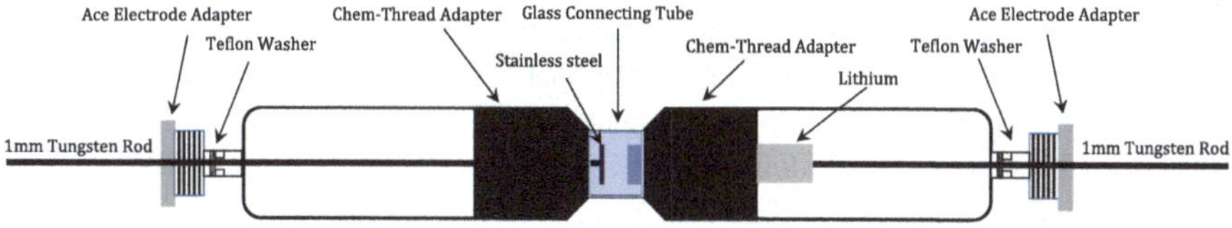

Supplementary Fig. 7.4 In-house built electrochemical setup for visualization experiments involving electrodeposition of lithium metal onto stainless steel electrode

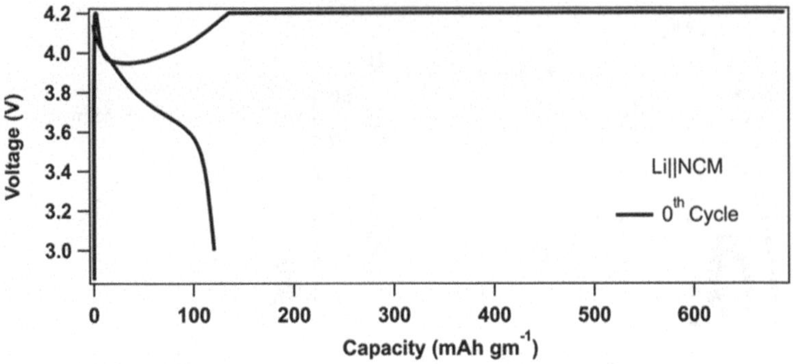

Supplementary Fig. 7.5 Initial cycle of Li||NCM cell using the electrolyte 1 M $LiPF_6$ in EC/DMC with additives 10% FEC and 12 mM *In(TFSI)*$_3$. The long charging curve is due to the formation of SEI by breakdown of TFSI anions

Methods

Computational Details

We perform density functional theory calculations using the open-source JDFTx software [43], employing a plane wave basis with ultrasoft pseudopotentials [44] at the recommended kinetic energy cutoffs of 20 and 100 Hartrees for wave functions and charge densities, respectively. We use the PBE generalized gradient approximation to the exchange-correlation functional [44] and the nonlinear PCM model [34] to describe solvation in acetonitrile [45] along with 1 M non-adsorbing electrolyte. (Electrolytes in continuum solvation models are non-adsorbing by definition.) We use k-point grids that correspond to an effective supercell size of at least 30 A in each periodic direction and Fermi smearing with a width of 0.01 Hartrees.

For the body-centered tetragonal lattice of bulk indium, we obtain lattice constants a = 3.27 A and c = 4.96 A, in excellent agreement with experiment (errors +0.5% and +0.4%, respectively). For the surfaces, we use inversion-symmetric five-layer slab models with a vacuum/solvent gap of 15 A, along with truncated Coulomb potentials to exactly eliminate interactions between periodic images along the slab normal [46]. For diffusion calculations, we use 3 × 3 supercells along the surface directions and map the energy landscape of a Li atom over a symmetry-irreducible wedge of one-unit cell. Specifically, we use (the irreducible wedges of) an 8 × 8 uniform grid of Li positions for the (001) surface and a 6 × 6 grid for (011). All ionic positions are optimized self-consistently, except for the central layer held at bulk geometry for surface calculations, and planar coordinates of Li constrained for mapping the energy landscape.

Experimental Details

Materials

Lithium discs were obtained from MTI corporation. Indium foil (0.1 μm), indium(III) tris(trifluoromethanesulfonimide), ethylene carbonate, diethylene carbonate, propylene carbonate, lithium hexafluorophosphate, and lithium bis(trifluoromethanesu lfonimide) were all purchased from Sigma Aldrich. Fluoroethylene carbonate was obtained from Alfa Aesar. Celgard 3501 separator was obtained from Celgard Inc. Lithium titanate was obtained from NEI Corporation. Nickel cobalt manganese oxide cathode was bought from Electrodes and More Co. All the chemicals were used as received in after rigorous drying in a ~0 ppm water level and <5 ppm oxygen glove box.

Scanning Electron Microscopy and EDX

Surface analysis of indium-coated lithium samples was done using SEM and EDX techniques using the LEO155FESEM instrument. The samples were prepared by 6-h treating of lithium disc in a solution of 12 mM $In(TFSI)_3$ in EC/DMC solvent, followed by 2-day drying in glove box antechamber. Morphology of electrodeposition was studied by a postmortem analysis of a Li||stainless steel battery using the electrolyte 1 M $LiPF_6$ EC/DMC with a Celgard separator. For the case of indium coating, 12 mM $In(TFSI)_3$ was added in the control electrolyte. The battery was discharged at a rate of 0.5 mA/cm^2 for 6 h.

X-Ray Diffraction

XRD was carried out on a Scintag Theta-Theta X-ray diffractometer using Cu K-α radiation at $\lambda = 1.5406$ Å. Samples were quickly transferred to the XRD chamber with minimal exposure to air. For XRD same indium-coated lithium samples were used as for SEM characterizations.

X-Ray Photoelectron Spectroscopy

XPS was conducted using Surface Science Instruments SSX-100 with operating pressure of $\sim 2 \times 10^{-9}$ torr. Monochromatic Al K-α x-rays (1486.6 eV) with beam diameter of 1 mm were used. Photoelectrons were collected at an emission angle of 55°. A hemispherical analyzer determined electron kinetic energy, using pass energy of 150 V for wide survey scans and 50 V for high-resolution scans. Samples were ion-etched using 4 kV Ar ions, which were rastered over an area of 2.25×4 mm with total ion beam current of 2 mA, to remove adventitious carbon. Spectra were referenced to adventitious C 1s at 284.5 eV. CasaXPS software was used for XPS data analysis with Shelby backgrounds. Samples were exposed to air only during the short transfer time to the XPS chamber (less than 10 s).

Impedance Spectroscopy

The impedance spectroscopy measurement was done using a Novocontrol N40 Broadband Dielectric instrument. Symmetric cells were prepared using two lithium discs using the electrolyte 1 M $LiPF_6$ EC/DMC with a Celgard separator. The indium-based cells were prepared by addition of 12 mM $In(TFSI)_3$ as the co-salt in the control electrolyte. The measurements were done in a frequency range from 10^{-3} to 10^7 Hz.

Cyclic Voltammetry

Cyclic voltammetry was performed in two-electrode setup. In one of the tests, the cell comprised of lithium as reference and counter electrode and indium foil as working electrode, while in another experiment, stainless steel was used as the working electrode. The electrolyte used in the former case was 1 M $LiPF_6$ EC/DMC, while in the latter the same electrolyte was used with 12 mM $In(TFSI)_3$. In both cases the separator used was Celgard, and the scanning rate was 1 mV/s operated between 0 and 2 V (vs. Li/Li^+).

Direct Visualization Experiments

The visualization experiment was carried out for understanding the electrodeposition process with and without indium coatings. The electrolyte utilized was 1 M $LiPF_6$-EC/DMC and the same electrolyte with 12 mM $In(TFSI)_3$ additive. The setup consists of a glass chamber with a lithium rod on one side and stainless steel disc on the other, connected using tungsten rods. The glass tube was tightly closed and sealed with Teflon film to ensure complete inert environment inside the cell. In these experiments, the stainless steel electrode was continuously charged with a current density of 8 mA/cm^2. The electrode, being charged, was monitored over time, and images of lithium deposition at different intervals were captured from an optical microscope.

Battery Performance

2032 type Li||Li coin cells with and without 12 mM $In(TFSI)_3$ were prepared inside an argon-filled glove box. The amount of electrolyte used for all battery testing was 60 µl. The cells were evaluated using galvanostatic (strip-plate) cycling in a Neware

CT-3008 battery tester. The batteries were repeatedly charged and discharged with each half cycle 1 h long. Coulombic efficiency test was performed in Li||stainless steel cell with a higher current density of 1 mA/cm^2 and capacity 1 mAh/cm^2; 10% (vol.) fluoroethylene carbonate was used in additive in both control and indium-based electrolytes. Half-cell test was performed in lithium versus lithium titanate cell at a C-rate of 1 C. The cathode loading was 3 mAh/cm^2 and the voltage range was between 1 and 3 V. For cycling NCM cells with 2 mAh cm^{-2}, the voltage range was chosen to be 4.2 to 3 V. A constant voltage step was applied at the end of the charge cycle at 4.2, until the current reduced to 10% of the current used in galvanostatic charging process. The charging was done at C/2 rate and discharge at 1 C.

References

1. Armand, M., Tarascon, J.-M. Building better batteries: Nature. **451**, 652–657 (2008)
2. Tarascon, J.-M., Armand, M.: Issues and challenges facing rechargeable lithium batteries. Nature. **414**, 359–367 (2001)
3. Tikekar, M.D., Choudhury, S., Tu, Z., Archer, L.A.: Design principles for electrolytes and interfaces for stable lithium-metal batteries. Nat. Energy. **1**, 16114 (2016)
4. Cheng, X., Zhang, R., Zhao, C., Wei, F., Zhang, J., Zhang, Q.:A Review of Solid Electrolyte Interphases on Lithium Metal Anode. Adv. Sci. **3**, 1–20 (2016)
5. Tu, Z., Nath, P., Lu, Y., Tikekar, M.D., Archer, L.A.: Nanostructured Electrolytes for Stable Lithium Electrodeposition in Secondary Batteries Acc. Chem. Res. **48**, 2947–2956 (2015)
6. Tikekar, M.D., Archer, L.A., Koch, D.L.: Stability Analysis of Electrodeposition across a Structured Electrolyte with Immobilized Anions. J. Electrochem. Soc. **161**, A847–A855 (2014)
7. Tikekar, M.D., Archer, L.A., Koch, D.L.: Stabilizing electrodeposition in elastic solid electrolytes containing immobilized anions. Sci. Adv. **2**, e1600320 (2016)
8. Bachman, J.C., Muy, S., Grimaud, A., Chang, H., Pour, N., Lux, S.F., Paschos, O., Maglia, F., Lupart, S., Lamp, P., et al.: Inorganic Solid-State Electrolytes for Lithium Batteries: Mechanisms and Properties Governing Ion Conduction. Chem. Rev. **116**, 140–162 (2016)
9. Tu, Z., Kambe, Y., Lu, Y., Archer, L.A.: Nanoporous Polymer-Ceramic Composite Electrolytes for Lithium Metal Batteries. Adv. Energy Mater. **4**, 1300654 (2014)
10. Tung, S.-O., Ho, S., Yang, M., Zhang, R., Kotov, N.A.: A dendrite-suppressing composite ion conductor from aramid nanofibres. Nat. Commun. **6**, 6152 (2015)
11. Tu, Z., Zachman, M.J., Choudhury, S., Wei, S., Ma, L., Yang, Y., Kourkoutis, L.F., Archer, L.A.: Rechargeable Batteries: Nanoporous Hybrid Electrolytes for High-Energy Batteries Based on Reactive Metal Anodes. Adv. Energy Mater. **7**, 1602367 (2017)
12. Choudhury, S., Mangal, R., Agrawal, A., Archer, L.A.: A highly reversible room-temperature lithium metal battery based on crosslinked hairy nanoparticles. Nat. Commun. **6**, 10101 (2015)
13. Khurana, R., Schaefer, J.L., Archer, L.A., Coates, G.W.: Suppression of lithium dendrite growth using cross-linked polyethylene/poly(ethylene oxide) electrolytes: a new approach for practical lithium-metal polymer batteries. J. Am. Chem. Soc. **136**, 7395–7402 (2014)
14. Cheng, X.B., Hou, T.Z., Zhang, R., Peng, H.J., Zhao, C.Z., Huang, J.Q., Zhang, Q.: Dendrite-Free Lithium Deposition Induced by Uniformly Distributed Lithium Ions for Efficient Lithium Metal Batteries. Adv. Mater. **28**, 2888–2895 (2016)
15. Lu, Y., Tikekar, M., Mohanty, R., Hendrickson, K., Ma, L., Archer, L.A.: Stable Cycling of Lithium Metal Batteries Using High Transference Number Electrolytes. Adv. Energy Mater. **5**, 1402073 (2015)

16. Schaefer, J.L., Yanga, D.A., Archer, L.A.: High Lithium Transference Number Electrolytes via Creation of 3-Dimensional, Charged, Nanoporous Networks from Dense Functionalized Nanoparticle Composites. Chem. Mater. **25**, 834–839 (2013)
17. Wu, H., Yu, G., Pan, L., Liu, N., McDowell, M.T., Bao, Z., Cui, Y.: Stable Li-ion battery anodes by in-situpolymerization of conducting hydrogel to conformally coat silicon nanoparticles. Nat. Commun. **4**, 1943 (2013)
18. Aurbach, D., Zinigrad, E., Cohen, Y., Teller, H.: A short review of failure mechanisms of lithium metal and lithiated graphite anodes in liquid electrolyte solutions. Solid State Ionics. **148**, 405–416 (2002)
19. Lin, D., Liu, Y., Cui, Y.: Reviving the lithium metal anode for high-energy batteries. Nat. Nanotechnol. **12**, 194–206 (2017)
20. Choudhury, S., Wan, C.T., Al Sadat, W.I., Tu, Z., Lau, S., Zachman, M.J., Kourkoutis, L.F., Archer, L.A.: Designer interphases for the lithium-oxygen electrochemical cell. Sci. Adv. **3**, e1602809 (2017)
21. Ding, F., Xu, W., Graff, G.L., Zhang, J., Sushko, M.L., Chen, X., Shao, Y., Engelhard, M.H., Nie, Z., Xiao, J., et al.: Dendrite-Free Lithium Deposition via Self-Healing Electrostatic Shield Mechanism. J. Am. Chem. Soc. **135**, 4450–4456 (2013)
22. Choudhury, S., Archer, L.A.: Lithium Fluoride Additives for Stable Cycling of Lithium Batteries at High Current Densities. Adv. Electron. Mater. **2**, 1500246 (2015)
23. Zheng, G., Lee, S.W., Liang, Z., Lee, H.-W., Yan, K., Yao, H., Wang, H., Li, W., Chu, S., Cui, Y.: Interconnected hollow carbon nanospheres for stable lithium metal anodes. Nat. Nanotechnol. **9**, 618–623 (2014)
24. Kozen, A.C., Lin, C., Pearse, A.J., Schroeder, M.A., Han, X., Hu, L., Lee, S., Rubloff, G.W., Noked, M.: Next-Generation Lithium Metal Anode Engineering via Atomic Layer Deposition. ACS Nano. **9**, 5884–5892 (2015)
25. Liang, Z., Zheng, G., Liu, C., Liu, N., Li, W., Yan, K., Yao, H., Hsu, P.-C., Chu, S., Cui, Y.: Polymer nanofiber-guided uniform lithium deposition for battery electrodes. Nano Lett. **15**, 2910–2916 (2015)
26. Zhang, X., Cheng, X., Chen, X., Yan, C., Zhang, Q.: Fluoroethylene Carbonate Additives to Render Uniform Li Deposits in Lithium Metal Batteries. Adv. Funct. Mater. **27**, 1605989 (2017)
27. Qian, J., Henderson, W.A., Xu, W., Bhattacharya, P., Engelhard, M., Borodin, O., Zhang, J.-G.: High rate and stable cycling of lithium metal anode. Nat. Commun. **6**, 6362 (2015)
28. Seh, Z.W., Sun, J., Sun, Y., Cui, Y.: A Highly Reversible Room-Temperature Sodium Metal Anode. ACS Cent. Sci. **1**, 449–455 (2015)
29. Lu, Y., Tu, Z., Archer, L.A.: Stable lithium electrodeposition in liquid and nanoporous solid electrolytes. Nat. Mater. **13**, 961–969 (2014)
30. Ozhabes, Y., Gunceler, D., Arias, T.A.: Stability and surface diffusion at lithium-electrolyte interphases with connections to dendrite suppression. arXiv. **1504.05799**, 1–7 (2015)
31. Choudhury, S., Wei, S., Ozhabes, Y., Gunceler, D., et al.: Designing solid-liquid interphases for sodium batteries. Nat. Commun. **8**, 898 (2017).
32. Jäckle, M., Groß, A.: Microscopic properties of lithium, sodium, and magnesium battery anode materials related to possible dendrite growth. J. Chem. Phys. **141**, 174710 (2014)
33. Gunceler, D., Schwarz, K.A., Sundararaman, K.L.R., Arias, T.A.: 16th International Workshop on Computational Physics and Materials Science: Total Energy and Force Methods (2013)
34. Gunceler, D., Letchworth-Weaver, K., Sundararaman, R., Schwarz, K.A., Arias, T.A.: The importance of nonlinear fluid response in joint density-functional theory studies of battery systems. Model. Simul. Mater. Sci. Eng. **21**, 074005 (2013)
35. Naumkin, A.V., Kraut-Vass, A., Powell, C.J.: NIST X-ray Photoelectron Spectroscopy Database, Version 4.1. National Institute of Standards and Technology, Gaithersburg (2012)
36. Aurbach, D., Markovsky, B., Shechter, A., Ein-Eli, Y., Cohen, H.: A Comparative Study of Synthetic Graphite and Li Electrodes in Electrolyte Solutions Based on Ethylene Carbonate-Dimethyl Carbonate Mixtures. J. Electrochem. Soc. **143**, 3809–3820 (1996)

37. Ota, H., Sakata, Y., Wang, X., Sasahara, J., Yasukawa, E.: Characterization of Lithium Electrode in Lithium Imides/Ethylene Carbonate and Cyclic Ether Electrolytes II. Surface Chemistry. J. Electrochem. Soc. **151**, A437–A446 (2004)
38. Verma, P., Maire, P., Novak, P.: A review of the features and analyses of the solid electrolyte interphase in Li-ion batteries. Electrochim. Acta. **55**, 6332–6341 (2010)
39. Ha, S., Lee, Y., Woo, S.W., Koo, B., Kim, J., Cho, J., Lee, K.T., Choi, N.: Magnesium(II) bis(trifluoromethane sulfonyl) imide-based electrolytes with wide electrochemical windows for rechargeable magnesium batteries. ACS Appl. Mater. Interfaces. **6**, 4063–4073 (2014)
40. Zhang, T., Liao, K., He, P., Zhou, H.: A self-defense redox mediator for efficient lithium–O2 batteries. Energy Environ. Sci. **9**, 1024–1030 (2015)
41. Li, N.-W., Yin, Y.-X., Yang, C.-P., Guo, Y.-G.: An Artificial Solid Electrolyte Interphase Layer for Stable Lithium Metal Anodes. Adv. Mater. **28**, 1853–1858 (2016)
42. Cheng, X.-B., Yan, C., Chen, X., Guan, C., Huang, J.-Q., Peng, H.-J., Zhang, R., Yang, S.-T., Zhang, Q.: Implantable Solid Electrolyte Interphase in Lithium-Metal Batteries. Chem. **2**, 258–270 (2017)
43. Sundararaman, R., Gunceler, D., Letchworth-Weaver, K., Schwarz, K.A., Arias, T.A.: JDFTx code. http://jdftx.sourceforge.net (2012)
44. Garrity, K.F., Bennett, J.W., Rabe, K.M., Vanderbilt, D.: Pseudopotentials for high-throughput DFT calculations. Comput. Mater. Sci. **81**, 446–452 (2014)
45. Gunceler, D., Arias, T.A.: Towards a generalized iso-density continuum model for molecular solvents in plane-wave DFT. Model. Simul. Mater. Sci. Eng. **25**, 015004 (2017)
46. Sundararaman, R., Arias, T.A.: Regularization of the Coulomb singularity in exact exchange by Wigner-Seitz truncated interactions: towards chemical accuracy in nontrivial systems. Phys. Rev. B. **87**, 165122 (2013)

Chapter 8
Designer Interphases for the Lithium-Oxygen Electrochemical Cell

8.1 Introduction

The rechargeable lithium-oxygen (Li-O_2) electrochemical cell is peerless among energy storage technologies for its high theoretical specific energy (3500 Wh/kg), which far exceeds that of current state-of-the-art Li-ion battery technology [1–4]. Li-O_2 cells are under intense study for applications in electrified transportation because they are viewed as the gateway to Li-Air storage technology able to offer competitive specific storage capacities to fossil fuels. A Li-O_2 cell consists of a Li metal anode, an electrolyte that conducts Li^+ ions and uses O_2 gas hosted in a porous carbon or metal support as the active material in the positive electrode (cathode). Ideally, the cell operates on the principle that Li_2O_2 is reversibly formed and decomposed in the cathode, with the net electrochemical reaction of $2(Li^+ + e^-) + O_2 \leftarrow\rightarrow Li_2O_2$ at an equilibrium potential of 2.96 V vs. Li/Li^+. In practice, however, the physicochemical processes in Li-O_2 cells rarely, if ever, live up to these ideals. The insoluble electrically insulating Li_2O_2 is, for example, difficult to oxidize, which leads to high charging overpotential (charge voltage ~ 4.3 V) on the oxygen evolution reaction (OER) and limits the cell efficiency to ~60–70% [5–7]. Additionally, decomposition reactions between the electrolyte and reactive oxygen species at the positive electrode and lithium metal at the negative electrode form undesirable products that further limit cell life and efficiency [8, 9]. Finally, insoluble Li_2O_2 produced upon discharge accumulates in the cathode, eventually clogging pores in the cathode support, which simultaneously limits the Li-O_2 cell discharge capacity and compromise rechargeability [10, 11]. A grand challenge in the field concerns the development of materials and cell running protocols that are able to overcome all of these limitations without compromising the favorable attributes of the Li-O_2 cell.

S. Choudhury, *Rational Design of Nanostructured Polymer Electrolytes and Solid–Liquid Interphases for Lithium Batteries*, Springer Theses,
https://doi.org/10.1007/978-3-030-28943-0_8

Several approaches have been proposed for overcoming each of the challenges with the Li-O_2 cell, but a frustrating observation is that promising solutions to one problem often come at the expense of others or in some cases create new problems [1, 4]. For example, significant theoretical and experimental efforts to lower the overpotential of the OER have resulted in the exploration of soluble redox mediators as electrolyte additives. Redox mediators are first oxidized electrochemically at a lower potential than Li_2O_2; the oxidized form of a soluble redox mediator can therefore diffuse to and oxidize otherwise electrochemically inaccessible Li_2O_2 deposited on the cathode surface, regenerating the mediator and allowing the Li-O_2 battery to be recharged at a lower overpotential. Since their introduction by Bruce and co-workers [12], multiple successful demonstrations of this concept have been reported using tetrathiafulvalene (TTF) [12], (2,2,6,6-tetramethylpiperidin-1-yl) oxidanyl (TEMPO) [13], nitroxides [13, 14], lithium iodide (LiI) [15–17], tris[4-(diethylamino) phenyl] amine (TDPA) [18], iron phthalocyanine (FePc) [19], and LiBr [20] as electrolyte additives. A drawback of this approach is that, with few exceptions [13, 21], the redox mediator is free to diffuse throughout the cell and is reduced by Li metal in a parasitic process that depletes both the anode and redox mediator. Likewise, efforts to improve the stability of electrolytes in the presence of the highly nucleophilic O_2^- species produced at the cathode and the Li metal anode [22, 23] have produced mixed results.

It is now known that electrolytes based on ethers, carbonates, ketones, and esters are all broken down at the cathode of a Li-O_2 cell by the highly nucleophilic Li_2O_2 discharge product. At the anode, no liquid electrolyte presently exists that can survive long-term contact with metallic Li and few form a stable solid electrolyte interphase (SEI) with Li [24]. Results from electrochemical mass spectrometry studies have shown that straight-chain alkyl amides N,N-dimethylformamide (DMF) and N,N-dimethylacetamide (DMA) are unique among electrolyte solvents for their stability against nucleophilic attack at the Li-O_2 cathode [25, 26]. Burke et al. [27] reported that the high donor number solvents (like, DMSO in their case) induce a solution-mediated reaction pathway at the cathode by stabilizing LiO_2 intermediates and the anion NO_3^-, which leads to higher cell discharge capacity [27–31].A perhaps obvious drawback is that these high donor number electrolytes undergo continuous chemical reaction with the Li anode, degrading the anode and electrolyte. $LiNO_3$ salt additives have been investigated for its ability to form stable coatings on Li metal in certain electrolytes, which passivate the metal against attack even by electrolytes containing oxidizing sulfur species [32–35]. In an important study, Walker et al. [36] showed that electrolytes that combine the beneficial effects of $LiNO_3$ and N,N-dimethylacetamide do in fact enable longer term cycling of Li-O_2 cells, underscoring the synergistic benefits of a high donor number electrolyte and anode protection in the Li-O_2 cell.

An unprotected Li metal anode can fail by other, more catastrophic processes than those precipitated by uncontrolled chemical reaction with a liquid electrolyte [24, 37, 38]. Electrodeposition of lithium metal during battery recharge is known to be physically unstable towards formation of rough/dendritic structures on the anode that ultimately grow to short-circuit the cell. The ohmic heat generated by this process can trigger thermal run-away of the cell in organic liquid electrolytes leading to cell failure by fire and/or explosions [39–41]. Furthermore, because rough electrodeposition increases the surface area of Li in contact with liquid electrolytes, physical instability of the Li anode exacerbates chemical instability at the anode/electrolyte interphase. Three recent reviews provide a comprehensive assessment of the strengths and shortcomings of practiced strategies for stabilizing rechargeable lithium batteries against failure by dendrite-induced short-circuits [24, 42, 43]. An important conclusion is that because Li deposition is fundamentally unstable, fundamentally based approaches that take advantage of multiple physical processes are likely to be the most successful in guaranteeing long-term stability of rechargeable batteries that use metallic lithium as anode.

Herein, we report on the stability of Li-O_2 cells employing liquid electrolytes containing an ionomer salt additive that spontaneously forms a multifunctional solid electrolyte interphase (SEI) at the anode. The additive and in situ-formed SEI forms are deliberately designed to take advantage of three, fundamentally based mechanisms for stabilizing electrochemical processes at the anode and cathode of the Li-O_2 cell. First, consistent with predictions from recent continuum [44, 45] and density functional analyses of lithium deposition [46], we report that ionomer electrolyte additives able to ensure low diffusion barriers and high cation fluxes in the SEI at the anode are highly effective in stabilizing deposition of Li. We demonstrate the success of these additives by means of electrochemical analysis and postmortem imaging. Second, we show that if the ionomer additives are designed to form thin conformal coatings at the Li surface, it is possible to passivate the anode surface against chemical attack by high donor number (DN = 27.8) liquid electrolytes capable of stabilizing oxide intermediates on the cathode. Finally, we report that the same material that stabilizes Li deposition on the anode also functions as an effective redox mediator that lowers the overpotential for the OER reaction at the Li-O_2 cathode.

8.2 Results and Discussion

8.2.1 Understanding the Anode Protection Mechanism

Characterization of the Anode

The electrolyte ionomer salt additive (2-bromoethanesulfonate lithium salt) investigated in the present study is illustrated in Fig. 8.1a. The material is chosen because of its ability to react with lithium to simultaneously anchor lithium ethanesulfonate at the anode/electrolyte interphase and to generate partially soluble lithium bromide (LiBr) in the electrolyte. The specific ionomer chemistry selected for the study is motivated by four fundamental considerations. First, recent continuum theoretical analysis [44, 45] and experiment [47–49] indicate tethering anions such as sulfonates at the anode/electrolyte interphase lowers the potential at the interface during

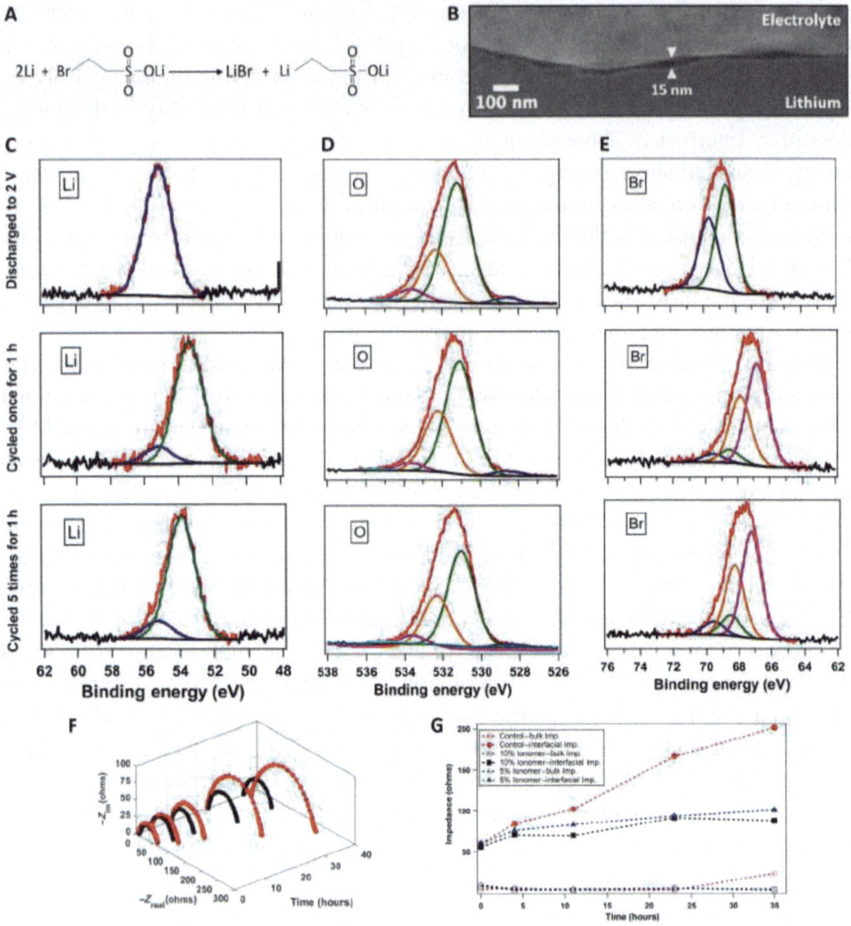

Fig. 8.1 Artificial SEI concept and experimental verification of its proposed operating mechanism. (**a**) Schematic for the reaction of lithium 2-bromoethanesulfonate with lithium metal forming LiBr and lithium-based organometallic. (**b**) SEM image of the interfacial layer between an intact electrolyte and a lithium electrode, revealed in a cross section produced by cryo-FIB milling. (**c**) Lithium 1s peak obtained from XPS of the lithium metal anode of a Li-O2 battery with the electrolyte ionomer [10% (by weight)] in 1 M LiNO3-DMA. (**d**) Oxygen 1s peak of the lithium anode. (**e**) Bromine 3d peak of the lithium anode. In (**c**) to (**e**), the first row shows the postmortem analysis after discharging until 2 V, the second row shows the result after cycling once with each half-cycle 5 h long, and the third row shows the result after cycling five times with each half-cycle 1 h long. (**f**) Three-dimensional diagram of Nyquist plots obtained by impedance measurements at different intervals of time using symmetric lithium cells, in which –Zim is the imaginary component of the impedance and – Zreal is the real component of the impedance. (**g**) Comparison of interfacial and bulk impedance values for ionomer-based and control electrolytes as a function of time. In (**f**) and (**g**), the red symbols denote results with the control electrolyte (1 M LiNO3-DMA), whereas the black and blue symbols represent batteries with 10 and 5% (by weight) ionomer additive, respectively, with the same electrolyte

Li deposition and in so doing stabilizes the deposition. Second, joint density-functional (JDFT) calculations [46] show that the energy barrier E_a for Li^+ diffusion at a Li anode coated with LiBr salt ($E_{a,LiBr} \approx 0.03$ eV) is much lower, by a factor of around 8, compared to Li_2CO_3 ($E_{a,Li2CO3} \approx 0.24$ eV), which forms naturally when aprotic solvents react with Li. This means that under isothermal conditions, stable deposition of Li in given electrolyte can occur at deposition rates more than three orders of magnitude higher on a LiBr-coated Li anode than on an anode with a spontaneously formed Li_2CO_3-rich SEI. Third, the short hydrocarbon stem that connects the tethered sulfonate groups to Li should allow a dense hydrocarbon brush to form at the interface to protect the Li electrode from chemical attack by a high DN electrolyte required for stability at the cathode. Finally, soluble LiBr undergoes electrochemical oxidation and reduction in an appropriate potential window to function as a soluble redox mediator.

Cryo-focused ion beam (cryo-FIB) was used to characterize the morphology and thickness of the ionomer-enriched electrode-electrolyte interphase with the liquid electrolyte intact but cryo-immobilized. In this technique, a symmetric lithium cell (with ionomer-based electrolyte) was opened manually, and the sample was snap-frozen by immediately plunging it into slush nitrogen to preserve the electrolyte and to avoid air exposure. The sample was then transferred under vacuum into an FEI Strata 400 FIB fitted with a quorum PP3010T cryo-FIB/SEM preparation system and maintained at −165 °C for the duration of the experiment. To produce a cross section of the interface, the focused gallium ion beam was used to mill through the frozen electrolyte and into the electrode. This interface was then examined by scanning electron microscopy (SEM) and energy dispersive X-ray spectroscopy (EDX) directly in the cryo-FIB. SEM images revealed an interfacial layer up to approximately 25 nm thick in most areas. An example of the observed layer is shown in Fig. 8.1b. EDX could not confirm an increased bromine concentration in this layer, owing to the similar atomic composition of the reactant and product (in this case bulk electrolyte and interface). Therefore, it is important to perform X-ray analysis of washed electrode surface (i.e., excluding bulk liquid) to understand chemical compositions of the interface.

The proposed reaction mechanism was evaluated by means of EDX and high-resolution XPS analytical measurements. The XPS measurements are performed using monochromatic Al K- α-X-rays (1489.6 eV) with a beam diameter of 1 mm, and the results originate from a surface layer on the electrodes approximately 15–25 nm thick. Supplementary Fig. 8.1 reports the 2-D EDX results on a lithium anode that was thoroughly washed after cycling. Sulfur and bromine signals are clear everywhere on the surface of the materials as expected. XPS analysis was performed using postmortem measurements on lithium anodes harvested from $Li-O_2$ cells subjected to different running conditions. High-resolution scans for anodes retrieved after cycling or after a single discharge with the ionomer additive in 1 M $LiNO_3$-DMA electrolyte are reported in Fig. 8.1c–e. The corresponding results without the ionomer are shown in Supplementary Fig. 8.2. In Fig. 8.1c, it is apparent that after the first discharge, a Li 1s peak at 55.2 eV is observed on anodes with/without the ionomer present in the electrolyte. The peak may be attributed to the presence of LiOH, Li_2O_2, and Li_2CO_3 [50–57]. A more prominent Li 1s peak is observed at 53.8 eV, accounting for about 85% of lithium, only in spectra of anodes cycled in the presence of the ionomer additive. This peak is indicative of the formation of a different SEI in electrolytes containing the ionomer; Li 1s peaks with comparable binding energy are reported for organometallics containing Li-C bonds (54.2 eV) [58, 59]. This observation is consistent with the hypothesis that the ionomer reacts at the Li anode surface to form a lithium ethanesulfonate-rich SEI at the interface. Also, the fact that this binding energy is observed in the cycled anodes confirms that the SEI layer is stable and present even after repeated insertion and extraction of lithium ions into the underlying electrode.

Further evidence that the ionomer additive forms a stable SEI on Li can be deduced from the O 1s (Fig. 8.1d) and Br 3d (Fig. 8.1e) high-resolution scans. The O 1s peak at 532.2 eV comprises approximately 18% of the oxygen signal in cells without the ionomer additive, whether the anodes originate from cells that were subjected to a single discharge or were cycled. The 532.2 eV peak has been previously reported to originate from sulfonates [64] which accounts for, respectively, 27% and 38% of the oxygen signal when the anode is discharged once or cycled in the presence of the ionomer additive. The corresponding sulfur atomic contribution for the same materials can be computed from the wide survey scans (Supplementary Table 8.1) to be about 2% for the once discharged anode and about twice as high for the cycled anodes. The high-resolution scans of Br 3d reveal the formation of a single bond (a $3d_{5/2}$ and $3d_{3/2}$ doublet) with a Br $3d_{5/2}$ peak at 68.5 eV when the anode is discharged once in the presence of the ionomer. We attribute this peak to the formation of Br-Li bond, which has been previously reported to occur at a binding energies between 68.8 and 69.5 [53, 60]. The same peak persists when the anode is cycled in the presence of the ionomer however with a contribution of only around 15%. The reduced LiBr species in the anodes of cycled cells is an indication of LiBr being solvated by the DMA electrolyte that can further participate in the redox mediation of oxygen cathode recharging. In fact, a more prominent Br 3d peak at 67.0 eV is observed only for the cycled anodes that is likely to originate from Br-C bonds (binding energies between 66.7 and 71.0 eV [60–65]) in the SEI originating from untethered ionomer. The untethered ionomer in the electrolyte can help in the regeneration of the SEI layer in repeated cycling. Our results based on XPS analysis thus show that the ionomer-added electrolyte forms a SEI layer of lithium ethanesulfonate and LiBr, in accordance with the proposed reaction mechanism.

The effectiveness of ionomer-based SEI on Li was analyzed using impedance spectroscopy measurements on symmetric lithium cells. The results are shown in Fig. 8.1f with Nyquist-type plots at progressive time periods for control cells and those that contain 10%(wt.) of ionomer additive. The Nyquist plots for 5% ionomer-added cells are shown in Supplementary Fig. 8.3. The experimental data points are fitted with the circuit model illustrated in Supplementary Fig. 8.4 to deduce the bulk and interfacial resistances (Fig. 8.1g) as a function of time for the control electrolyte as well as with 10% and 5% (by wt.) ionomer additive. It is seen that the bulk resistance for all cells remains essentially constant for approximately 20 h, beyond which the bulk resistance of the control diverges (the increase is much larger for the results for the control cells after 48 and 56 hours as seen Supplementary Fig. 8.5). The time-dependent interfacial impedance provides an even more sensitive indicator of the stability of the anode-electrolyte interphase in a high donor number solvent. It is seen that the initial interfacial resistances for control and ionomer-SEI stabilized Li electrodes are approximately equal (~50 Ω). However, there is an exponential rise in the interfacial resistance of the control cell over time consistent with rapid reaction between Li and DMA. It is important to note that this reaction is observed even though $LiNO_3$ is present at large concentration in the electrolyte. These results therefore challenge the view that $LiNO_3$ provides an effective means of passivating Li metal anodes against reactive liquid electrolytes. In contrast, the results in Fig. 8.1g show that the interfacial resistance remains constant (see also Supplementary Fig. 8.4) when the ionomer-based SEI is present. It is seen that the stabilization with 10% ionomer additive is marginally better than 5% ionomer. Together these findings demonstrate that a SEI based on bromide ionomers has a large stabilizing effect on Li anodes in DMA-based electrolyte solvents.

Lithium-Electrolyte Stability

Figure 8.2a, b reports on the quality of lithium ion deposition on stainless steel substrates mediated by control and ionomer-containing 1MLiNO$_3$-DMA electrolytes. For these experiments, cells were assembled with lithium as anode and stainless steel as a virtual cathode. Lithium of capacity 10mAh/cm^2 was deposited at a rate of 1 mA/cm^2 onto stainless steel after which the cell was rested for a period of 10 h and the voltage monitored over time. Figure 8.2b shows that in case of a control electrolyte, Li deposition takes place at a higher voltage compared to the ionomer-containing electrolyte. Also, it can be observed that after the rest period, the voltage measured in the control cells immediately rises to approximately 0.5 volts. Such a high open circuit potential after Li deposition is a reflection of the complete decomposition of Li deposits on stainless steel due to corrosion by the electrolyte. It is again worth noting that despite using the Li-passivating salt $LiNO_3$ at high concentrations in the electrolyte, the freshly deposited lithium reacts completely with the electrolyte solvent. Figure 8.2b also reports the corresponding voltage profiles observed in rested cells containing the ionomer as an electrolyte additive. It is seen that the cell voltage remains close to 0 volts (vs. Li/Li$^+$), i.e., near the open circuit potential of a symmetric lithium cell, which means that the Li electrode is chemically stable in the reactive DMA electrolyte solvent.

To further examine the morphology of Li deposits, postmortem analysis was performed, wherein the surface features of the electrodes were visualized under a scanning electron microscope (SEM). Figure 8.2a shows the SEM image of the surface of stainless steel in the control and ionomer-based electrolytes. For the control, there are few patches of Li observed, and large sections of bare stainless steel are clearly visible. In contrast, in electrolytes containing the ionomer, the stainless steel surface is covered with a thick layer of lithium. It is also seen that Li electrodeposits

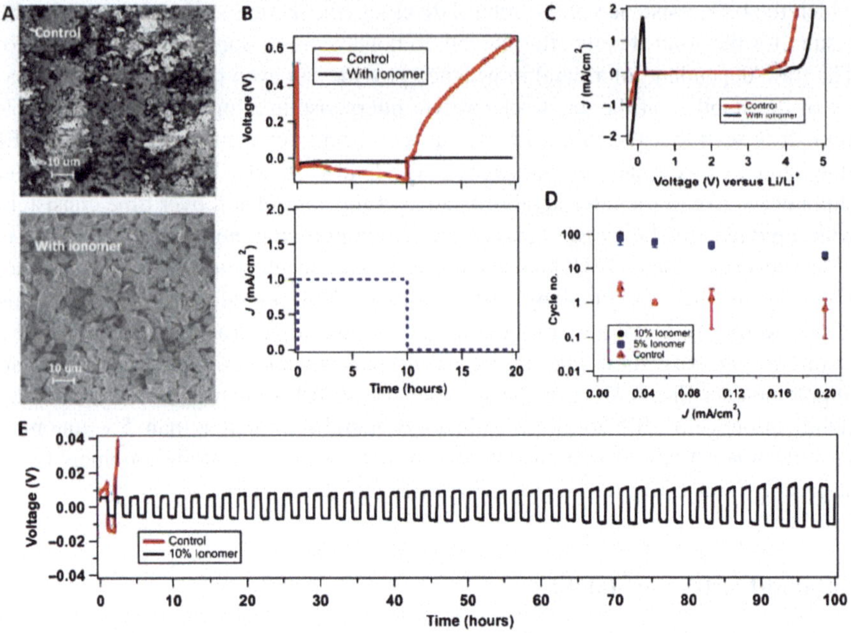

Fig. 8.2 Stabilizing the lithium-electrolyte interphase. (**a**) SEM images of stainless steel (SS) electrode after depositing lithium (10 mAh/cm^2) in a Li||SS cell with and without the ionomer additive using the same electrolyte of 1 M LiNO3-DMA. (**b**) Voltage profile of the Li||SS cell plotted over time. In this experiment, Li + ions were deposited onto the stainless steel side at a current density of 1 mA/cm^2 for 10 h, after which the cell was kept at rest for an additional 10 h, as shown in the current-versus-time curve. In the voltage-versus-time graph, the red line represents the profile of the control electrolyte (1 M $LiNO_3$-DMA), whereas the black line is for the same electrolyte enriched with 10% (by weight) ionomer additive. The dashed blue line in the current-versus-time graph is the applied current for both cases. (**c**) Linear scan voltammetry showing current as a function of voltage versus Li/Li+, with Li as both working and reference electrode and SS being the counter electrode. (**d**) In a Li||SS cell, lithium with 10 mAh/cm^2 capacity is deposited onto SS, and the battery was charged and discharged consecutively at various current densities. The cycle number associated with the divergence of voltage is plotted against the respective current densities. (**e**) Voltage profile for the strip-and-plate experiment under the abovementioned condition using a current density of 0.05 mA/cm^2. In all figures, red indicates the control electrolyte (1 M $LiNO_3$-DMA), and black represents the addition of 10% (by weight) ionomer additive, whereas blue denotes 5% (by weight) addition

formed in the latter electrolytes are evenly sized and spherical in shape, even at a relatively high current density of 1 mA/cm^2. This observation is consistent with previous reports of more compact electrodeposition of Li in electrolytes with halide salt-enriched SEIs and single ion conducting features [24, 42].

To fundamentally understand the basis of these observations, electrochemical stability of the electrolytes was characterized by means of linear scan voltammetry in the range −0.2 to 5 V vs. Li/Li$^+$, at a fixed scan rate of 1 mV/s. Figure 8.2c shows current as a function of voltage in a two-electrode setup of Li||stainless steel. It is seen that for the control (indicated by red curve), the current diverges at a value around 4 V vs. Li/Li^+, while for electrolytes containing ionomer additives, the current diverges at a higher voltage, around 4.3 V vs. Li/Li^+. This improved stability is consistent with previous reports of electrolyte composites with tethered anions [49], wherein anions fixed at/near the electrode surface limit access to and chemical reaction of anions in an electrolyte with the negative electrode. Another important feature of the results can be seen at a potential close to 0 volts vs. Li/Li^+. The significant current peak apparent at approx. −0.2 V vs. Li/Li^+ for both control and ionomer-containing electrolytes is a characteristic of lithium plating onto stainless steel. However, as the voltage is progressively increased, the corresponding Li stripping peak is not seen in the control cell but is readily apparent in cells with ionomer-containing electrolyte. This behavior is indicative of the complete consumption of lithium deposits on stainless steel in the control cells and is consistent with previous results of SEM.

Figure 8.2d, e report results from so-called galvanostatic "plating-stripping" experiments. These experiments are used to evaluate the stability of Li electrodeposition and to assess the propensity of the material to electrodeposit as rough, dendritic structures. In contrast to previous studies [38], where thick (~0.75 mm) Li foil is used on both electrodes employed in pate-strip protocols, we performed these experiments using asymmetric Li/Li cells comprised of one thick Li and one Li-lean (10mAh/cm^2 of Li deposited on stainless steel at 1 mA/cm^2) electrode. The stability of the Li deposition reaction is normally assessed using three criteria: (1) The overpotential of lithium deposition. It can be seen from Fig. 8.2e that at a fixed current density (0.05 mA/cm^2), the voltage response for cells with ionomer-based SEI is low (approx. 6 mV), while the corresponding value for the control is much higher (approx. 150 mV). This difference is indicative of formation of insulating products on the surface of the Li electrodes. (2) Steep decrease of the cell voltage to zero with continuous charge-discharge. This is an indication of short-circuiting of the cell when dendritic lithium formed at one or both electrodes bridges the two electrodes. It is apparent that this phenomenon is not observed in either the control or for the ionomer-SEI stabilized electrodes. (3) A steady increase of the voltage over extended cycles of charge and discharge. This observation is indicative of an unstable SEI that grows continuously, eventually consuming the Li deposited on the stainless steel substrate. As seen in Supplementary Fig. 8.6, after only two cycles at both current densities studied, the control cell fails after a steep rise in voltage. This is quite different from what is observed for cells in which Li is stabilized by an ionomer SEI, which are stable for over 150 cycles. Figure 8.2d reports the number of cycles at which the cell voltage diverges as a function of current density (J). The ionomer-based SEI are seen to improve cell lifetime at a fixed current density by nearly two orders of magnitude. These results underscore the effectiveness of the ionomer-based SEI in stabilizing electrodeposition of Li in amide-based electrolytes, which were previously thought to be unfeasible for lithium metal batteries due to their high reactivity with and ready decomposition by Li.

Anode Protection Mechanism

We hypothesize that the stability of Li anode in DMA originates from two fundamental sources: (i) accumulation of LiBr salt at the Li/electrolyte interphase, which facilitates Li-ion transport to the Li electrode during charging and (ii) the existence of tethered sulfonate anions at the interface, which lowers the electric field at the electrode. Previous joint density-functional theoretical (JDFT) analysis revealed that the presence of lithium halides in the SEI of Li metal anode lowers the activation energy barrier by an order of magnitude or more for lateral Li diffusion at a Li/electrolyte interphase, thereby increasing the tendency of Li to form smooth deposits [46]. Comparing the surface diffusion barriers for various constituents of a typical SEI layer, Arias et al. [46] found that Li_2CO_3, a common SEI constituent in carbonate electrolytes, has an energy barrier of 0.23 eV, while the barrier for a SEI composed of LiF is 0.17 eV. This difference has been argued previously to explain the much greater tendency of Li to form flat, compact deposits during battery recharge as revealed by experiments in which weakly soluble LiF salts are enriched in the SEI by precipitating out of liquid electrolytes [37]. Interestingly, the JDFT analysis shows that the activation energy barrier for Li-ion diffusion at a LiBr/Li interface is much lower (0.062 eV) and comparable to that of magnesium [46, 66], which is known in the literature to electrodeposit without formation of dendrites [67]. Thus, the LiBr created during the formation of the SEI should provide an even more powerful (than LiF) stabilizing effect on Li deposition.

In addition to the presence of LiBr, the SEI created by the ionomer contains bound anionic groups in form of lithium ethanesulfonate ($Li\text{-}CH_2CH_2\text{-}SO_3^-$). Thus, the electrolyte consists of a combination of free and tethered anion. In the past, researchers have realized the importance of single ion conducting electrolytes [42, 68], as they prevent the formation of ion concentration regions within a cell, leading to stable ion transport even at high charge rate. Interestingly, recent linear stability analysis of electrodeposition by Tikekar et al. [24, 44, 45] showed that the stability of an electrolyte can be significantly enhanced by immobilizing only a small fraction (10%) of the anions. The design of our electrolyte comprising of a fraction of anions near the anodic surface, with $LiNO_3$ as the free salt, is explicitly motivated by this theoretical framework. Thus, a modified SEI based on bromide ionomers tethered to the Li anode provides a powerful combination of processes that stabilize the anode against unstable electrodeposition.

8.2.2 Understanding the Cathode Stabilization Mechanism

Characterizing Cathode Products

Figure 8.3a shows a representative voltage profile for the galvanostatic discharge and charge for a $Li\text{-}O_2$ cell with 1 M $LiNO_3$ in an ionomer-enriched DMA electrolyte. Cutoff voltages of 2.2 V and 4.3 V, respectively, were used for the discharge

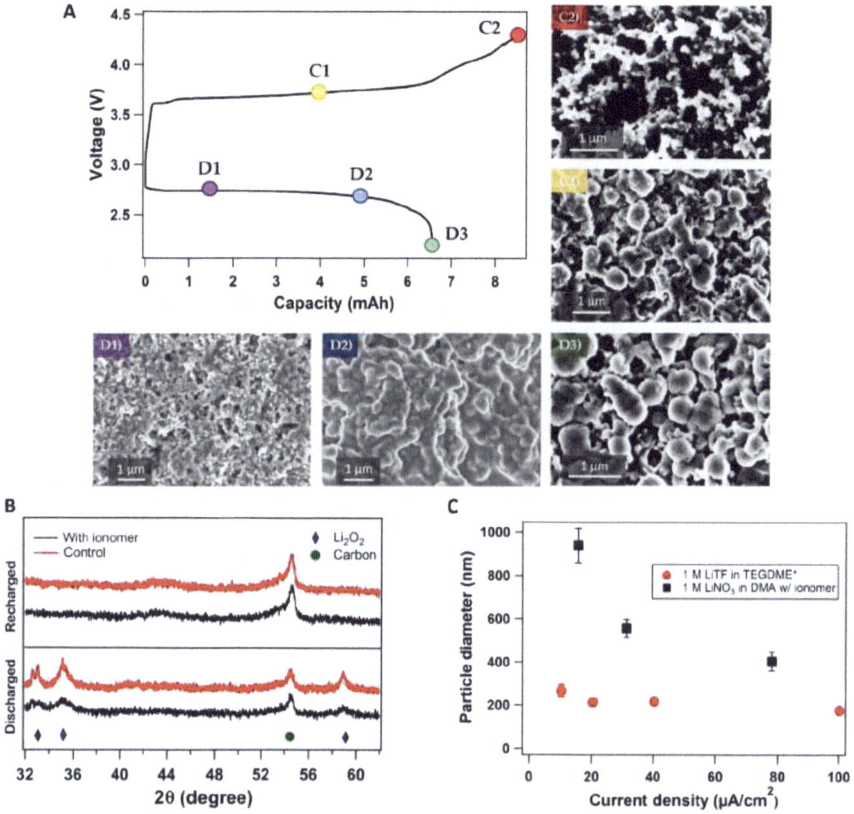

Fig. 8.3 Characterization and electrochemical analysis of oxygen cathode. (**a**) Full charge-discharge cycle of a Li||O_2 cell using ionomer-enriched 1MLiNO$_3$-DMA electrolyte operated at a current density of 31.25 mA/cm^2. The different points on the voltage profile indicate various stages at which the same-type cells were stopped for ex situ analysis. The images below the voltage profile show the surface of a carbon cathode at the D1, D2, and D3 discharge phases. The size of the Li_2O_2 is seen to be increasing over the course of discharge. C1 and C2 show the stages of recharge; it is seen in C1 that the cathode is absent of Li_2O_2 particles. (**b**) XRD analysis showing various characteristic peaks for a fully discharged and a recharged Li||O_2 battery. Here, diamonds denote Li_2O_2 peak and circles represent carbon. The red lines refer to the control electrolyte (1 M LiNO3-DMA), whereas black lines show the result for the same electrolyte with the ionomer additive. (**c**) The diameter of Li_2O_2 particles obtained by fully discharging a Li||O_2 cell is plotted as a function of current density. Here, black indicates the electrolyte (1MLiNO$_3$-DMA) with the ionomer additive, whereas red represents data from Lau and Archer's paper that used the electrolyte 1MLiTF in TEGDME. (∗From Lau and Archer [10])

and charge cycles, and both processes were performed at a fixed current density of 31.25 μA/cm^2. Postmortem SEM analysis was employed to study the evolution of discharge products on the cathode at three stages of discharge (D1, D2, D3) and two stages of charge (C1, C2). The SEM images show that the reversible formation and decomposition of an insoluble solid product on the cathode. Complementary XRD analysis (Fig. 8.3b) shows that the cathode product is exclusively Li_2O_2 and no other products such as LiOH are observed. The SEM analysis shows that Li_2O_2 particles grow increasingly larger as the discharge progresses and nucleation sites for growth are filled, and the full discharge capacity of the cell is reached. Analysis of the particle sizes on discharge (see Supplementary Fig. 8.7) reveals that at low current densities (e.g., 15 μA/cm^2), large Li_2O_2 particles (1 μm and higher) are formed. Comparing these results to those reported by Lau et al. [10] for Li-O2 cells discharged in a 1 M LiTF in TEGDME (a low donor no. solvent), the Li_2O_2 particles formed in DMA are at least four times larger (see Fig. 8.3c). These findings are consistent with expectation for the high donor number of DMA, which solvates Li^+ cations and enables a solution-mediated mechanism, circumventing capacity limitations from the passivation layer formed at the cathode, which enables deep discharge [31]. At higher current densities, the particle size at the voltage cutoff decreases drastically, consistent with the idea that kinetic diffusion limitations [27] set the maximum particle size. The SEM images (R1, R2) shows that the cathode upon charge, closely resembles that of the pristine electrode prior to discharge. Redox mediation from lithium 2-bromoethanesulfonate is thought to aid in the electrochemical decomposition of the large, insulating Li_2O_2 particles formed on the cathode. Support for this hypothesis comes from the effectiveness of the recharge process as well as from the flat charge profile observed until the full capacity of the discharge is reached; the voltage ultimately begins to rise because of the set voltage limit of 4.3 V. Thus, Fig. 8.3 shows that a Li-O_2 cell with 1 M $LiNO_3$ in DMA with an ionomer-based SEI on Li is able to reach a high capacity through LiO_2 disproportionation; can fully utilize the formed Li_2O_2 during the recharge; and cycles with features indicative of the presence of a redox mediator.

Cycling Performance

To evaluate the hypothesis that a high donor number electrolyte solvent and redox mediator provide significant synergistic benefits for Li-O_2 cells, we compare the voltage profiles for fully discharged cells without and with these attributes (see Fig. 8.4a). It is seen that the discharge capacity of Li-O_2 cells with a 1 M $LiNO_3$ DMA + ionomer electrolyte is noticeably higher (~6.5mAh) than with a conventional 1 M LiTFSI-diglyme (~5.1mAh) with same cathode loading. This finding is consistent with the observation of large-sized lithium peroxide structures owing to the solution-mediated nucleation of peroxides. Comparison of the charge cycle shows that with the diglyme electrolyte, the voltage diverges to >4.2 V in ~3.5mAh capacity, which is believed to be an indication of Li_2CO_3 formation and its effect on the charging process, whereas with ionomer-based electrolyte, the voltage diverges at ~6.5mAh (same as discharge). Figure 8.4b shows the cyclic voltammetry experiment for a lithium-oxygen cell in a two-electrode setup with lithium as both reference and counter electrode. The measurements were performed

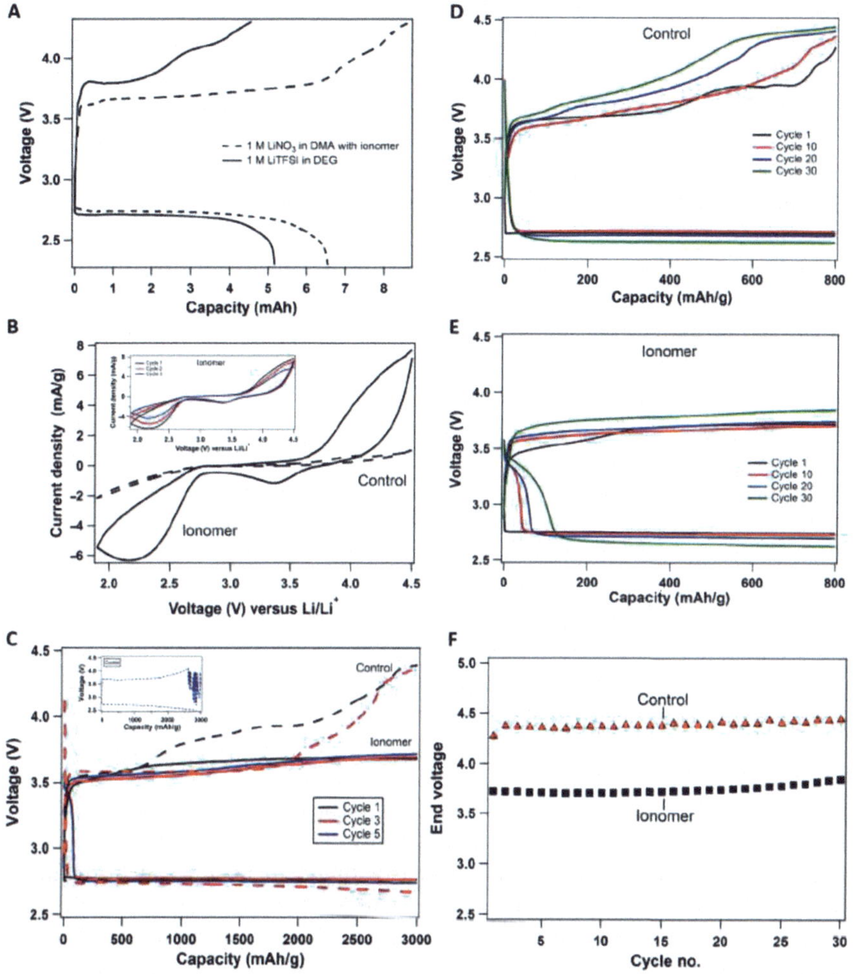

Fig. 8.4 Galvanostatic cycling performance of lithium-oxygen electrochemical cell. (**a**) Voltage profile for batteries fully discharged and recharged with 1 M $LiNO_3$-DMA + ionomer electrolyte (shown with a solid black line) and a low donor number electrolyte, 1 M LiTFSI-diglyme (shown with a dashed black line), at a current density of 31.25 mA/cm^2. (**b**) Comparison of cycling voltammetry results for the control electrolyte (1 M $LiNO_3$-DMA; shown with dashed lines) and the same electrolyte with the ionomer additive (shown with solid lines). The inset shows three cycles of cyclic voltammetry for the ionomer case. (**c**) Voltage profile of the Li||O_2 battery with a cutoff capacity of 3000 mAh/g and a current density of 0.04 mA/cm^2. The solid lines indicate ionomer-based electrolytes, whereas the control is shown with dashed lines. The inset shows the noisy profile of the fifth cycle with the control electrolyte. (**d**) Voltage profile with a capacity cutoff of 800 mAh/g and a current density of 0.08 mA/cm^2 for a Li||O_2 cell using the control electrolyte (1 M $LiNO_3$-DMA). (**e**) Voltage-versus-capacity curve with the same cutoff of 800 mAh/g using the ionomer additive in the electrolyte. (**f**) End voltage of charging cycle for the control and the ionomer-added electrolyte is plotted as function of cycle number

between 1.9 V and 4.5 V (vs. Li/Li^+) at a scan rate of 1 mV/s, and normalized current is plotted against voltage. The current peaks for the ionomer-based electrolyte are an order of magnitude higher than the control electrolyte. Thus, it can be inferred that there is higher electrochemical activity with owing to higher stability of the electrolyte and redox mediation due to presence of LiBr. The peak seen at ~3.5 V can be attributed to Br_3^-/Br^- redox couple. The inset shows three cycles with ionomer-added electrolytes, where there is slight shift of the current peaks to lower values.

Discharge and charge profiles for cells having the electrolyte 1 M $LiNO_3$-DMA with and without ionomer with a capacity cutoff of 3000mAh/gm and current density of 0.04 mA/cm^2 are displayed in Fig. 8.4c. It is seen that both discharge and charge voltage curves tend to diverge to lower and higher values, respectively. Further it can be seen from the inset of Fig. 8.4c that the voltage profile becomes extremely noisy in the fifth cycle of the control electrolyte, while that with ionomer additive is stable. This instability without ionomer can be attributed to the degradation of the electrolyte by reaction with the unprotected lithium metal. One major benefit of cells cycled with ionomer is reduced overpotential during charge relative to that of the control cell, thus increasing cycling efficiency. This is studied in a $Li-O_2$ battery with a lower capacity cutoff of 800mAh/gm at a current density of 0.08 mA/cm^2 for control (Fig. 8.4d) and ionomer-added electrolyte (Fig. 8.4e). As demonstrated in Fig. 8.4e, the highest voltage on charge for cells with ionomer is approximately 3.7 V, close to the Br^-/Br_3^- redox reaction at 3.48 V [1]. Control cells with solely 1 M $LiNO_3$ in DMA reach voltages of around 4.45 V as seen in Fig. 8.4d. This suggests similar action to a redox mediator, in which Li_2O_2 is oxidized by Br_3^- to reform Br^- in a cycle that lowers charge overpotential. The discharge and charge profiles remain similar over 30 cycles for cells with additive, while the charge profile in untreated cells increases more drastically. The distinct gentle slope of the initial portion of the discharge profile in cells with ionomer can be attributed to the presence of bromine species in the system. Figure 8.4f compared the end voltage of recharge with and without the ionomer additive. The ~1 V improvement in the round-trip efficiency not only saves loss of input energy but also ensures long life cycling by preventing electrolyte decomposition [4].

Cathode Stabilization Mechanism

At the cathode surface, LiBr is thought to participate in the redox mediation that promotes the OER reaction. In this process, the Li_2O_2 can be co-reduced with Br^- to form O_2 and Br_3^-. The potential for $Br^- \rightarrow Br_3^-$ is known to be 3.48 V, thus the charging of a $Li-O_2$ cell can be limited to this voltage. DMA's ability to dissolve peroxides also aids in the effective electrolyte-side redox mediation. Support for the uniqueness of these ideas comes from recent experiments which demonstrate the efficacy of LiI and LiBr as redox mediators in $Li-O_2$ cells based on glymes [20]. In the absence of water in the electrolyte, LiI was reported to produce a gradual rise in the discharge voltage due to formation of iodine and similar products. LiBr was found to be ineffective in maintaining a steady charge voltage. In electrolytes with high water contents and LiI, LiOH has been shown to be the primary discharge product, which has been reported to be thermodynamically impossible to undergo OER. Our results therefore clearly show that protecting the Li anode in a 1 M $LiNO_3$-DMA electrolyte with a SEI based on bromide ionomer overcomes fundamental limitations of the anode, cathode, and electrolyte in previously studied systems and enables stable cycling of these cells.

8.3 Conclusions

In summary, we demonstrate that addition of lithium 2-bromoethanesulfonate (ionomer) to 1 M $LiNO_3$ in DMA electrolytes produces a SEI at lithium surface that stabilizes the anode in Li-O_2 cells by at least two powerful processes. Compared to control, cells with the ionomer SEI, Li-O_2 cells based on lithium 2-bromoethanesulfonate exhibit flatter, more stable charge profiles and are able to withstand deeper cycling. Furthermore, we show that electrochemical charge-discharge processes in the cells coincide with formation and decomposition of large Li_2O_2 particles as the principal OER product in the cathode. Analysis by linear scan voltammetry and "plate-strip" cycling analysis of the Li anode show that a SEI based on lithium 2-bromoethanesulfonate ionomer on the anode provides chemical stability to Li against attack by DMA, as well as physical stability against rough, dendritic electrodeposition. Although we expect the "perfect" electrolyte for Li-O_2 cells significant additional work, by addressing fundamental issues that limit performance of the anode and cathode, we predict that multifunctional SEIs of the sort discussed in this study will emerge as critical to further progress.

Appendix: Supplementary Information

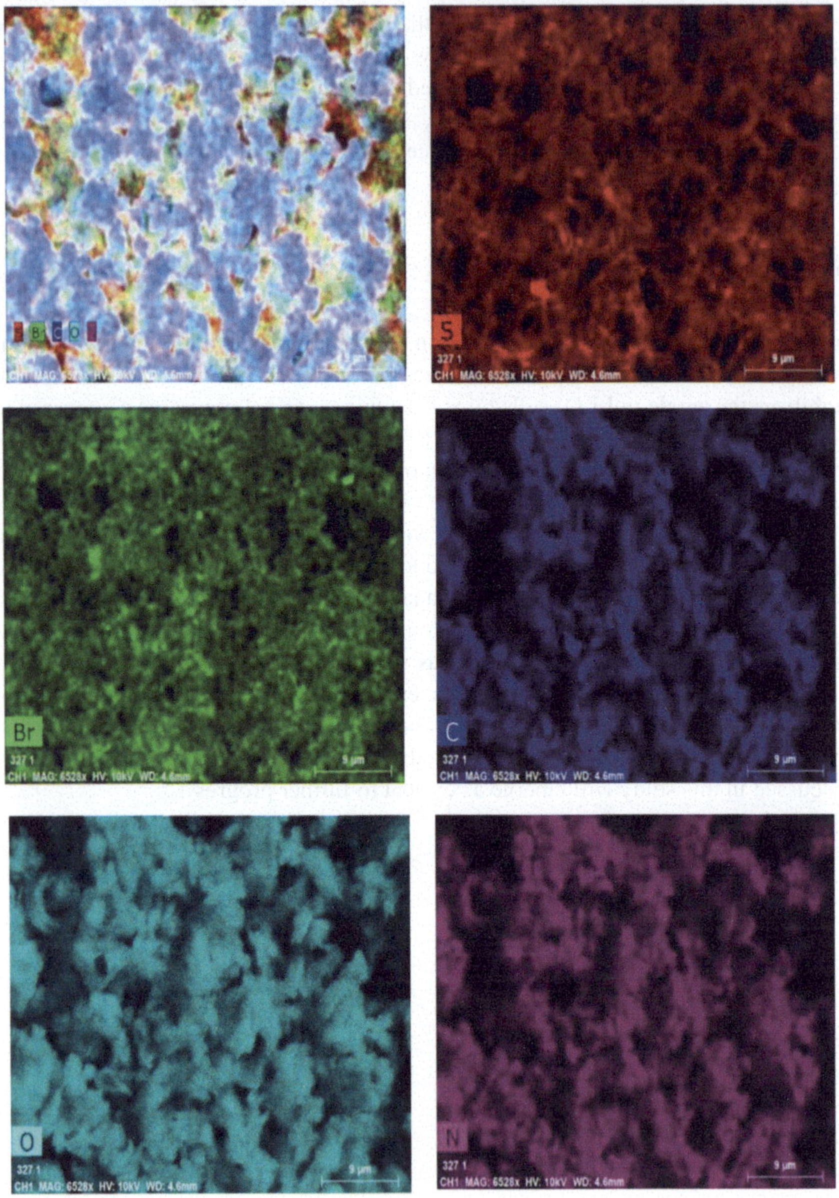

Supplementary Fig. 8.1 2D EDAX mapping of lithium-deposited stainless steel substrate with 1 M $LiNO_3$-DMAc electrolyte and 10% ionomer additive. The atoms taken into consideration are sulfur, bromine, carbon, oxygen, and nitrogen

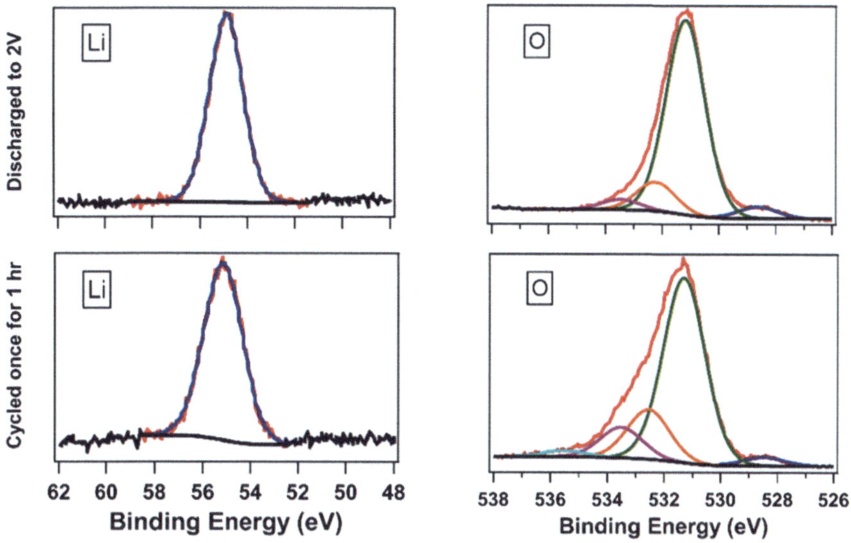

Supplementary Fig. 8.2 XPS results showing the binding energy of Li and O atom with control electrolyte of 1 M $LiNO_3$-DMAc electrolyte. The first row shows results when the battery is discharged to 2 V; the second row shows results when the Li-O_2 battery is cycled once for 1 h

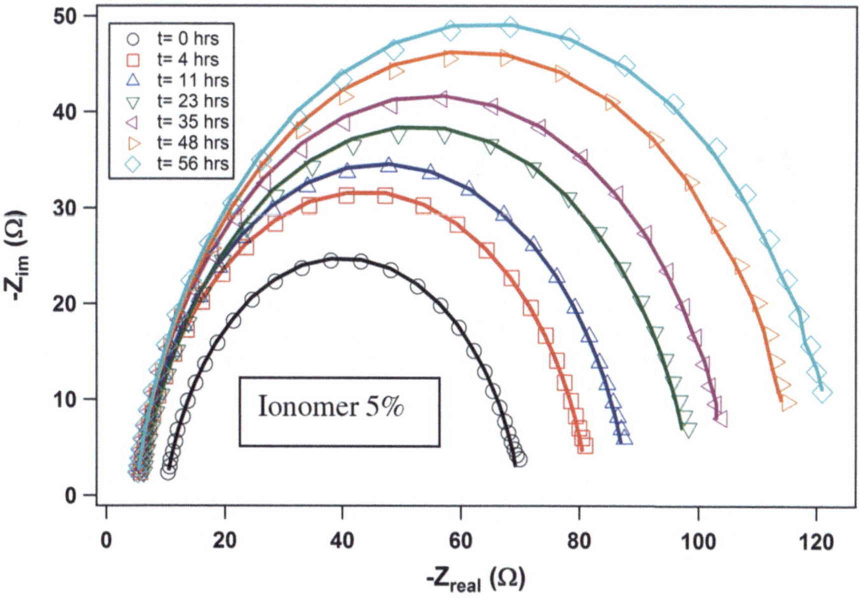

Supplementary Fig. 8.3 Nyquist plots of 1 M $LiNO_3$-DMAc enriched with 5% (by wt.) of ionomer additive, showing impedance for different storage times of the battery

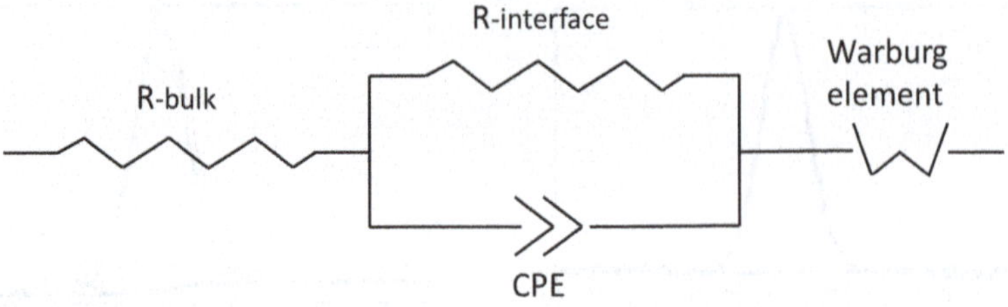

Supplementary Fig. 8.4 Equivalent circuit model to fit the Nyquist plot obtained from impedance spectroscopy measurement comprising of bulk resistance, interfacial resistance parallel to a constant phase element, and a solid-state diffusion element

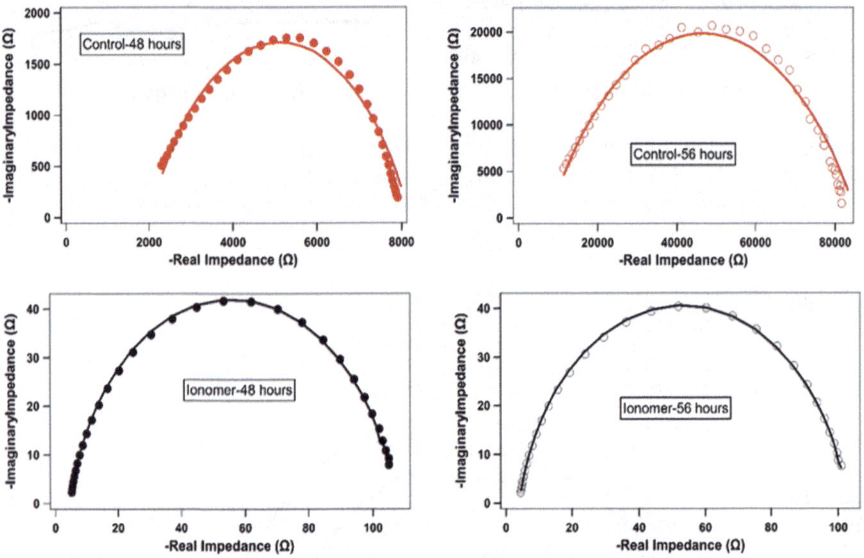

Supplementary Fig. 8.5 Nyquist plots showing experimental as well as circuit model fitted results of impedance measurements with symmetric cells for control electrolyte and ionomer-added batteries after 48 hrs and 56 hrs of storage. The red plot represents control and black shows data for ionomer-added electrolyte

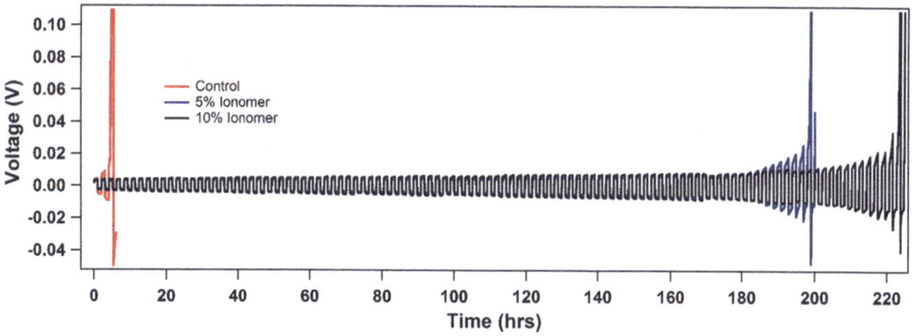

Supplementary Fig. 8.6 Stripping and plating of Li vs. SS cell after depositing 10 mAh/cm^2 of lithium onto stainless steel. It is seen that for all cells the voltage diverges for all cells however at different point of times

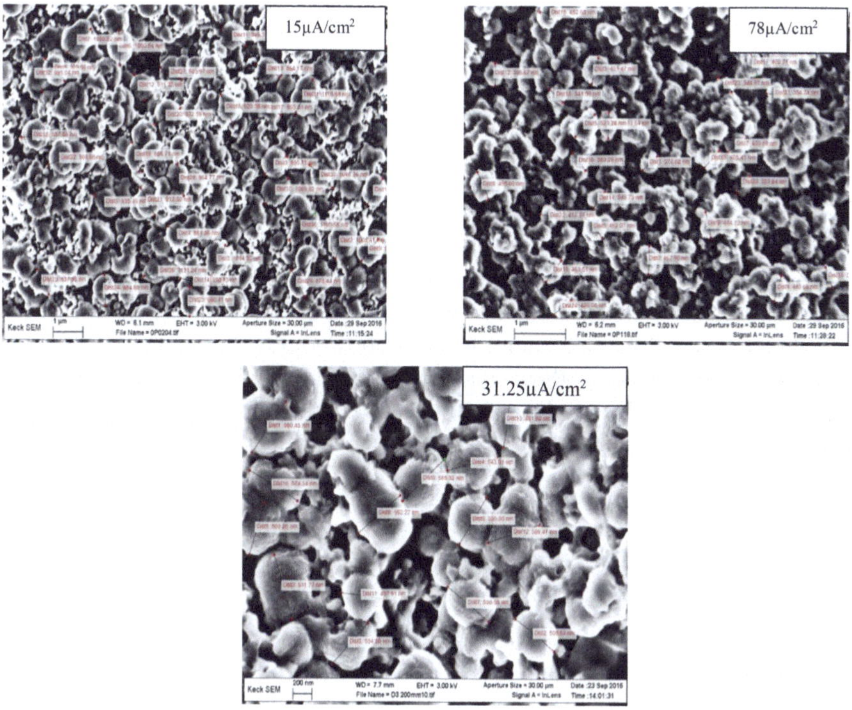

Supplementary Fig. 8.7 Size analysis of lithium peroxide particles after discharging a Li-O_2 cell with 1 M $LiNO_3$-DMAc electrolyte and ionomer additive at different current densities as indicated in the box

Supplementary Table 8.1 Atomic percentage of detected elements on lithium anodes

	Anode				
Component	1	2	3	4	5
O 1s	44.93	38.42	42.35	40.11	40.99
C 1s	10.52	15.23	12.88	16.4	16.42
N 1s	2.02	3.61	2.61	2.84	2.92
F 1s	1.12	5.12	1.25	1.73	2.35
Br 3p			1.01	1.74	1.71
S 2p			2.11	3.80	4.09
Li 1s	41.42	37.62	37.8	33.38	31.52

Samples (1) without ionomer discharged to 2 V, (2) without ionomer discharged and recharged for 1 h, (3) with ionomer discharged to 2 V, (4) with ionomer discharged and recharged for 1 h, and (5) with ionomer cycled five times for 1 h each

Experimental

Li-O_2 Battery Methods and Materials

Cathode Preparation

A cathode slurry was prepared by mixing 180 mg of Super P carbon (TIMCAL), 20 mg of polyvinylidene fluoride (PVDF; Aldrich), and 2000 mg of N-Methyl-2-pyrrolidone (NMP; Aldrich) in a ball mill at 50 Hz for 1 h. Toray TGP-H-030 carbon paper was coated with an 80 μm thick layer of carbon slurry using a doctor blade. The resulting coated carbon paper was dried at 100 °C overnight under vacuum and transferred into an argon-filled glovebox ($O_2 < 0.2$ ppm, $H_2O < 1.0$ ppm; Innovative Technology) without exposure to air. 5/8-inch diameter disks were punched and weighed from the carbon paper to yield individual carbon cathodes. The weight of the active carbon layer (not including the carbon paper) averaged 1.0 mg ± 0.1 mg.

Electrolyte Preparation

$LiNO_3$ and LiTFSI were heated under vacuum overnight at 100 °C to remove all traces of water and transferred directly into the glovebox. N,N-dimethylacetamide (DMA; Aldrich) and bis(2-methoxyethyl) ether (diglyme; Aldrich) solvents were dried over 3 Å molecular sieves (Aldrich). Lithium 2-bromoethanesulfonate was obtained through ion exchange with sodium 2-bromoethanesulfonate (Aldrich).

Coin Cell Assembly

First, a 1/2-inch (12.7 mm) diameter hole was punched in the top (cathode) side of each CR2032 case. Then, a stainless steel wire cloth disk, 3/4-inch (19 mm) disk diameter, and 0.0055-inch (0.140 mm) wire diameter from McMaster-Carr were added, followed by a cathode disk, ¾-inch diameter separator (either Whatman GF/D glass fiber or Celgard 3501), 100 μL of desired electrolyte, ½-inch diameter lithium metal, 15.5 mm diameter stainless steel spacer disk, stainless steel wave spring (MTI Corporation), and anode cap of the CR2032 case. The assembly was crimped to a pressure of 14 MPa with a hydraulic coin cell crimple (BT Innovations).

Testing Environment

Cells were tested at a regulated pure O_2 environment of 1.3 atm and allowed to equilibrate for 6 h prior to electrochemical testing. Galvanostatic measurements were conducted using a Neware CT-3008 battery tester.

Cyclic Voltammetry

The cyclic voltammetry test was done in a two-electrode setup of Li||air cathode. The batteries were cycled between 1.9 V and 4.5 V at a scan rate of 1 mV/sec several times.

Anode Stability Methods and Materials

Impedance Spectroscopy

Cells in the symmetric configuration were assembled in an Ar glovebox. Measurements were done using a Solatron frequency analyzer at a frequency range of 10^{-3} to 10^7 Hz. The data was fitted into Nyquist-type plots using the equivalent circuit shown in Supplementary Fig. 10.2 with the software zsimpwin. Impedance was conducted at room temperature at various time intervals.

Linear Scan Voltammetry

Linear scan voltammetry was done in a Li||SS cell. The batteries were first swept to −0.2 V vs. Li/Li^+; then they were swept in reverse direction until the voltage diverges.

Lithium Versus Stainless Steel Cycling

For cycling tests, lithium vs. stainless steel cells were prepared and were cycled at 0.01 mA/cm^2 between 0 and 0.5 V ten times in order to form a stable SEI layer. Then different tests were done as given in the manuscript.

Characterization Techniques

Scanning Electron Microscopy and EDAX

Discharged cells were disassembled inside the glovebox, and the cathodes were removed and transported to the scanning electron microscope (Zeiss LEO 1550 Field Emission SEM) within an airtight container. The cathodes were loaded onto the stage in the presence of a nitrogen stream. Images were taken with a single pass after focusing on a nearby region. EDAX measurements were done by taking multiple counts on a small section of sample.

X-Ray Diffraction

Cathodes were mounted on a glass microscope slide inside an argon-filled glovebox and coated with paraffin oil to protect them from air during the X-ray diffraction (XRD) measurements. Measurements were done on a Scintag Theta-Theta X-ray diffractometer using Cu K-α radiation at $\lambda = 1.5406$ Å and fitted with a 2-dimensional detector. Frames were captured with an exposure time of 10 min, after which they were integrated along χ (the polar angle orthogonal to 2θ to yield an intensity) vs. 2θ plot.

X-Ray Photoelectron Spectroscopy

XPS was conducted surface science instruments SSX-100 with operating pressure of ~ 2×10^{-9} torr. Monochromatic Al K-α X-rays (1486.6 eV) with beam diameter of 1 mm were used. Photoelectrons were collected at an emission angle of 55°. A hemispherical analyzer determined electron kinetic energy, using a pass energy of 150 V for wide survey scans and 50 V for high-resolution scans. Samples were ion-etched using 4 kV Ar ions, which were rastered over an area of 2.25 × 4 mm with total ion beam current of 2 mA, to remove adventitious carbon. Spectra were referenced to adventitious C 1s at 284.5 eV. CasaXPS software was used for XPS data analysis with Shelby backgrounds. Li 1s and O 1s were assigned to single peaks for each bond, whereas Br 3d was assigned to double peaks ($3d_{5/2}$ and $3d_{3/2}$) for each bond with 1.05 eV separation. Residual SD was maintained close to 1.0 for the calculated fits. Samples were exposed to air only during the short transfer time to the XPS chamber (less than 5 s).

References

1. Bruce, P.G., Freunberger, S.A., Hardwick, L.J., Tarascon, J.-M.: Li-O2 and Li-S batteries with high energy storage. Nat. Mater. **11**, 19–29 (2012)
2. Luntz, A.C., McCloskey, B.D.: Nonaqueous Li-air batteries: a status report. Chem. Rev. **114**, 11721–11750 (2014)
3. Girishkumar, G., McCloskey, B., Luntz, A.C., Swanson, S., Wilcke, W.: Lithium-air battery: promise and challenges. J. Phys. Chem. Lett. **1**, 2193–2203 (2010)
4. Aurbach, D., McCloskey, B.D., Nazar, L.F., Bruce, P.G.: Advances in understanding mechanisms underpinning lithium–air batteries. Nat. Energy. **1**, 16128 (2016)
5. Lu, Y.-C., Gasteiger, H.A., Parent, M.C., Chiloyan, V., Shao-Horn, Y.: The influence of catalysts on discharge and charge voltages of rechargeable Li–oxygen batteries. Electrochem. Solid-State Lett. **13**, A69 (2010)
6. Mccloskey, B.D., et al.: Combining accurate O2 and Li2O2 assays to separate discharge and charge stability limitations in nonaqueous Li-O2 batteries. J. Phys. Chem. Lett. **4**, 2989–2993 (2013)
7. McCloskey, B.D., et al.: Limitations in rechargeability of Li-O2 batteries and possible origins. J. Phys. Chem. Lett. **3**, 3043–3047 (2012)
8. McCloskey, B.D., et al.: Twin problems of interfacial carbonate formation in nonaqueous Li-O2 batteries. J. Phys. Chem. Lett. **3**, 997–1001 (2012)
9. Ottakam Thotiyl, M.M., et al.: A stable cathode for the aprotic Li-O2 battery. Nat. Mater. **12**, 1050–1056 (2013)
10. Lau, S., Archer, L.A.: Nucleation and growth of lithium peroxide in the Li-O2 battery. Nano Lett. **15**, 5995–6002 (2015)
11. Højberg, J., et al.: An electrochemical impedance spectroscopy investigation of the overpotentials in Li-O2 batteries. ACS Appl. Mater. Interfaces. **7**, 4039–4047 (2015)
12. Chen, Y., Freunberger, S.A., Peng, Z., Fontaine, O., Bruce, P.G.: Charging a Li-O2 battery using a redox mediator. Nat. Chem. **5**, 489–494 (2013)
13. Lee, D.J., Lee, H., Kim, Y.J., Park, J.K., Kim, H.T.: Sustainable redox mediation for lithium-oxygen batteries by a composite protective layer on the lithium-metal anode. Adv. Mater. **28**, 857–863 (2016)
14. Bergner, B.J., et al.: Understanding the fundamentals of redox mediators in Li-O2 batteries: a case study on nitroxides. Phys. Chem. Chem. Phys. **17**, 31769–31779 (2015)
15. Liu, T., et al.: Cycling Li-O2 batteries via LiOH formation and decomposition. Science (80-.). **350**, 3–6 (2015)
16. Kwak, W.-J., et al.: Understanding the behavior of Li–oxygen cells containing LiI. J. Mater. Chem. A. **3**, 8855–8864 (2015)
17. Lim, H.D., et al.: Superior rechargeability and efficiency of lithium-oxygen batteries: hierarchical air electrode architecture combined with a soluble catalyst. Angew. Chem. Int. Ed. Engl. **53**, 3926–3931 (2014)
18. Kundu, D., Black, R., Adams, B., Nazar, L.F.: A highly active low voltage redox mediator for enhanced rechargeability of lithium-oxygen batteries. ACS Cent. Sci. **1**, 510–515 (2015)
19. Sun, D., et al.: A solution-phase bifunctional catalyst for lithium – oxygen batteries. J. Am. Chem. Soc. **136**, 8941–8946 (2014)
20. Hirshberg, D., et al.: Li-O_2 cells with LiBr as an electrolyte and redox mediator. Energy Environ. Sci. **9**, 2334–2345 (2016)
21. Zhang, T., Liao, K., He, P., Zhou, H.: A self-defense redox mediator for efficient lithium-O2 batteries. Energy Environ. Sci. **9**, 1024–1030 (2015)
22. Veith, G.M., Nanda, J., Delmau, L.H., Dudney, N.J.: Influence of lithium salts on the discharge chemistry of Li – air cells. J. Phys. Chem. Lett. **3**, 1242–1247 (2012)
23. Black, R., et al.: Screening for superoxide reactivity in Li-O2 batteries: effect on Li2O2/LiOH crystallization. J. Am. Chem. Soc. **134**, 2902–2905 (2012)

24. Tikekar, M.D., Choudhury, S., Tu, Z., Archer, L.A.: Design principles for electrolytes and interfaces for stable lithium-metal batteries. Nat. Energy. **1**, 16114 (2016)
25. Freunberger, S.A., et al.: Reactions in the rechargeable lithium-O2 battery with alkyl carbonate electrolytes. J. Am. Chem. Soc. **133**, 8040–8047 (2011)
26. Bryantsev, V.S., et al.: The identification of stable solvents for nonaqueous rechargeable Li-air batteries. J. Electrochem. Soc. **160**, A160–A171 (2012)
27. Burke, C.M., Pande, V., Khetan, A., Viswanathan, V., Mccloskey, B.D.: Enhancing electrochemical intermediate solvation through electrolyte anion selection to increase nonaqueous Li – O 2 battery capacity. Proc. Natl. Acad. Sci. **2**, 201505728 (2015)
28. Johnson, L., et al.: The role of LiO2 solubility in O2 reduction in aprotic solvents and its consequences for Li-O2 batteries. Nat. Chem. **6**, 1091–1099 (2014)
29. Khetan, A., Luntz, A., Viswanathan, V.: Trade-offs in capacity and rechargeability in nonaqueous Li-O2 batteries: solution-driven growth versus nucleophilic stability. J. Phys. Chem. Lett. **6**, 1254–1259 (2015)
30. Knudsen, K.B., Vegge, T., McCloskey, B.D., Hjelm, J.: An electrochemical impedance spectroscopy study on the effects of the surface- and solution-based mechanisms in Li-O2 cells. J. Electrochem. Soc. **163**, A2065–A2071 (2016)
31. Aetukuri, N.B., et al.: Solvating additives drive solution-mediated electrochemistry and enhance toroid growth in non-aqueous Li–O2 batteries. Nat. Chem. **7**, 50–56 (2015)
32. Rosenman, A., et al.: The effect of interactions and reduction products of LiNO3, the anti-shuttle agent, in Li-S battery systems. J. Electrochem. Soc. **162**, A470–A473 (2015)
33. Zhang, S.S.: Role of LiNO3 in rechargeable lithium/sulfur battery. Electrochim. Acta. **70**, 344–348 (2012)
34. Liang, X., et al.: Improved cycling performances of lithium sulfur batteries with LiNO 3-modified electrolyte. J. Power Sources. **196**, 9839–9843 (2011)
35. Aurbach, D., et al.: On the surface chemical aspects of very high energy density, rechargeable Li–sulfur batteries. J. Electrochem. Soc. **156**, A694–A702 (2009)
36. Walker, W., et al.: A rechargeable Li-O-2 battery using a lithium nitrate/N, N-dimethylacetamide electrolyte. J. Am. Chem. Soc. **135**, 2076–2079 (2013)
37. Choudhury, S., Archer, L.A.: Lithium fluoride additives for stable cycling of lithium batteries at high current densities. Adv. Electron. Mater. 1–6 (2015). https://doi.org/10.1002/aelm.201500246
38. Choudhury, S., Mangal, R., Agrawal, A., Archer, L.A.: A highly reversible room-temperature lithium metal battery based on crosslinked hairy nanoparticles. Nat. Commun. 1–9 (2015). https://doi.org/10.1038/ncomms10101
39. Agrawal, A., Choudhury, S., Archer, L.A.: A highly conductive, non-flammable polymer–nanoparticle hybrid electrolyte. RSC Adv. **5**, 20800–20809 (2015)
40. Tarascon, J.M., Armand, M.: Issues and challenges facing rechargeable lithium batteries. Nature. **414**, 359–367 (2001)
41. Kashiwagi, T., et al.: Nanoparticle networks reduce the flammability of polymer nanocomposites. Nat. Mater. **4**, 928–933 (2005)
42. Tu, Z., Nath, P., Lu, Y., Tikekar, M.D., Archer, L.A.: Nanostructured electrolytes for stable lithium electrodeposition in secondary batteries. Acc. Chem. Res. acs.accounts. **5b00427**, (2015). https://doi.org/10.1021/acs.accounts.5b00427
43. Cheng, X., et al.: A review of solid electrolyte interphases on lithium metal anode. Adv. Sci. **3**, 1–20 (2016)
44. Tikekar, M.D., Archer, L.A., Koch, D.L.: Stability analysis of electrodeposition across a structured electrolyte with immobilized anions. J. Electrochem. Soc. **161**, A847–A855 (2014)
45. Tikekar, M.D., Archer, L.A., Koch, D.L.: Stabilizing electrodeposition in elastic solid electrolytes containing immobilized anions. Sci. Adv. **2**, e1600320 (2016)
46. Ozhabes, Y., Gunceler, D., Arias, T.A.: Stability and surface diffusion at lithium-electrolyte interphases with connections to dendrite suppression. arXiv. **1504.05799**, 1–7 (2015)
47. Lu, Y., et al.: Stable cycling of lithium metal batteries using high transference number electrolytes. Adv. Energy Mater. **5**, n/a–n/a (2015)

48. Schaefer, J.L., Yanga, D.A., Archer, L.A.: High lithium transference number electrolytes via creation of 3-dimensional, charged, nanoporous networks from dense functionalized nanoparticle composites. Chem. Mater. **25**, 834–839 (2013)
49. Bouchet, R., et al.: Efficient electrolytes for lithium-metal batteries. Nat. Mater. **12**, 452–457 (2013)
50. Yao, K.P.C., et al.: Thermal stability of Li2O2 and Li2O for Li-air batteries: in situ XRD and XPS studies. J. Electrochem. Soc. **160**, A824–A831 (2013)
51. Lu, Y.-C., et al.: In situ ambient pressure X-ray photoelectron spectroscopy studies of lithium-oxygen redox reactions. Sci. Rep. **2**, 715 (2012)
52. Kundu, D., Black, R., Berg, E.J., Nazar, L.F.: A highly active nanostructured metallic oxide cathode for aprotic Li–O_2 batteries. Energy Environ. Sci. **8**, 1292–1298 (2015)
53. Moulder, J., Stickle, W., Sobol, P., Bomben, K.: Handbook of X-ray Photoelectron Spectroscopy. PerkinElmer Corp. Eden Prairie, Eden Prairie (1992)
54. Zhang, Z., et al.: Increased stability toward oxygen reduction products for lithium-air batteries with oligoether-functionalized silane electrolytes. J. Phys. Chem. C. **115**, 25535–25542 (2011)
55. Lu, J., et al.: Magnetism in lithium-oxygen discharge product. ChemSusChem. **6**, 1196–1202 (2013)
56. Basile, A., Bhatt, A.I., O'Mullane, A.P.: Stabilizing lithium metal using ionic liquids for long-lived batteries. Nat. Commun. **7**, 1–11 (2016)
57. Hausbrand, R., et al.: Fundamental degradation mechanisms of layered oxide Li-ion battery cathode materials: methodology, insights and novel approaches. Mater. Sci. Eng. B. **192**, 3–25 (2015)
58. Xiong, S., Xie, K., Diao, Y., Hong, X.: Properties of surface film on lithium anode with LiNO 3 as lithium salt in electrolyte solution for lithium-sulfur batteries. Electrochim. Acta. **83**, 78–86 (2012)
59. Wu, Y., Fang, S., Jiang, Y.: Effects of nitrogen on the carbon anode of a lithium secondary battery. Sci. York. **120**, 117–123 (1999)
60. NIST X-ray photoelectron spectroscopy database, Version 4.1. Natl. Inst. Stand. Technol. Gaithersbg. **1**, (2012)
61. Ferrighi, L., et al.: Control of the intermolecular coupling of dibromotetracene on Cu(110) by the sequential activation of C-Br and C-H bonds. Chem. Eur. J. **21**, 5826–5834 (2015)
62. Basagni, A., et al.: On-surface photo-dissociation of C–Br bonds: towards room temperature Ullmann coupling. Chem. Commun. **51**, 12593–12596 (2015)
63. Gutzler, R., et al.: Ullmann-type coupling of brominated tetrathienoanthracene on copper and silver. Nanoscale. **6**, 2660–2668 (2014)
64. Desai, S.M., Solanky, S.S., Mandale, A.B., Rathore, K., Singh, R.P.: Controlled grafting of N-isopropyl acrylamide brushes onto self-standing isotactic polypropylene thin films: surface initiated atom transfer radical polymerization. Polymer. **44**, 7645–7649 (2003)
65. Di Giovannantonio, M., et al.: Insight into organometallic intermediate and its evolution to covalent bonding in surface-confined ullmann polymerization. ACS Nano. **7**, 8190–8198 (2013)
66. Jäckle, M., Groß, A.: Microscopic properties of lithium, sodium, and magnesium battery anode materials related to possible dendrite growth. J. Chem. Phys. **141**, 174710 (2014)
67. Ha, S., et al.: Magnesium (II) Bis (trifluoromethane sulfonyl) Imide-based electrolytes with wide electrochemical windows for rechargeable magnesium batteries. ACS Appl. Mater. Interfaces. **6**, 4063–4073 (2014)
68. Feng, S., et al.: Single lithium-ion conducting polymer electrolytes based on poly[(4-styrenesulfonyl)(trifluoromethanesulfonyl)imide] anions. Electrochim. Acta. **93**, 254–263 (2013)

Chapter 9
Soft Colloidal Glasses as Solid-State Electrolytes

9.1 Introduction

Rechargeable batteries that utilize a metallic lithium anode simultaneously offer opportunities and challenges as reversible electrochemical storage systems [1–5]. Lithium has the highest electronegativity and lowest atomic radius among other metals; however a major drawback is the propensity of the metal to form unstable, dendritic deposits during battery recharge, which produce premature battery failure by internal short circuits or by voltage runaway when the deposited lithium reacts with electrolyte to form a thick, ion-retarding interphase. The ohmic heat generated by Li short circuits in a conventional liquid electrolyte poses a serious impediment to progress for at least three interrelated reasons [6–8]. First a consequence of the flammability of the liquid electrolytes in current use is that release of such heat in a closed electrochemical cell would invariable end in fire, explosion, or both. Second, the relatively low melting point (T_m = 180 °C) of metallic lithium means that heat produced from a localized short can quickly destabilize the structural integrity of the electrode, causing catastrophic cell failure by thermal runaway. Finally, the intrinsic high reactivity with and exothermic reactions of metallic Li with commonly used fire-fighting reagents means that highly specialized procedures would be required to successfully intervene to stop a lithium metal battery fire.

Solid-state electrolytes are attractive candidates for lithium metal cells both because of their non-flammable characteristics and the potential to eliminate leakage, meaning that cells in a wider range of form factors are possible [9–11]. Additionally, a solid-state electrolyte can limit growth of Li dendrites and may limit transport of reactive species to a thin boundary region near the interface, localizing parasitic reactions between a Li anode and electrolyte components. The work by Li et al. [12] provides the most compelling demonstration of how these features of a solid-state electrolyte can be used to advantage. Specifically, the authors show that a micro-lithium battery comprised of metallic lithium anode, a nickel manganese cobalt oxide (NCM) cathode, and a glassy lithium phosphorus

S. Choudhury, *Rational Design of Nanostructured Polymer Electrolytes and Solid–Liquid Interphases for Lithium Batteries*, Springer Theses,
https://doi.org/10.1007/978-3-030-28943-0_9

oxynitride (LiPON) solid electrode can be cycled stably for at least 10,000 cycles with minimal loss of capacity. Notwithstanding intense research by research teams worldwide, it has been a daunting challenge to achieve similar stability in larger-scale versions of these cells. Myriad challenges ranging from the low bulk ionic conductivity of LiPON, its low mechanical toughness, and poor interfacial contact have been reported [13–20]. Recently, Han et al. [21] have demonstrated that some of these challenges can be overcome in solid-state batteries in which alumina-coated lithium is used in tandem with a $Li_7La_{2.75}Ca_{0.25}Zr_{1.75}Nb_{0.25}O_{12}$ (LLCZN) garnet-type solid electrolyte, but these cells face other challenges associated with cost of the electrolyte, environmental stability of the electrolyte under conditions typically used for battery assembly, mechanical instability of the alumina coating layer, poor interfacial ion transport at phase boundaries between the solid-state electrolyte and intercalating cathodes, and the tendency of Li to form three-dimensional, dendritic deposits that appear to grow along the grain boundaries of the solid-state electrolyte.

Solid electrolytes based on amorphous or semi-crystalline polymers, most notably poly(ethylene oxide) (PEO), have been investigated for decades because of their flexible mechanics, straight forward manufacturability, and ability to transport lithium ions in both amorphous and crystalline forms [22, 23]. Perhaps, the greatest challenge in solid electrolytes common in both ceramics and polymers is the poor interfacial contact between the solid electrode and electrolyte, and many recent articles have focused on developing novel strategies to ensure unrestricted electron and ion transport in such interfaces [24–26]. Solid polymer electrolytes are problematic, however, because of their poor oxidative stability, strong coordination with Li, and low molecular mobility, which make them unsuitable for either high-voltage or high-power cells capable of room-temperature operation. Infusing inorganic fillers in the PEO electrolyte has attracted significant interest as a facile strategy for simultaneously increasing electrolyte modulus and fraction of high-conductivity amorphous phases in the polymer, potentially breaking the usual modulus-conductivity trade-off [7, 11, 27, 28]. However, at high loading of fillers, new challenges associated with aggregation and phase separation of the polymer and particle phases obscure these benefits [7, 29, 30].

In this work, we report what is to our knowledge the first example of a solid-state polymer electrolyte that offers the combination of oxidative stability, excellent mechanical properties, and high-enough ion mobility to enable stable operation of a lithium metal battery at 25 °C. Composed of short (M_w = 5 KDa) PEO chains covalently grafted to SiO_2 nanostructures, the electrolytes exhibit soft glassy behaviors, including existence of a yield stress that allows them to flow in response to an external load and to vitrify when the load is removed. We demonstrate the practical utility of these soft glassy electrolytes in electrochemical cells in which a metallic lithium anode is paired with a Nickel Cobalt Manganese Oxide (NCM) cathode. Further, it is shown that the oxidative instability of ether-based electrolytes at a high-voltage cathode and morphological instability of Li during battery recharge can be simultaneously addressed in a simple solid-state electrolyte design.

9.2 Results and Discussion

9.2.1 *Synthesis and Chemical Analysis*

Figure 9.1 shows the schematic of the solid-state electrolyte design used in the study. Specifically, we synthesize silanized poly(ethylene oxide) by the reaction of amine-functionalized PEO (5 KDa) with 3-(triethoxysilyl)propyl isocyanate (in 1:1 molar basis) in anhydrous chloroform [31]. Thereafter, silane-PEO is covalently grafted on the surface of silica nanoparticles (25 nm) in aqueous media (Fig. 9.1a). A rigorous washing process is carried out to expel the free polymer chains from the grafted nanoparticles. The presence of free chains in the composite is a concern because it can trigger similar adverse effects to those reported in plasticized solid polymer electrolytes, where free molecules migrate to the electrode-electrolyte interface to participate in parasitic reactions, similar to what is observed in liquid electrolytes. The self-suspended materials produced by this synthesis protocol are mixed with bis(trifluoromethane) sulfonimide lithium salt (LiTFSI) at different ratios to provide a cation source in the composite (Fig. 9.1b). The SiO_2-PEO/LiTFSI composite forms the entire soft glassy electrolyte (SGE) composition (see Fig. 9.1c). To understand the relationship between salt composition and physicochemical and transport properties of the SGE, we created SiO_2-PEO/LiTFSI electrolytes with different salt concentrations. In so doing it is possible to vary the ratio of Li^+ cation and ethylene oxide (EO) units in the composite from $r = 0$ to $r = 0.2$.

Figure 9.2 reports results from Fourier transform infrared spectroscopy (FTIR) measurements on the soft glassy electrolytes. The major differences in the FTIR spectra occur in the "fingerprint" region ranging from wavelength 900 to 1500 cm^{-1}. Among the most obvious observations from these spectra is that absence of contamination associated with water absorption in the materials [32, 33]. It is also seen that there are several IR bands corresponding to similar vibrations in pure PEO ($r = 0$) and LiTFSI. The intensity of the –CF bond stretch at 1175 cm^{-1} can be utilized to understand molecular structuring in the salt in the SGE. At ratios, $r = 0$, 0.00625, and 0.0125, the peak at 1175 cm^{-1} is absent. This observation is consistent

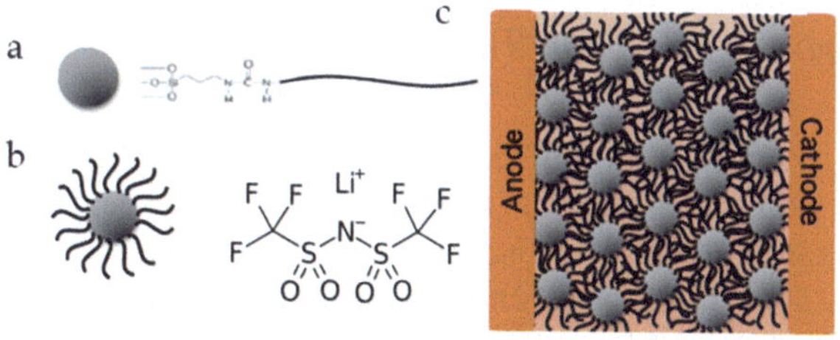

Fig. 9.1 (**a**) Schematic of silica nanoparticle (25 nm) and silane-functionalized polyethylene oxide (5000 Da) involved in the covalent grafting process; (**b**) hairy nanoparticles blended with LiTFSI salt to form electrolytes; (**c**) cartoon showing the soft glassy electrolyte sandwiched between two electrodes

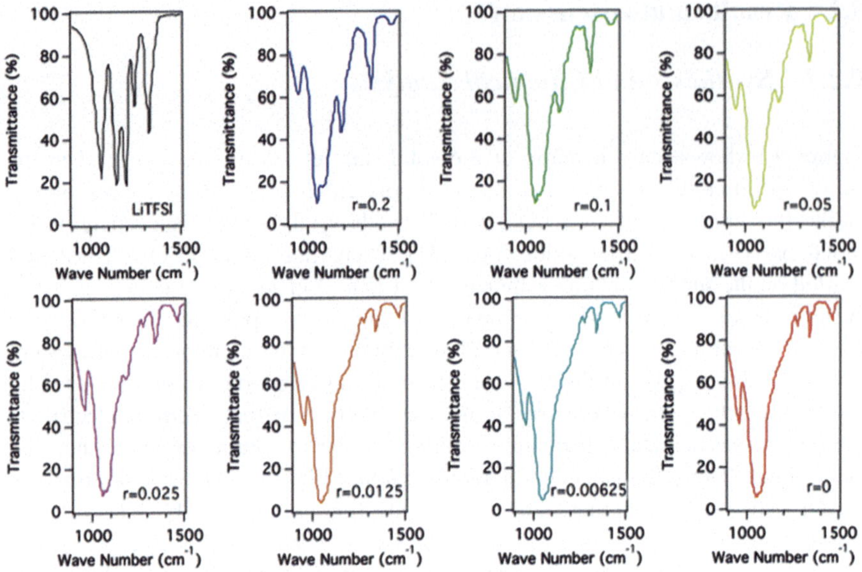

Fig. 9.2 Transmitted infrared wave as a function of wave number obtained using FTIR. The different *r*'s represent the ratio between lithium ions and ethylene oxide monomers in the glassy electrolyte

with a model in which all Li^+ cations and $TFSI^-$ anions are dissociated and coordinated with PEO moieties in the composite. Further complexation of LiTFSI ($r = 0.025$ and 0.05) leads to the appearance of the 1175 cm^{-1} peak. At higher salt contents ($r = 0.10$ and 0.20), the intensity of the peak rises, implying that the LiTFSI forms aggregates in the composite with a low degree of dissociation and consequently low fractions of mobile ions for charge transport in the electrolyte. These observations suggest that $r = 0.025$ and 0.05 are close to the optimum salt concentrations for the studied SGE electrolytes as nearly complete salt dissociation is promoted by the particle-tethered PEO chains.

9.2.2 *Calorimetry and Ion Transport*

The molecular structure of the SGE can be further analyzed using differential scanning calorimetry (DSC). Figure 9.3a reports the gravimetric heat flow as a function of temperature for SGE with salt concentration ranging from $r = 0$ to 0.2. The sharp singlet peak at ~54 °C for the $r = 0$ sample is an indication of a lone crystallite structure in contrast to three melting peaks reported for free PEO polymer [31, 32]. It can be seen that for samples with LiTFSI salt, the thermogram still maintains a singular peak, however, with differing intensities. Supplementary Fig. 9.1 reports the melting temperatures (T_m) for the different LiTFSI:EO ratios. Interestingly, it is seen that

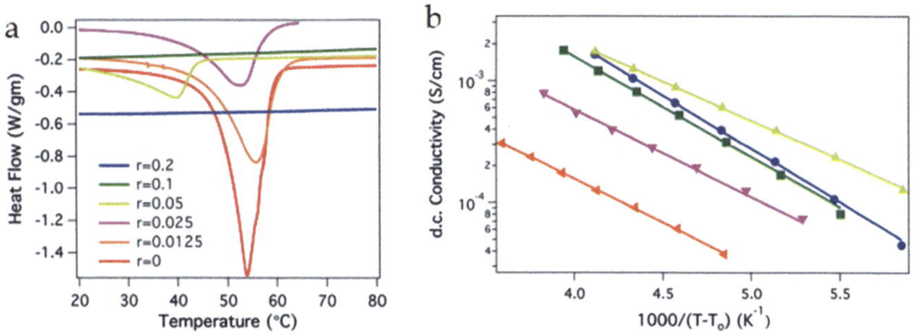

Fig. 9.3 (**a**) Thermogram showing gravimetric heat flow as a function of temperature; (**b**) VFT-fitted experimental data of conductivity in the temperature range above melting

the T_m is maintained very close to ~54 °C for SPE samples ranging from $r = 0$ to $r = 0.025$, although the intensity is seen to go down due to the decrease in the overall content of PEO moieties. The low degree of variation in the melting temperature in this range is thought to reflect the minimum influence of the salt in disrupting the crystallite structures of PEO. In other words, the interaction of LiTFSI with PEO in this range is ionic and effective in dissociating Li and TFSI ions in the salt but appears to have no effect on the PEO crystallite size. At higher salt contents, there is a stiff drop in the T_m at $r = 0.05$ to $T_m = \sim 40$ °C; corresponding to this change, the crystallization peak in Fig. 9.3a also significantly broadens in comparison to the lower or zero salt concentration. The decrease in T_m provides evidence of molecular interaction between LiTFSI and PEO groups. At $r = 0.10$ and $r = 0.20$, no melting transition is observed, implying that interactions with the salt completely disrupt crystallization of PEG chains tethered to the SiO_2 nanocores. Taken together, these observations imply that $r = 0.05$ is a critical point in that it heralds a transition from less disruptive ionic to more disrupting molecular interactions between ions in the salt and tethered PEO chains.

The molecular structure and ionic transport in an electric field can be further inferred using conductivity measurements of these composite materials. Supplementary Fig. 9.2 reports the d.c. conductivity obtained using dielectric spectroscopy performed over a wide temperature range, plotted in Arrhenius form, for SGE ranging from $r = 0.0125$ to $r = 0.2$. It is known that although PEO molecules can transport ions in amorphous as well as crystalline states, the transport timescales are considerably different. The conductivity values for $r = 0.0125$, 0.025, and 0.05 are consistent with this understanding and reveal an abrupt change in slope at 48 °C, 36 °C, and 24 °C, respectively. This observation is also in agreement with the DSC results, which reveal a crystallization transition in the materials. We isolated the temperature range from 48 to 120 °C for the measurements and fitted the measured conductivities with a Vogel-Fulcher-Tammann (VFT) model, $\sigma = A\exp(-E_a/R(T - T_o))$, where A is the prefactor, E_a is the apparent activation energy for ion transport, R is the universal gas constant, and T_o is the shift temperature. That the VFT model provides a good fit to the data points in this range (see Fig. 9.3b) indicates the absence of any thermal degradation of the material or temperature-induced abnormalities in ion transport. The fitting parameters are given in Supplementary Table 9.1. It is important to note that the previously observed temperature-induced jamming in the self-suspended hairy nanoparticles appears to have no noticeable effect on ionic motion [34–37]. Figure 9.3b further shows that the ionic conductivity at $r = 0.05$ is higher than the values measured at all other Li:EO ratios in the measured temperature range (see Supplementary Fig. 9.3). The maxima in the conductivity in a material that is evidently still semi-crystalline at low temperature provide support to our earlier suggestion that at this salt concentration, the tethered PEO chains provide maximum dissociation of ion pairs in the LiTFSI salt. On this basis, one could further conclude that the salt concentration at lower ratios is insufficient to produce full complexation with all the available ether oxygens, while at higher than $r = 0.05$, LiTFSI partially exist as undissociated and non-conducting ion pairs. For this reason, we chose $r = 0.05$ as the optimum electrolyte composition for the electrochemical studies discussed next.

9.2.3 *Structure Analysis and Rheology*

The bulk scale (nm-μm) characteristics of SPEs are dominated by structural contributions from the SiO_2 nanoparticle cores. For this reason we used small-angle X-ray scattering and oscillatory shear rheology to analyze the electrolytes.

Supplementary Fig. 9.4 reports the scattered X-ray intensities plotted against the wave vector q, for measurements performed at 90 °C. Several features of the intensity profile can be used to understand the structure of these materials. At high q, the $I(q)$ decay as the fourth power of the wave vector ($I(q) \sim q^{-4}$) with repeated oscillations, indicating that the particles are spherical in shape. Further at low q, the

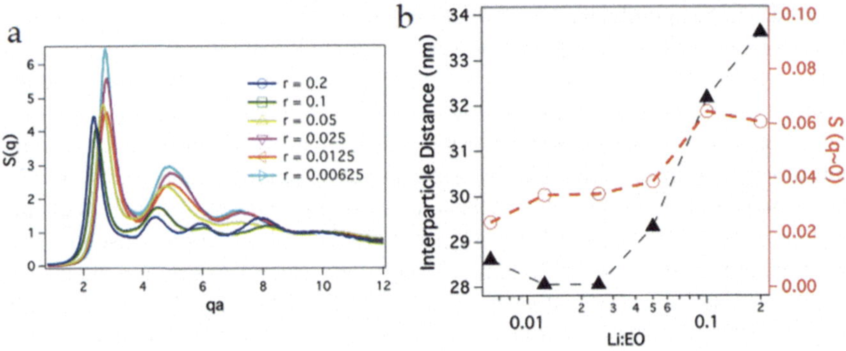

Fig. 9.4 (**a**) Structure factor obtained from SAXS analysis as a function of radius normalized wave vector; (**b**) interparticle distance obtained from the first peak as well as value of $S(q)$ as $q \to 0$ for different salt compositions

$I(q)$ is independent of q, denoting the absence of long-range density fluctuations and structure in the materials [38–40]. Both characteristics are indicative of well-dispersed particles. Figure 9.4a reports the structure factor ($S(q)$) plotted as a function of wave vector normalized by the particle radius ~12.5 nm. Remarkably, in the limit as $q \rightarrow 0$, $S(q)$ is seen to be significantly lower than previously obtained results for hard sphere suspensions. This behavior has been reported previously and reflects the effect of space-filling constrains on the tethered polymer chains which drive hyperuniformity in the materials, such that $S(q = 0) \rightarrow 0$ [41, 42]. Figure 9.4b reports $S(q \sim 0)$ for the different salt concentrations investigated. The results show there are no noticeable differences in $S(0)$ until $r = 0.05$; however, upon increasing the salt concentration beyond this value, there is a jump in $S(0)$, reminiscence of long-range ordering. This finding lends support to our earlier inference that above the critical salt concentration LiTFSI is no longer associated with PEO chains and instead occupies space between the silica nanoparticles, reducing the strength of the space-filling constraint on grafted polymer chains. The location of the peaks in the $S(q)$ (see Fig. 9.4a) confirms this point. Specifically, the center-to-center distance between the silica nanoparticles can be estimated from the location of the first peak S_1, plotted in Fig. 9.4b. It is seen to rise steeply beyond $r = 0.05$ consistent with the existence of LiTFSI as undissociated salt clusters in the materials. It has been previously reported that the first peak of the structure factor (S_1) signifies the steric repulsions and the second peak (S_2) reflects the entropic attractions in these materials [31]. Consistent with previous results using covalently grafted particles, the S_1 peak height is much larger than that of S_2, in contrast to their ionic counterparts, also signifying that the ionic linkages formed due to the salts do not significantly alter the macroscopic distribution of the nanoparticles.

We performed oscillatory shear rheology on these materials to understand the relationships between their dynamics and bulk transport properties. Results from oscillatory shear measurements at a fixed shear strain (γ = 0.1%) and variable dynamic frequency (ω) and at a fixed frequency (ω = 10 s^{-1}) and variable strain amplitude are reported in Fig. 9.5a and b, respectively. All measurements were performed at 90 °C. Interestingly, for all salt compositions, the storage modulus (G') dominates the loss modulus (G'') in the low-strain, linear viscoelastic regime, indicating the materials possess solid-like, elastic consistency. Large amplitude oscillatory shear (LAOS) measurements (Fig. 9.5b) show that the materials are in fact soft glasses [43–45]. At low shear strain, $G' \gg G''$ and nearly independent of ω (Fig. 9.5a) and γ (Fig. 9.5b). In contrast at higher shear strain, G' decreases with increasing strain, while G'' initially rises and then falls less rapidly than G'. As a result G'' displays a local maximum, crosses G', and ultimately becomes larger than G' at high shear strains. This transition of the materials from solid-like (G' dominant) to liquid-like (G'' dominant) consistencies at higher strains and the appearance of the G'' maximum at an intermediate shear strain are all well-known traits of soft glasses. They are known to arise from arrested motion or caging of the SiO_2 cores by the interdigitated tethered PEO molecules followed by strain-induced breakdown of the cages, yielding particles that slide past each other dissipating energy as a result of frictional contacts between the dislocated corona polymer chains [31, 40].

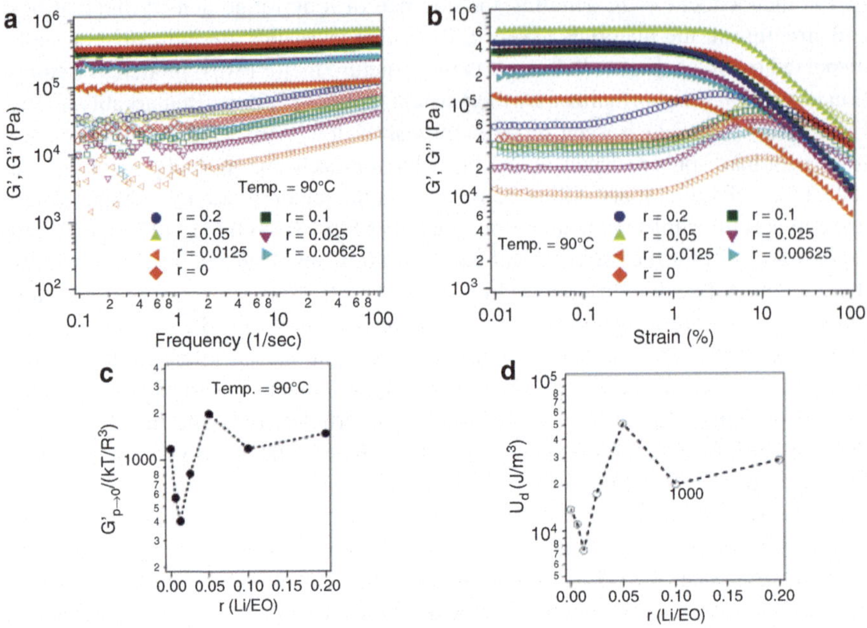

Fig. 9.5 (**a**) and (**b**) Frequency sweep and amplitude sweep measurements obtained from oscillatory shear measurements, respectively, at 90 °C. The frequency sweep was done at strain = 0.1% and amplitude sweep at **ω** = 10%; (**c**) plateau modulus as strain → 0 from different compositions of glassy electrolytes; (**d**) dissipation energy obtained by calculating the area under the curve of G'' maxima, obtained by lognormal fitting

Figure 9.5c reports the normalized elastic modulus obtained from the results in Fig. 9.5a at different salt concentrations. Here G' is normalized by the Brownian stress kT/R^3, such that values above unity imply that the stresses produced by caging are sufficient to prohibit uncorrelated, random motion of the cores. The results show that at all salt concentrations, $<G'/(kT/R^3)>$ is significantly larger than unity meaning that the particle motions are completely arrested by the interdigitated PEO corona. The PEO chains can therefore be thought of effective cross-links that lock the SiO_2 cores in place to create a tortuous nanoporous medium in which ions must move in these electrolytes.

At low salt concentrations, results in Fig. 9.5c show that addition of salt to the SiO2-PEO material causes $<G'/(kT/R^3)>$ to decrease. The decrease continues until $r = 0.0125$; whereafter it begins to rise, reaching a maximum value at $r = 0.05$. It is known that Li^+ cations are able coordinate with multiple EO moieties in an amorphous polymer, which would enhance the bridging effect produced by the interdigitated PEO chains [46, 47]. The saturation of the elastic modulus beyond $r = 0.05$ is consistent with our designation of $r = 0.05$ as the critical salt concentration. The specific energy dissipated (U_d) (shown in Fig. 9.5d) during the cage breakage transition can be calculated from the area under the $G''(\gamma)$ curve, obtained by fitting the experimental results with a normal distribution function. The effect of

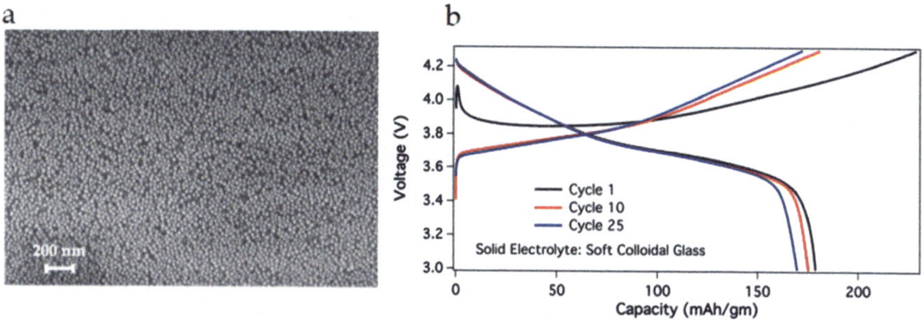

Fig. 9.6 (**a**) SEM image of the surface of lithium metal electrode covered with multiple stacks of colloidal soft glassy electrolytes; (**b**) voltage profile for the Li‖NCM cell at a current density of 0.20 mA/cm^2 for cycles 1, 10, and 25

salt concentration on U_d tracks closely the $<G'/(kT/R^3)>$ data, indicating that the two effects originate from the same source and consistent with what one would expect from a cage breakage event arising from breakage of PEO cross-links.

To investigate electrochemical properties of SGE materials, we designed a lithium metal battery using lithium metal electrode and SGE with $r = 0.05$ as the solid electrolyte. Figure 9.6a reports the scanning electron microcopy (SEM) image of the surface of lithium metal laminated with the SGE. The particles are seen to be well dispersed without any visible aggregates. It is noteworthy than even after multiple layers, the particles are essentially randomly distributed in space, consistent with the idea that the materials can be conceptualized as nanoporous media, with pore size set by the interparticle distance, which is of the order 4 nm for $r = 0.05$ (Fig. 9.4b). On the basis of linear stability analysis [48, 49] and experiment [11, 50, 51], we previously reported that electrolytes with such nanoporous morphology are effective in suppressing growth of dendrites during metal electrodeposition.

9.2.4 Analysis of Electrochemical Performance

As discussed earlier, the poor oxidative stability of PEO-based electrolytes has traditionally limited use of such electrolytes to batteries in which metallic lithium is paired with relatively low voltage (<3.8 V vs. Li/Li$^+$) cathode chemistries, including lithium titanate $LiTi_4O_7$ or lithium iron phosphate $LiFePO_4$. This has in turn reduced practical interest in lithium batteries that utilize PEO-based polymers as solid-state electrolytes. We analyzed the voltage stability window of the soft glassy electrolyte using cyclic voltammetry using a lithium versus stainless steel electrochemical cell as shown in Supplementary Fig. 9.5. It was observed that the oxidative potential for the SGE is ~4.2 V vs. Li/Li$^+$. This extended stability window can be asserted to the immobilization of the PEO groups by the surface grafting on silica nanoparticles. Similar improvement has been also reported in block-copolymer electrolytes based on PS-PEO with LiTFSI salt [52]. However, in addition to the physical approach, it is important to chemically inhibit the electrochemical oxidation of ether-oxide groups below 4.3 V vs. Li/Li^+ to enable stable cycling in a high-voltage Li||NCM battery. Recently, we performed ab initio calculations and inferred that a salt additive lithium bis(oxalate) borate (LiBOB) can be employed as a sacrificial agent in a liquid electrolyte to produce negatively charged ion clusters at the cathode-electrolyte interphase (CEI) [53]. A negatively charged SEI is hypothesized to create a rectifying mechanism that restricts transport of oxidation products formed during electrochemical breakdown of PEO-based electrolytes. Here, we evaluate the effectiveness of this concept for enabling high-voltage operation of a solid polymer electrolyte based on PEO. Specifically, we create a CEI on the surface of a NCM cathode (areal loading = 2 mAh/cm^2) by first wetting the cathode with a LIBOB-containing liquid electrolyte (0.4 M LiBOB, 0.6 M LiTFSI, 0.05 M $LiPF_6$-EC/DMC) [54] and use the wetted cathodes in electrochemical cells. $LiPF_6$ is included in the formulation because it is thought to be important for preventing corrosion of the Al current collector used for the cathode [55]. LiTFSI is the common ion carrier for transport at the CEI and in the bulk SGE phase.

Supplementary Fig. 9.6 shows the impedance spectra of the Li||SGE | liq.||NCM plotted in the form of a Nyquist plot at 30 °C. The experimental data was fitted to the equivalent circuit model shown in the inset of Fig. 9.6, which comprises of a bulk resistor in series with two other resistors representing the anodic and cathodic interfaces parallel to constant phase elements representing the static charges in the Debye layer. The bulk resistance is calculated to be 19 Ω, while the anodic and cathodic interfacial resistances are 127 Ω and 435 Ω, respectively. It can be inferred from these resistances that the liquid electrolyte partially solvates the SGE at the anode side, thus offering low overall battery resistance for operation at room temperature at low to moderate currents. The ionic conductivity of the solvated SGE electrolyte can be calculated as 0.5 mS/cm. Furthermore, we analyzed the impedance spectra of the battery after 100 cycles shown in the inset of Supplementary Fig. 9.6. It can be seen that, while the bulk impedance (20 Ω) remains similar to the original value, the interfacial resistance drops from 127 to 40 Ω, which implies that formation of ionic pathways sue to wetting of the PEO chains by the liquid electrolyte. Although we operate the battery in this work at ambient conditions in the current work, it will be interesting in the future to observe the physical behavior of the soft glassy electrolyte as well as robustness of the interfacial layer when operated at elevated temperature. However, it is important to note that the hairy nanoparticles maintain their viscoelastic nature at high temperatures and also the borate salts are thermally stable.

The CEI layer formed during cycling was analyzed using Fourier transform infrared spectroscopy measurements. The IR spectra are shown in Supplementary Fig. 9.7 after the battery was cycled five times and the relevant peaks are marked using dashed lines. The peaks at 1060 cm^{-1} represent O–B–O bond, while that at 1260 and 1360 cm^{-1} indicate C–O and B–O bonds [56]. Thus, it can be argued that the CEI is dominated by carboxylic and boro-oxalate species. Figure 9.6b shows the voltage profiles for the Li||NCM cell operated at 0.20 mA/cm^2, cycled between 4.3 and 3 V. The first charging involves a long step associated with the formation of an interfacial phase. It is observed that the nominal discharge voltage is ~3.7 V as expected for a NCM cathode in absence of any IR loss. The initial discharge capacity is obtained to be ~178 mAh/g, and in 25 cycles the capacity retention is over 97%. The coulombic efficiency of the first cycle is seen to be ~78%; thereafter it increases to more than 99% in the course of cycling as seen in Supplementary Fig. 9.8. The low C.E. in the initial cycle can be arise from the irreversible reactions to form the interfacial layer on the cathode. We further performed postmortem analysis of the electrodes at different stages of cycling to fundamentally understand the cycling behavior. Supplementary Fig. 9.9 displays the surface of lithium metal anode and the NCM cathode after five cycles. It can be seen that the lithium metal anode is covered with the multilayers of PEO-grafted silica nanoparticles similar to the anode surface before cycling shown in Fig. 9.6a. This supports many previous observations in nanocomposite electrolytes, where it is argued the hairy nanoparticles adsorb to the surface of the anode forming a mechanically stable artificial SEI layer [57, 58]. At the end of 100 cycles of stable charging and discharging the Li||NCM cell, it is seen from Supplementary Fig. 9.10 that the surface of lithium metal anode is cracked with multiple wide openings, presumably due to the large volume expansion and contraction (10 μm) during lithium plating and stripping. Unlike previous reports of battery cycling using liquid electrolytes, there is no sign of dendritic growth even after prolonged cycling due to the mechanical reinforcement by the soft glassy electrolytes. The SEM analysis of the NCM cathode after 5 and 100 cycles (as shown in Supplementary Figs. 9.9 and 9.10) reveals that there is minimal disintegration of the cathode particles upon cycling.

9.3 Conclusion

In conclusion, we designed a glass-type solid polymer by covalent linkage between silica nanoparticles and polyethylene oxide polymer and blending with a lithium salt. Variation of the salt content in the solid electrolyte leads to number of interesting structural and transport properties. The molar ratio of 0.05 for Li^+ and ethylene oxide molecules was seen to be the optimum, such that above this value, lithium salt remains as aggregates while at low content the number of ions inadequately complex with the tethered polymers. The amount of salt in the composite also leads to macroscopic changes in the modulus and nanoparticle arrangements. We further characterized the novel material in a lithium metal battery, where the colloidal particles were seen to arrange in a uniform multilayered fashion. As a first example of a safe, high-voltage solid-state lithium metal battery, we demonstrated stable cycling of a nickel manganese cobalt oxide cathode after interfacial engineering with ionic additives. We believe this work will open up a new avenue in the solid polymer electrolyte literature with ether-oxide-based chemistries that often fail at high oxidation potentials.

Acknowledgments The authors acknowledge support of the National Science Foundation, Division of Materials Research, through Award No. DMR-1609125. This work made use of the Cornell Center for Materials Research Shared Facilities, which are supported through the NSF MRSEC program (DMR-1719875).

Appendix: Supplementary Information

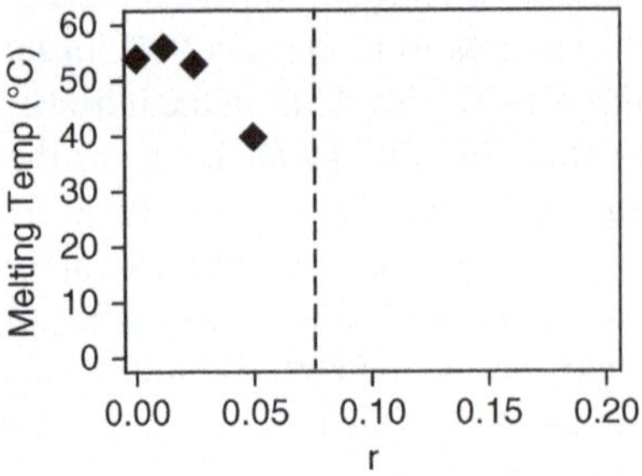

Supplementary Fig. 9.1 Melting temperature obtained from DSC plotted with Li:EO ratio

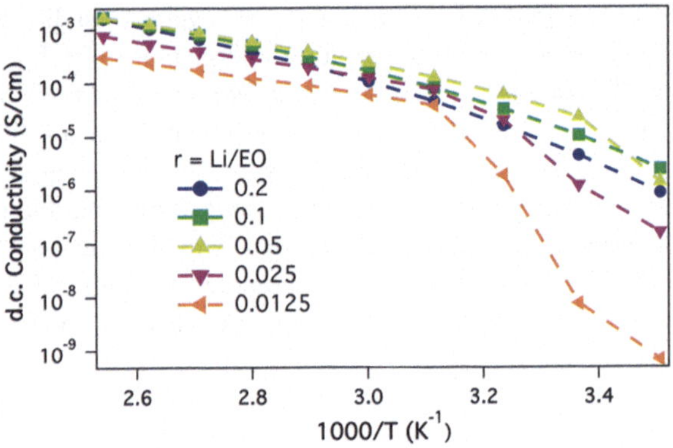

Supplementary Fig. 9.2 d.c. conductivity plotted a function of Arrhenius temperature

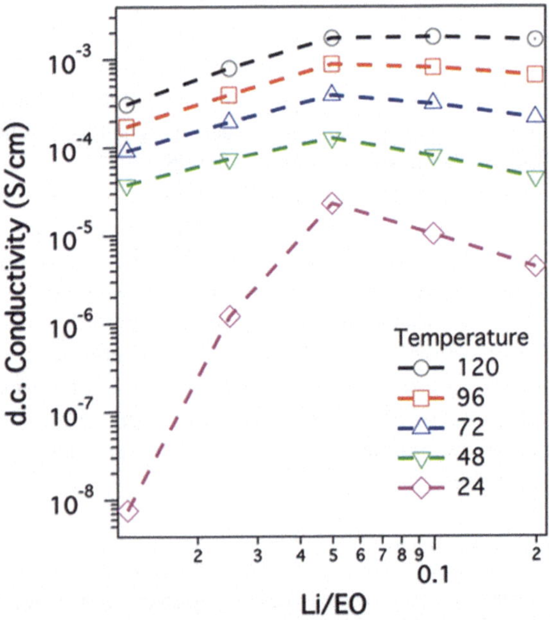

Supplementary Fig. 9.3 Variation of d.c. conductivity with the Li/EO ratios at various temperatures

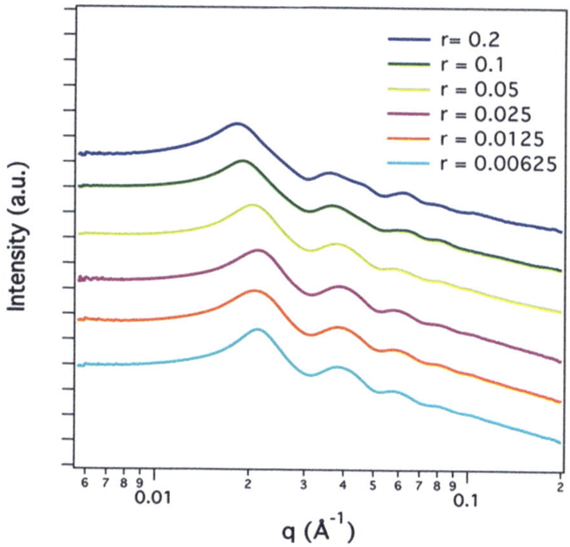

Supplementary Fig. 9.4 Intensity of scattered X-ray as a function of wave vector for different glassy electrolytes with varying salt concentrations

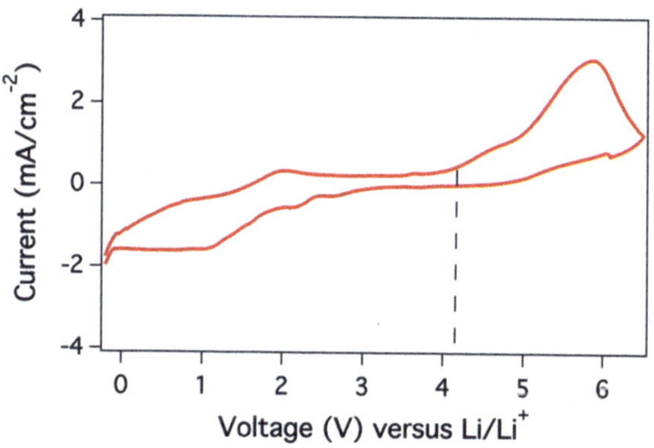

Supplementary Fig. 9.5 Cyclic voltammetry at 90 °C for a coin cell comprising of lithium vs. stainless steel battery with the soft glassy electrolyte with Li/EO ratio being 0.05. The dotted line shows the oxidative breakdown potential at ~4.2 V vs. Li/Li^+

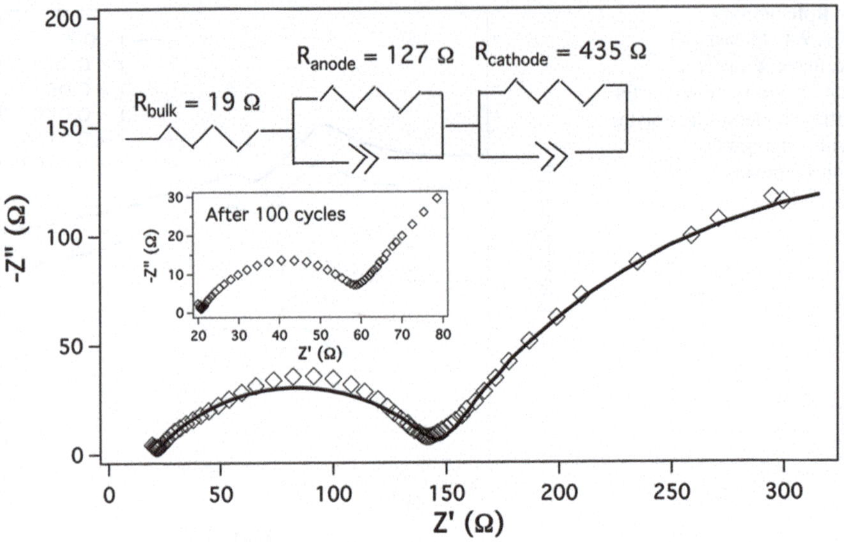

Supplementary Fig. 9.6 Impedance spectra obtained at 30 °C for a coin cell comprising of lithium vs. NCM battery with the soft glassy electrolyte with Li/EO ratio being 0.05 on the anode side along with a layer of liquid electrolyte on the cathode surface that comprises of 0.4 M LiBOB, 0.6 M LiTFSI, and 0.05 M $LiPF_6$ in EC/DMC. The inset shows the equivalent circuit model along with the corresponding resistances. The second inset shows the impedance profile of the same battery after 100 cycles

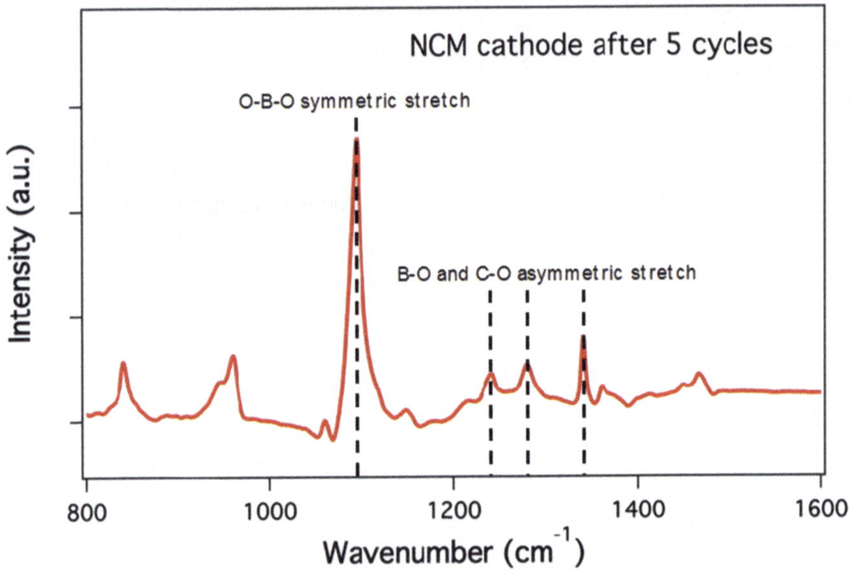

Supplementary Fig. 9.7 FTIR analysis of the NCM cathode after cycling five times. The dashed lines represent the location of the peaks for the carboxylic and boro-oxalate bonds signifying the chemistry of the cathode-electrolyte interphase

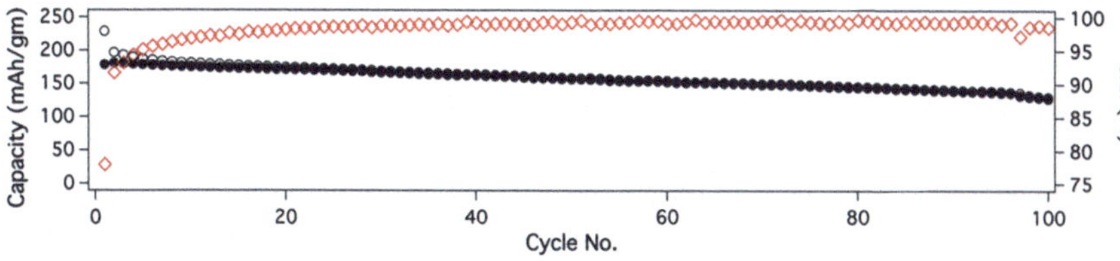

Supplementary Fig. 9.8 Cycling performance of lithium vs. NCM battery with the soft glassy electrolyte with Li/EO ratio being 0.05 on the anode side along with a layer of liquid electrolyte on the cathode surface that comprises of 0.4 M LiBOB, 0.6 M LiTFSI, and 0.05 M $LiPF_6$ in EC/DMC. The inset shows the equivalent circuit model along with the corresponding resistances

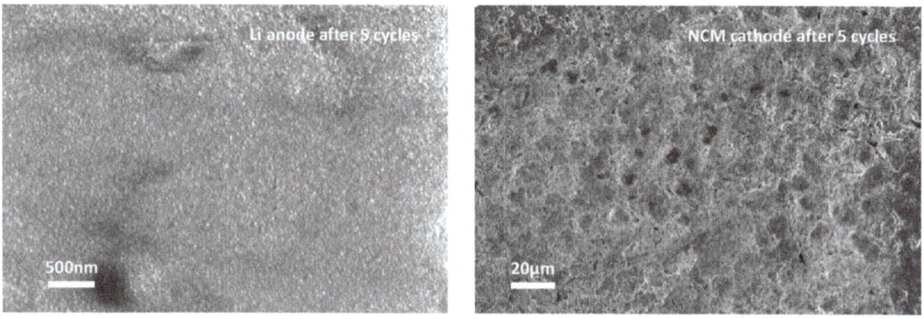

Supplementary Fig. 9.9 SEM micrographs of the lithium metal anode and NCM cathode after five cycles

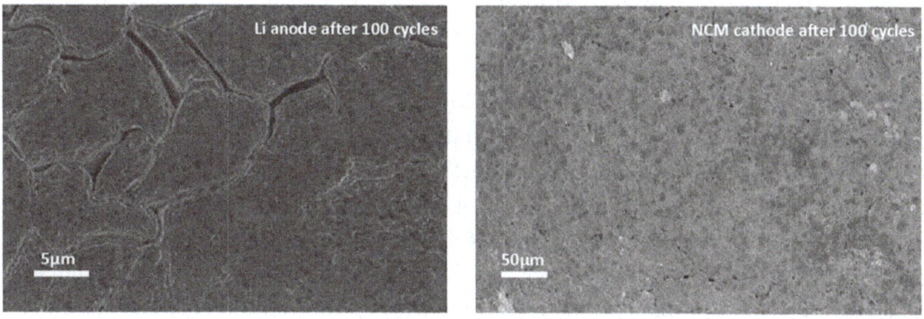

Supplementary Fig. 9.10 SEM micrographs of the lithium metal anode and NCM cathode after 100 cycles

Supplementary Table 9.1 VFT parameters for different samples obtained by fitting the experimental conductivity values using least squares method

Sample	*A* (S/cm)	E_a (kJ/mole)	*T* (K)
$r = 0.20$	6.38	16.7	150.2
$r = 0.10$	3.11	15.8	139.3
$r = 0.05$	0.73	12.2	150.4
$r = 0.025$	0.44	13.8	131.8
$r = 0.0125$	0.12	13.7	114.8

Materials and Methods

Synthesis of Self-Suspended Covalently Grafted Hairy Nanoparticles

Self-suspended covalently grafted nanoparticles were prepared using a previously described method. Briefly, these nanoparticles are prepared using a two-step process, in which polyethylene oxide was first functionalized with a silane group by reacting the isocyanate end in 3-(triethoxysilyl) propyl isocyanate (Sigma-Aldrich) with the amine end in amino-polyethylene oxide (MW ≈ *5000 Da*, PDI ≈ 1.1, purchased from Laysan Bio) in a stoichiometric ratio, creating a urethane bond in the process. The prepared silane functionalized polymer was condensed onto silica nanoparticles with diameter 25 ± 2 nm by reaction with the hydroxyl groups on the surface of the particles (TM-50, Sigma-Aldrich). Excess unreacted polymer chains were removed by repeated centrifugation in a chloroform-hexane mixture. The inorganic content of these hairy nanoparticles was analyzed using thermogravimetric analysis (TGA) on a TGA Q1000 (TA Instruments). The TGA for different samples revealed inorganic content of 51% corresponding to grafting density of approximately 1.47 chains/nm^2.

Characterization

Glassy electrolytes were prepared by mixing the hairy nanoparticles with predetermined amount of LiTFSI salt (Sigma-Aldrich). The molecular structuring in the glassy electrolytes was studied using attenuated total reflectance-Fourier transform infrared spectroscopy (ATR-FTIR) on a Nicolet iS10 FTIR spectrometer (Thermo Fisher Scientific) equipped with a deuterated triglycine sulfate (DTGS) detector and a SMART iTR diamond ATR accessory. Melting transitions were then investigated using differential scanning calorimetry on a DSC Q2000 (TA Instruments) at a scan rate of 10 °C/min. Morphology of the electrolytes at the electrode-electrolyte interface was studied using scanning electron microscope (Zeiss Gemini 500) at the Cornell Centre for Materials Research (CCMR). Fourier transform infrared spectroscopy (ATR-FTIR) was performed on a Nicolet iS10 FTIR spectrometer (Thermo Fisher Scientific) equipped with a deuterated triglycine sulfate (DTGS) detector and a SMART iTR diamond ATR accessory.

Electrochemical Measurements

2030 coin-type cells were assembled in a glovebox (MBraun Labmaster) with Nickel Cobalt Manganese Oxide (NCM) Cathode (2 mA/cm^2) as the cathode and lithium foil (Alfa Aesar) as the anode. The NCM cathode has 80% active material (NCM 622), 10% binder (PTFE), and 10% carbon. The active material loading was

~11 mg/cm^2. The glassy electrolyte was sandwiched between the two electrodes, after wetting the cathode in a liquid electrolyte comprised of LiBOB (Oakwood Chemicals), LiTFSI (Sigma-Aldrich), and $LiPF_6$ (Sigma-Aldrich) salts in an ethylene carbonate/dimethyl carbonate (Sigma-Aldrich) mixture.

Ionic transport in the bulk and at the interface in this system was studied using conductivity and impedance measurements using a Novocontrol N40 broadband spectrometer fitted with a Quarto temperature control system. The samples were sandwiched between two gold-plated blocking electrodes. The cyclic voltammetry was done in lithium versus stainless steel configuration using a CHI600 potentiostat. The cycling characteristics of the cells were evaluated under galvanostatic conditions using Neware CT-3008 battery testers.

Small-Angle X-Ray Scattering Measurements

X-ray scattering (SAXS) measurements were performed on the solventless SiO2-PEGME nanoparticles at sector 12-ID-B of Argonne National Laboratory, using a point collimated X-ray beam. The sample was smeared on a thermal cell, and the measurements were performed at different temperatures, all above melting point of PEG. The measured scattering intensity, $I(q)$, depends on wave vector q and particle volume fraction φ as

$$I(q,\varphi) = P(q)\, S(q,\varphi)$$

where $P(q)$ and $S(q, \varphi)$ represent the particle form factor and the interparticle structure factor, respectively. Because in the limit of infinite dilution $S(q, \varphi \rightarrow 0) \approx 1$, the particle form factor can thus be obtained from the scattering intensities of dilute aqueous suspensions of particles. The structure factor can then be obtained by normalizing the scattering intensity with the form factor.

Rheology Measurements

Rheology measurements. Oscillatory shear rheology measurements were performed at a temperature of 90 °C using a Physica MCR 501 rheometer (Anton Paar at Cornell Energy Systems Institute (CESI)), outfitted with a cone and plate geometry (10 mm diameter, 2° cone angle). To study the linear and nonlinear viscoelastic properties of the materials, variable strain amplitude measurements at a fixed angular frequency of $\omega = 10$ rad/s as well as variable frequency measurements at a fixed strain amplitude $\gamma = 0.5\%$ were employed.

References

1. Cheng, X., Zhang, R., Zhao, C., Wei, F., Zhang, J., Zhang, Q.: A review of solid electrolyte interphases on lithium metal anode. Adv. Sci. **3**, 1–20 (2016)
2. Tikekar, M.D., Choudhury, S., Tu, Z., Archer, L.A.: Design principles for electrolytes and interfaces for stable lithium-metal batteries. Nat. Energy. **1**, 16114 (2016)
3. Lin, D., Liu, Y., Cui, Y.: Reviving the lithium metal anode for high-energy batteries. Nat. Nanotechnol. **12**, 194–206 (2017)
4. Cheng, X.-B., Zhang, R., Zhao, C.-Z., Zhang, Q.: Toward safe lithium metal anode in rechargeable batteries: a review. Chem. Rev. **117**, 10403–10473 (2017)
5. Wei, S., Choudhury, S., Tu, Z., Zhang, K., Archer, L.A.: Electrochemical interphases for high-energy storage using reactive metal anodes. Acc. Chem. Res. **51**, 80–88 (2018)
6. Stalin, S., Choudhury, S., Zhang, K., Archer, L.A.: Multifunctional cross-linked polymeric membranes for safe, high-performance lithium batteries. Chem. Mater. **30**, 2058–2066 (2018)
7. Agrawal, A., Choudhury, S., Archer, L.A.: A highly conductive, non-flammable polymer-nanoparticle hybrid electrolyte. RSC Adv. **5**, 20800 (2015)
8. Guo, Q., Han, Y., Wang, H., Hong, X., Zheng, C., Liu, S., Xie, K.: Safer lithium metal battery based on advanced ionic liquid gel polymer nonflammable electrolytes. RSC Adv. **6**, 101638–101644 (2016)
9. Hu, J., Wang, W., Peng, H., Guo, M., Feng, Y., Xue, Z., Ye, Y., Xie, X.: Flexible organic–inorganic hybrid solid electrolytes formed via thiol–acrylate photopolymerization. Macromolecules. **50**, 1970–1980 (2017)
10. Khurana, R., Schaefer, J.L., Archer, L.A., Coates, G.W.: Suppression of lithium dendrite growth using cross-linked polyethylene/poly(ethylene oxide) electrolytes: a new approach for practical lithium-metal polymer batteries. J. Am. Chem. Soc. **136**, 7395–7402 (2014)
11. Choudhury, S., Mangal, R., Agrawal, A., Archer, L.A.: A highly reversible room-temperature lithium metal battery based on crosslinked hairy nanoparticles. Nat. Commun. **6**, 10101 (2015)
12. Li, J., Ma, C., Chi, M., Liang, C., Dudney, N.J.: Solid electrolyte: the key for high-voltage lithium batteries. Adv. Energy Mater. **5**, 1401408 (2015)
13. Takada, K.: Progress in solid electrolytes toward realizing solid-state lithium batteries. J. Power Sources. **394**, 74–85 (2018)
14. Zheng, F., Kotobuki, M., Song, S., Lai, M.O., Lu, L.: Review on solid electrolytes for all-solid-state lithium-ion batteries. J. Power Sources. **389**, 198–213 (2018)
15. Put, B., Vereecken, P., Stesmans, A.: On the chemistry and electrochemistry of LiPON breakdown. J. Mater. Chem. A. **6**, 4848–4859 (2018)
16. Le Van-Jodin, L., Claudel, A., Secouard, C., Sabary, F., Barnes, J.-P., Martin, S.: Role of the chemical composition and structure on the electrical properties of a solid state electrolyte: case of a highly conductive LiPON. Electrochim. Acta. **259**, 742–751 (2018)
17. Bai, L., Xue, W., Li, Y., Liu, X., Li, Y., Sun, J.: The interfacial behaviours of all-solid-state lithium ion batteries. Ceram. Int. **44**, 7319–7328 (2018)
18. Lei, F., Shuya, W., Siyuan, L., Qi, L., Yingying, L.: Recent progress of the solid-state electrolytes for high-energy metal-based batteries. Adv. Energy Mater. **8**, 1702657 (2018)
19. Yu, X., Bates, J.B., Jellison, G.E., Hart, F.X.: A stable thin-film lithium electrolyte: lithium phosphorus oxynitride. J. Electrochem. Soc. **144**, 524–532 (1997)
20. Chih-Jung, C., Tatsuhiro, M., Anirudha, J., Hung-Yu, L., Nai-Hsuan, Y., Nae-Lih, W., Ho, C., Shu-Fen, H., Ru-Shi, L.: Optimizing the lithium phosphorus oxynitride protective layer thickness on low-grade composite Si-based anodes for lithium-ion batteries. ChemistrySelect. **3**, 729–735 (2018)
21. Han, X., Gong, Y., Fu, K.K., He, X., Hitz, G.T., Dai, J., Pearse, A., Liu, B., Wang, H., Rubloff, G., et al.: Negating interfacial impedance in garnet-based solid-state Li metal batteries. Nat. Mater. **16**, 572 (2016)
22. Zhang, W., Tu, Z., Qian, J., Choudhury, S., Archer, L.A., Lu, Y.: Design principles of functional polymer separators for high-energy, metal-based batteries. Small. **14**, 1703001 (2018)

23. Stephan, A.M.: Review on gel polymer electrolytes for lithium batteries. Eur. Polym. J. **42**, 21–42 (2006)
24. Zhang, W., Zhuang, H.L., Fan, L., Gao, L., Lu, Y.: A "cation-anion regulation" synergistic anode host for dendrite-free lithium metal batteries. Sci. Adv. **4**, eaar4410 (2018)
25. Cheng, X.-B., Yan, C., Zhang, X.-Q., Liu, H., Zhang, Q.: Electronic and ionic channels in working interfaces of lithium metal anodes. ACS Energy Lett. **3**, 1564–1570 (2018)
26. Zhao, C.-Z., Zhang, X.-Q., Cheng, X.-B., Zhang, R., Xu, R., Chen, P.-Y., Peng, H.-J., Huang, J.-Q., Zhang, Q.: An anion-immobilized composite electrolyte for dendrite-free lithium metal anodes. Proc. Natl. Acad. Sci. **114**, 11069–11074 (2017)
27. Baker, G.L., Colsons, S.: Composite polymer electrolytes using fumed silica fillers: rheology and ionic conductivity. Chem. Matter. **6**, 2359–2363 (1994)
28. Lu, Y., Korf, K., Kambe, Y., Tu, Z., Archer, L.A.: Ionic-liquid-nanoparticle hybrid electrolytes: applications in lithium metal batteries. Angew. Chem. **126**, 498–502 (2014)
29. Stone, G.M., Mullin, S.A., Teran, A.A., Hallinan, D.T., Minor, A.M., Hexemer, A., Balsara, N.P.: Resolution of the modulus versus adhesion dilemma in solid polymer electrolytes for rechargeable lithium metal batteries. J. Electrochem. Soc. **159**, A222–A227 (2012)
30. Bruce, P.: Conductivity and transference number measurements on polymer electrolytes. Solid State Ionics. **28–30**, 918–922 (1988)
31. Choudhury, S., Agrawal, A., Kim, S.A., Archer, L.A.: Self-suspended suspensions of covalently grafted hairy nanoparticles. Langmuir. **31**, 3222–3231 (2015)
32. Kim, S.A., Archer, L.A.: Hierarchical structure in semicrystalline polymers tethered to nanospheres. Macromolecules. **47**, 687–694 (2014)
33. Wong, D.H.C., Thelen, J.L., Fu, Y., Devaux, D., Pandya, A.A., Battaglia, V.S., Balsara, N.P., DeSimone, J.M.: Nonflammable perfluoropolyether-based electrolytes for lithium batteries. Proc. Natl. Acad. Sci. U. S. A. **111**, 3327–3331 (2014)
34. Agarwal, P., Srivastava, S., Archer, L.A.: Thermal jamming of a colloidal glass. Phys. Rev. Lett. **107**, 268302 (2011)
35. Agrawal, A., Wenning, B.M., Choudhury, S., Archer, L.A.: Interactions, structure, and dynamics of polymer-tethered nanoparticle blends. Langmuir. **32**, 8698–8708 (2016)
36. Agrawal, A., Yu, H.-Y., Sagar, A., Choudhury, S., Archer, L.A.: Molecular origins of temperature induced jamming in self-suspended hairy nanoparticles. Macromolecules. **49**, 8738 (2016)
37. Agrawal, A., Yu, H.-Y., Srivastava, S., Choudhury, S., Narayanan, S., Archer, L.: Dynamics and yielding of binary self-suspended nanoparticle fluids. Soft Matter. **11**, 5224–5234 (2015)
38. Yu, H.-Y., Srivastava, S., Archer, L.a., Koch, D.L.: Structure factor of blends of solvent-free nanoparticle-organic hybrid materials: density-functional theory and small angle X-ray scattering. Soft Matter. **10**, 9120–9135 (2014)
39. Ashcroft, N.W., Langreth, D.C.: Structure of binary liquid mixtures. I. Phys. Rev. **16**, 685–692 (1967)
40. Srivastava, S., Choudhury, S., Agrawal, A., Archer, L.A.: Self-suspended polymer grafted nanoparticles. Curr. Opin. Chem. Eng. **16**, 92–101 (2017)
41. Chremos, A., Panagiotopoulos, A.Z., Koch, D.L.: Dynamics of solvent-free grafted nanoparticles. J. Chem. Phys. **136**, 44902 (2012)
42. Chremos, A., Panagiotopoulos, A.Z., Yu, H.-Y., Koch, D.L.: Structure of solvent-free grafted nanoparticles: molecular dynamics and density-functional theory. J. Chem. Phys. **135**, 114901 (2011)
43. Sollich, P.: Rheological constitutive equation for a model of soft glassy materials. Phys. Rev. E. **58**, 738–759 (1998)
44. Sollich, P., Lequeux, F., Hébraud, P., Cates, M.: Rheology of soft glassy materials. Phys. Rev. Lett. **78**, 2020–2023 (1997)
45. Zaccarelli, E., Mayer, C., Asteriadi, A., Likos, C., Sciortino, F., Roovers, J., Iatrou, H., Hadjichristidis, N., Tartaglia, P., Löwen, H., et al.: Tailoring the flow of soft glasses by soft additives. Phys. Rev. Lett. **95**, 268301 (2005)

46. Nugent, J.L., Moganty, S.S., Archer, L.A.: Nanoscale organic hybrid electrolytes. Adv. Mater. **22**, 3677–3680 (2010)
47. Schaefer, J.L., Moganty, S.S., Yanga, D.A., Archer, L.A.: Nanoporous hybrid electrolytes. J. Mater. Chem. **21**, 10094 (2011)
48. Tikekar, M.D., Archer, L.A., Koch, D.L.: Stabilizing electrodeposition in elastic solid electrolytes containing immobilized anions. Sci. Adv. **2**, e1600320 (2016)
49. Tikekar, M.D., Archer, L.A., Koch, D.L.: Stability analysis of electrodeposition across a structured electrolyte with immobilized anions. J. Electrochem. Soc. **161**, A847–A855 (2014)
50. Tu, Z., Zachman, M.J., Choudhury, S., Wei, S., Ma, L., Yang, Y., Kourkoutis, L.F., Archer, L.A.: Nanoporous hybrid electrolytes for high-energy batteries based on reactive metal anodes. Adv. Energy Mater. **7**, 1602367 (2017)
51. Tu, Z., Nath, P., Lu, Y., Tikekar, M.D., Archer, L.A.: Nanostructured electrolytes for stable lithium electrodeposition in secondary batteries. Acc. Chem. Res. **48**, 2947 (2015)
52. Bouchet, R., Maria, S., Meziane, R., Aboulaich, A., Lienafa, L., Bonnet, J., Phan, T.N.T., Bertin, D., Gigmes, D., Devaux, D., et al.: Single-ion BAB triblock copolymers as efficient electrolytes for lithium-metal batteries. Nat. Mater. **12**, 452–457 (2013)
53. Choudhury, S., Tu, Z., Nijamudheen, A., Zachman, M.J., Stalin, S., Deng, Y., Zhao, Q., Vu, D., Kourkoutis, L.F., Mendoza-Cortes, J.L., Archer, L.A.: Stabilizing polymer electrolytes in high-voltage lithium batteries. Nat. Commun. **10**, 3091 (2019)
54. Zheng, J., Engelhard, M.H., Mei, D., Jiao, S., Polzin, B.J., Zhang, J.-G., Xu, W.: Electrolyte additive enabled fast charging and stable cycling lithium metal batteries. Nat. Energy. **2**, 17012 (2017)
55. Zhang, S.S.: A review on electrolyte additives for lithium-ion batteries. J. Power Sources. **162**, 1379–1394 (2006)
56. Beaucage, G.: Small-angle scattering from polymeric mass fractals of arbitrary mass-fractal dimension. J. Appl. Crystallogr. **29**, 134–146 (1996)
57. Choudhury, S., Agrawal, A., Wei, S., Jeng, E., Archer, L.A.: Hybrid hairy nanoparticle electrolytes stabilizing lithium metal batteries. Chem. Mater. **28**, 2147–2157 (2016)
58. Wei, S., Xu, S., Agrawral, A., Choudhury, S., Lu, Y., Tu, Z., Ma, L., Archer, L.A.: A stable room-temperature sodium–sulfur battery. Nat. Commun. **7**, 11722 (2016)

Chapter 10
Solid Polymer Interphases for Lithium Metal Batteries

10.1 Introduction

Powerful and long-lasting electrical energy storage devices are essential in modern times. One highly sought-after pathway to such devices is evolving today's ubiquitous lithium-ion batteries to so-called metal batteries that replace the graphitic anode with an alkali metal block, including lithium, sodium, and potassium [1, 2]. These batteries not only augment the anodic capacity by factors as high as ten times but also enable use of higher-energy cathodes, including sulfur and oxygen [3–7]. However, fundamental issues involving morphological, chemical, and hydrodynamic instabilities at the alkali metal anode must be addressed. First, rough electrodeposit structures grow on the metallic anode to eventually short-circuit the cell, inducing early battery failure and potential thermal runaway processes [8, 9]. Second, parasitic side reactions between the electrolyte and the expanded anode promote electrolyte loss, ultimately lowering the cell efficiency over time [10–13]. These effects are exacerbated both by the large volume change experienced by alkali metal electrodes during cycles of charge and discharge and by the tendency of associated interface strain to damage/destroy the passivating but fragile interfacial material phases formed spontaneously on the electrode. A third mode of failure arises from unstable electroconvective flows in liquid electrolytes at current densities above the diffusion limit. This latter process is a consequence of ion depletion at the electrode interface and has been studied extensively in the context of electroplating processes [14].

Conventional wisdom holds that solid-state electrolytes composed of mechanically strong and chemically inert materials may offer a unified strategy for mitigating all three sources of instability. The successes and failings of such all solid-state alkali metal batteries are beginning to emerge from fundamental [9, 15–17] and application-focused studies [11, 12, 18, 19]. Among the most important challenges that have emerged from such studies include (i) the difficulty of finding materials

S. Choudhury, *Rational Design of Nanostructured Polymer Electrolytes and Solid–Liquid Interphases for Lithium Batteries*, Springer Theses,
https://doi.org/10.1007/978-3-030-28943-0_10

that simultaneously offer sufficient mechanical rigidity to slow the growth kinetics of nonplanar metal deposits and fast room-temperature bulk and interfacial ion transport – particularly when the alkali metal electrode is used in tandem with high-capacity intercalating cathodes such as $LiCoO_2$ (LCO), $LiNiMnCoO_2$ (NMC), and $LiNiCoAlO_2$ (NCA); (ii) the inability for many of the best electrolyte candidates to reversibly deform and flex to accommodate volume change at the anode; and (iii) the complex, insulating interphases formed by solid-state electrolytes in contact with the reactive alkali metal electrode. Dense polyether-based networks with high cross-link densities have been reported in multiple recent studies [18, 20, 21] to be effective in overcoming some of these challenges at low current densities. More recently, Wei, et al. [22] reported that liquid electrolytes that incorporate high molar mass polymers to form molecular entanglements in the liquid and thereby impart viscoelasticity are effective in stabilizing deposition of metals at intermediate current densities, particularly at electrodes composed of softer alkali metals such as sodium.

We report on the physical and electrochemical characteristics of thin, cross-linked polymer electrolyte interphases formed directly on the surface of Li metal anodes. We find that the interactions and composition of such interfaces can be easily tuned in the presence of a compatible liquid electrolyte host to create single-component, elastic membranes on the surface of a lithium metal anode. It is further shown that intrinsic chemical and physical features of the membrane design impart high levels of reversibility to electrochemical processes at the anode, including elimination of hydrodynamic instability by indefinitely extending the diffusion-limited ion migration regime. Deployment of the cross-linked polymer membranes in electrochemical cells composed of metallic Li anodes and commercial-grade nickel manganese cobalt oxide (NMC) cathodes reveals that the membranes facilitate high coulombic efficiency stable cell operations in cell configurations where the membranes function either as artificial solid electrolyte interfaces or as single-component solid-state electrolytes. Direct visualization studies of Li electrodeposition reveal that the ability of the elastic interfaces formed in contact with the Li anode to promote compact deposition is an important source of the electrochemical stability and versatility of the materials.

10.2 Results and Discussion

Figure 10.1a shows a schematic of the synthesis scheme used to create the polymer membranes used in the present study. It is a facile, single-step polymerization process and does not involve incorporation of any undesired solvent. Specifically, poly(ethylene glycol) dimethacrylate (PEGDMA) of molecular weight *750 Da* is added to Bis(2-methoxyethyl) ether (diglyme) at varying volume fractions. In the presence of a suitable free-radical initiator, such as methyl benzoylformate (MBF), the DMA end groups form up to two linkages with other PEGDMA chains, rapidly cross-linking the material and trapping the mobile diglyme component in the pores. The ratio of lithium ions (Li^+) to ethylene oxide (EO) was maintained at 0.10 via

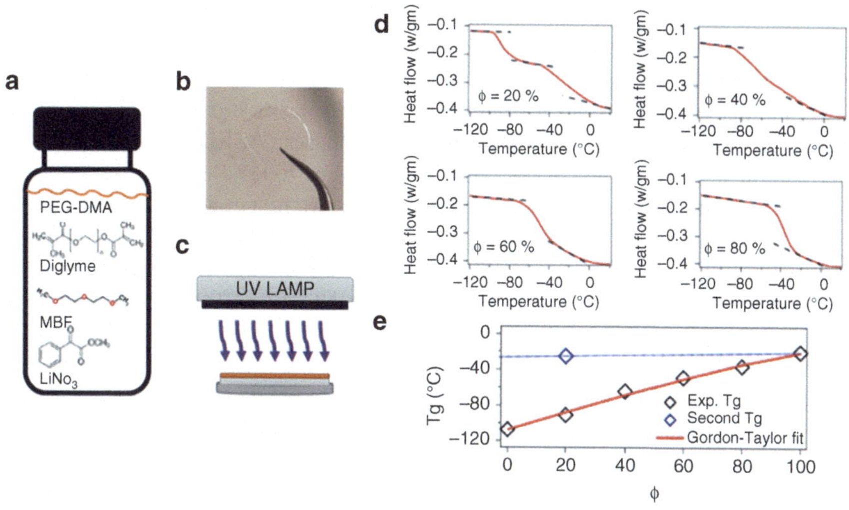

Fig. 10.1 Physical structure and thermodynamic characterizations: (**a**) Chemical structures of the precursors involved in the synthesis of the polymer network; (**b**) photograph of the polymer network. (**c**) Schematic demonstrating the concept of in situ cross-linking on metal electrode; (**d**) thermograms obtained from differential scanning calorimetry for various fractions of PEGDMA (Φ) in the membranes. The dotted lines mark the step-change in the heat-flow; (**e**) glass transition temperature as a function of PEGDMA fraction in the membranes (Φ). The red line represents Gordon-Taylor fit

addition of a simple, low-cost lithium salt, Lithium Nitrate ($LiNO_3$). Cross-linking was achieved in the present work through photopolymerization via exposure of the interfaces to ultraviolet (UV) light (*320 nm*) irradiation. The result of the photopolymerization is a soft membrane, tightly bound to the underlying substrate. The degree of softness is dependent on the diglyme content in the overall mixture because only PEGDMA participates in cross-linking. Figure 10.1b shows a membrane with 40% PEGDMA content (ϕ) that is transparent and homogenous, exhibiting no observable aggregates or signs of crystallite formation. Further it can be seen in Supplementary Fig. 10.1 that it has a rubbery texture and is able to recover fully even after macroscopic deformation. The curing process can be carried out on lithium or other substrates of lower surface energy (represented in the schematic of Fig. 10.1c). This versatility is important as it allows the materials to be studied in detail either in the form of intact solid electrolyte interphases on Li, as single-component solid-state electrolytes, or as freestanding films.

Evolution of the chemical bonding chemistry in the membranes with varying PEGDMA content in the precursor solution (ϕ) was studied via Fourier transform infrared spectroscopy (FTIR). The results illustrated in Supplementary Fig. 10.2 show that the $C{=}O$ bond at ~1700 cm^{-1} associated with PEGDMA increases in intensity and shifts to lower wave number with increasing PEGDMA content. This shift is an indication of a reduction in the effective moment of inertia of the absorbers, which correlates with a reduction in the spacing between network points in the cross-linked material. Thus, increasing PEGDMA content results in higher cross-link density, which leads to membranes that are macroscopically more elastic and mechanically stronger.

Supplementary Fig. 10.3 illustrates how such an in situ cross-linked membrane might be used as artificial solid electrolyte interphases (ASEI) to inhibit physical and chemical instabilities at an alkali metal electrode. Before assessing the electrochemical consequences of this ASEI design, we first consider the fundamental physical features of the materials and on that basis elucidate their versatility. Results from differential scanning calorimetry (DSC) of the of pure diglyme-$LiNO_3$ ($\phi = 0\%$) as well as pure PEGDMA membrane-$LiNO_3$ ($\phi = 100\%$) are shown in in Supplementary Fig. 10.4, while the DSC thermograms for intermediate ϕ's are provided in Fig. 10.1d. In the specific ranges of temperatures displayed in these figures, a second-order phase transition is observed that is associated with the glass transition of the polymer membranes. This transition reflects the phase change from melt to a glassy state due to kinetic entrapment of the molecules due to free volume reduction. The glass transition temperature (T_g) observed for $\phi = 0$, corresponding to pure diglyme-$LiNO_3$ and $\phi = 100\%$ (pure PEGDMA – $LiNO_3$), are −109.58 °C and −20.37 °C, respectively, as shown in Supplementary Fig. 10.4. It is known that compatible polymeric mixtures show distinct T_g's intermediate of the respective materials, while that of a fully mixed blend result in a single T_g dependent on the molecular dispersion and relative strength of intermolecular forces in the materials. The membrane with 20% PEGDMA content shows two distinct T_g values of −91.01 °C and −25 °C (see Fig. 10.1d), which can be asserted to two-phases that are rich in diglyme and PEGDMA, respectively. At this particular ϕ, the membrane exists in two separate phases, which implies that the PEGDMA in the original mixture is not adequate to form a fully connected network after cross-linking. This means that only a small fraction of the added diglyme interacts with the cross-linked phase, while the rest exists as a second phase that is essentially a free liquid.

At $\phi = 40\%$, the glass transition event occurs over a visibly broadened range of temperature, indicating that the PEGDMA network in the membrane is at that point at the limits where it exists as a one single component. At this composition, the separate T_g values begin to merge, indicating a chemically homogenous material such that the PEG chains are permeated evenly throughout the membrane and it inherits the mechanical toughness of a fully cross-linked network while maintaining conductivity close to that of a neat (liquid) electrolyte. As the PEGDMA composition is increased further, the T_g values fully merge to a sharp transition (shown at $\phi = 60\%$ and above), indicating the formation of a single-phase percolated network. At this point, the cross-linked networks exhibit characteristics of so-called solid-state electrolytes: mechanically resilient yet limited by low ionic conductivity.

Figure 10.1e compares T_g values for the synthesized material at all PEGDMA content. The measured T_g values were fitted to the classical Gordon-Taylor relation: $T_g = (w_1 T_{g1} + K_o w_2 T_{g2})/(w_1 + K_o w_2)$, where w_1 and w_2 are weight fractions of diglyme-$LiNO_3$ and PEGDMA-$LiNO_3$, while T_{g1} and T_{g2} are the glass transition temperature of the same. K_o here is 0.35 that is obtained from the least square fitting of the experimental values, which is indicative of the relative change in heat capacities compared to the pure components. The Gordon-Taylor relationships holds for non-interacting ($\chi = 0$) polymer mixtures. It was noted that the second glass transition temperature at $\phi = 20\%$ that doesn't follow the Gordon-Taylor relation was recorded to be close to the value at $\phi = 100\%$. Thus, it can be concluded that as the PEGDMA content in the overall mixture increases, the first phase (rich in PEG network) continues to grow and essentially engulf the diglyme molecules to form a dense and swollen percolated network of polymer chains. Frequency-dependent conductivity can reveal important information regarding the ion transport mechanisms as well as structural arrangements. Figure 10.2a reports the normalized real component of a.c. conductivity as a function of frequency for different PEGDMA content obtained using dielectric spectroscopy at 30 °C. The range of frequency values reported in Fig. 10.2a only captures the high-to-mid values below which polarization effects at electrode-electrolyte interface start to dominate. It is seen that at lower frequencies

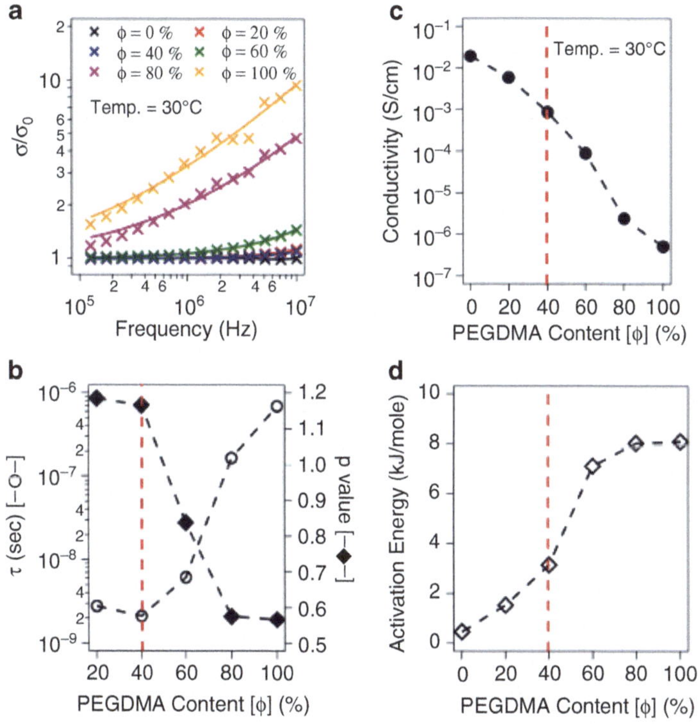

Fig. 10.2 Structure-dependent ion transport properties: (**a**) Normalized ac conductivity as a function of frequency. Measured values are shown as markers, while lines are the data fitted to Cramer's equation. (**b**) Results from frequency-dependent conductivity measurements. Left axis shows the ion transport relaxation time, and right axis describes ratio of back hop-rate over forward hop-rate obtained from the power-law model. (**c**) Conductivity as a function of PEGDMA content at fixed temperature. (**d**) Activation energy of ion transport as a function of PEGDMA content obtained by VFT model

the conductivity is independent of frequency denoting the bulk or d.c. conductivity (σ_0). In the high-frequency region, the conductivity values progressively rise beyond a critical frequency of ω_c due to correlated ionic motions and depend on the host media microstructure as well as temperature. Such behavior has also observed universally in several glassy and solid polymer electrolytes reported in the literature. The solid line fits in Fig. 10.2a represent the jump-relaxation model proposed by Jonscher [23] for thermally activated processes that was being mathematically reproduced by Cramer et al. [24] as a power-law expression: $\sigma = \sigma_0\,[1 + (\tau\omega)^p]$. Here τ is the timescale for dielectric relaxation that depends upon the coulombic interactions between the polymer chains and mobile ions. The power-law exponent p denotes the ratio between the forward hop to backward hop relaxation time. The variation of τ and p with different PEGDMA content (ϕ) is represented in Fig. 10.2b. It is seen that the ion hopping relaxation time (τ) increases rapidly beyond $\phi = 40\%$. This can be attributed to a transition of conduction mechanism from bulk motion of oligomers to ion hopping processes along the EO links in the percolated network. The evidence of this is the transition from a mixed diglyme-network to a single-phase solid electrolyte observed from the glass transition behavior at and beyond $\phi = 40\%$. Furthermore, the p-value is seen to changes from >1 to less than unity at the same fraction. Previously in the literature, solid polymer electrolytes based on PEO have been reported to exhibit $p < 1$ due to the cationic nature of the ion transport where Li^+ ions hop through coordination with EO molecules, however, at elevated temperatures as well as for liquid electrolytes p is seen be higher than 1. This argument similarly validates our proposed idea that all diglyme molecules are "tightly" bound by the percolated network beyond $\phi = 40\%$.

Supplementary Fig. 10.5 reports the d.c. conductivity obtained from the plateau region of the frequency-dependent conductivity as a function of inverse temperature. The continuous lines here represent the Vogel-Fulcher-Tammann (VFT) model given by, $\sigma_0 = A\exp(-E_a/R(T - T_0))$, which provides relationship between the ion transport rate and temperature. Here, A is a pre-factor related to the overall number of charge carrier, E_a is the activation energy, and T_0 denote the characteristic temperature below which the ion transport ceases to take place. The excellent fitting of the VFT model confirms the absence of any temperature-induced chemical or morphological changes in the measured range. The comparison of the conductivity values at 30 °C is plotted as a function of ϕ in Fig. 10.2c. It is seen that the conductivity from ϕ = 0–40% only decreases by an order of magnitude, and thereafter, as PEGDMA becomes the majority in mixture, the conductivity of the membrane quickly drops off by several orders of magnitude. Operation of cells created using the membrane of above 40% PEGDMA content as an electrolyte composite content would require increased operating temperatures similar to solid polymer electrolytes previously reported in the literature [20, 25, 26]. Figure 10.2d reports the activation energy obtained from the VFT fittings. It is evident that the activation energy for ion hopping increases sharply beyond $\phi = 40\%$. This implies that ion transport through the membrane occurs at more easily at lower PEGDMA content. Thus, the molecular interactions between the oligomers and polymer network ultimately regulate the bulk-scale ion transport processes.

In absence of forced convections, electrodeposition is a diffusion-limited process such that ion transport rate at every potential differences should be a function of ionic conductivity. However, at intermediate voltage differences, the ion transport rate should reach a maximum value, known as the limited current density, which can be calculated as $J_{\mathrm{lim}} = zFDc/\delta$, where z is the charge, F is Faraday constant, c is bulk concentration, D is ion diffusivity, and δ is the diffusion layer thickness. At even higher voltages, the ion migration rate cannot exceed the diffusion rate; thus the electroneutrality fails to be maintained at the electrode-electrolyte interfaces resulting in creation of a space charge region. It has been experimentally observed as well as predicted from Nernst-Planck Theory that a large electric field near the electrodes generates a convective flow field driving the ions across the space charge region, thereby exceeding the limiting current to an "over-limiting" value. Since the current flow in these conditions occurs not by the electroconvective flows rather than ion diffusion, it results in several unwanted events like electrolyte degradation and rampant dendritic growth on the electrodes. It can be safely argued that these instabilities are absent in all solid-state electrolytes but can occur in Newtonian fluids. Thus, such experiments are important with context of the present work to classify the cross-linked polymer networks.

Supplementary Fig. 10.6 reports current density as a function of voltage for our membranes of different PEGDMA content. The measurements were performed in symmetric (Li||Li) two-electrode cells. The voltage was scanned in staircase progression from 0 to 5 V vs. Li/Li^{+}, and the resultant steady-state current was recorded. In $\phi = 0$, 20% PEGDMA content electrolytes, quick divergence of current can be observed, without any obvious plateau region in the current density as opposed to the higher PEGDMA contents. Figure 10.3a shows the data reported in Supplementary Fig. 10.6 after normalization by the limiting current of ion transport. J_{lim} for $\phi = 0$, 20% were calculated from the theoretical expression using δ as the interelectrode spacing, while J_{lim} for the $\phi < 40\%$ was calculated from the observed plateau region. The data reports that for PEGDMA content $\phi < 40\%$ quickly rises to regions of transport instability, over the limiting current. Past the diffusion-limited current, electroconvection dominates ion transport. However, for $\phi \geq 40\%$, the current through the electrochemical systems is limited at the critical current, signaling stable ion transport. As the PEGDMA polymer network architecture exists as a single percolating solid at higher (40–100%) PEGDMA content, electroconvection is negated.

A cartoon summarizing the observed effect is represented in Fig. 10.3b. The architecture of the cross-linked PEGDMA chain network in the solid polymer interphase membrane is shown. With diglyme in excess, the physically wet membrane exhibits many free diglyme chains, and some interacting in a thinly populated PEGDMA supramolecules. As the PEGDMA content was incrementally increased, the percolated network exhibited more diglyme chains interactions. It is interesting that oligomer-polymer interactions not only regulate the thermodynamic behavior, but it is also manifested in ion transport behavior and macroscale hydrodynamics. Based on this, we can select PEGDMA content $\phi = 40\%$, as the optimum candidate for designing our solid polymer interphase (*SPI*) that simultaneously exhibit liquid and solid-like characteristics.

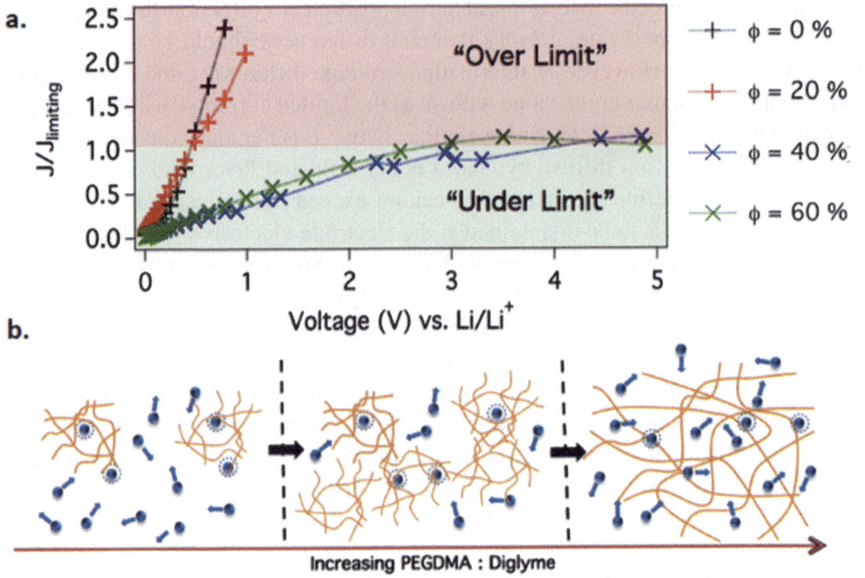

Fig. 10.3 Hydrodynamic stability during electrodeposition: (**a**) Normalized *I–V* curves for various membranes of different PEGDMA fractions (Φ). Here, the *y*-axis is normalized by the limiting current density of ion transport. (**b**) Cartoon showing the evolution of the cross-linked polymer architecture with increasing PEGDMA content. As Φ increases, free diglyme decreases in the network, and PEGDMA becomes cross-linked in a singular percolated structure

To characterize macroscopic morphological evolution during electrodeposition, an in-house built visualization setup was utilized. The setup contained dual-lithium metal rods as the electrodes coated with the *SPI* and liquid electrolyte of 1 M EC:DMC $LiPF_6$ in the interelectrode space, filling the center of the tube. Electrodeposition was visually recorded under an optical microscope using a current density of 4 mA cm^{-2}. Electrodeposit morphology for both coated and uncoated electrodes was observed at regular intervals up to 1 h, as shown in Fig. 10.4a. For the control electrolyte, the electrodeposit morphology is "mossy" and rough as compared to the case with the solid polymer interphase. This effect is the result of three mechanisms. The physical barrier provided by the *SPI* provides the electrolyte with the mechanical modulus to resist uneven deposition. Further, the negation of electroconvection due to the permeating polymer network encourages stable ion transport towards the electrode surface. Lastly, at the electrode surface, the presence of the polymer composites prevents continuous side reactions with the carbonate electrolyte. Quantitative analysis of electrodeposit morphology was also reported. This was done by plotting the height of deposit at various times, shown in Fig. 10.4b. In a linear regression of the data, the growth rate of the pristine lithium deposits with the control electrolyte was found to be ~90 nm s^{-1}, while that with the *SPI* was found to be ~20 nm s^{-1}. Supplementary Fig. 10.7 reports cell coulombic efficiency (CE) as a function of cycle number. Results were obtained using an antisymmetric

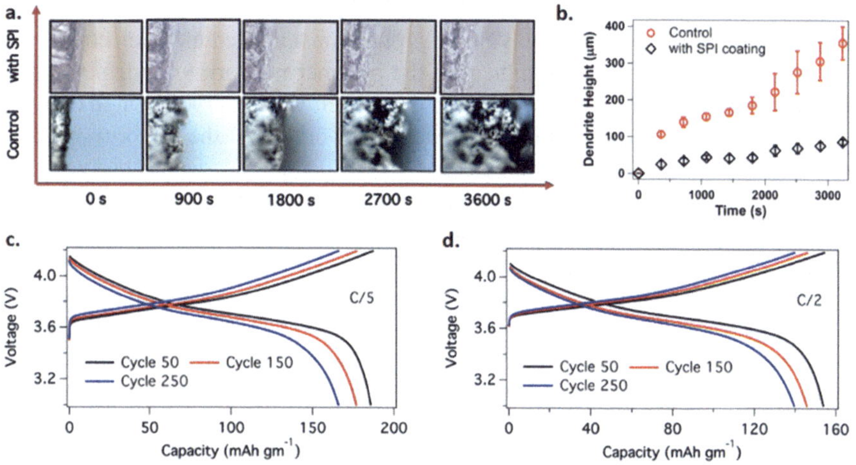

Fig. 10.4 Morphological stability and electrochemical performances: (**a**) Snapshots from direct visualization. The first row shows results lithium coated with the synthesized solid polymer interphase; the second row shows results with the uncoated lithium electrode. In both cases the electrolyte of 1 M $LiPF_6$ in EC/DMC was used and current density was 4 mA/cm^2. (**b**) Average height of dendrite as a function of time. (**c**) and (**d**) Voltage profiles for lithium vs. NCM batteries where the lithium anode is coated with the solid polymer interphase for C-rates of C/5 and C/2, respectively. The electrolyte used is 0.6 M LiTFSI, 0.4 M LiBOB 0.05 $LiPF_6$ in EC/DMC

coin cell, with pristine Li electrode and stainless steel counter electrode with and without the *SPI* coating. In the control case, the cell cycles stably at above 80% CE for around 30 cycles, before the CE drops and the cell fails. In contrast, the coin cells fabricated with the *SPI* coating exhibited >80% CE for at least 100 cycles. It is clear that the SPI can promote long-term electrodeposition stability.

Finally, we paired the SPI-coated lithium metal anode with high-voltage cathode of nickel manganese cobalt oxide (NMC). The electrolyte used is 0.6 M LiTFSI and 0.4 M LiBOB 0.05 $LiPF_6$ in EC/DMC (1/1 by wt.), which has been reported to be stable in high potential operation. Figure 10.4c and d reports the voltage profiles at a rate of C/5 and C/2, respectively. The capacity and coulombic efficiencies as a function of cycle numbers are reported in Supplementary Fig. 10.8a and b. Here, where 1 C corresponds to a current density of 2 mA cm^{-2}. It can be seen in both the C-rates that the initial capacity is ~100 mAh g^{-1} that eventually increases to ~185 mAh g^{-1} and ~155 mAh g^{-1} for C/5 and C/2, respectively. However, the coulombic efficiency is seen to be remained close to 100% throughout the battery operation; this indicates that initial rise in the capacity is due to the increasing interfacial conductivity as the liquid electrolyte wets the *SPI*. It can be seen at both C-rates that the discharge capacity retention even after 250 cycles of operation is more than 90%. We also demonsrate Li||NMC battery using the solid-state polymer ($\phi = 40\%$) as the electrolyte (Supplementary Fig. 10.9), which shows stable performance for at least 25 cycles.

In conclusion, we designed cross-linked solid polymer networks based on PEG chemistry for using as a solid polymer interphase on a lithium metal electrode. We systematically varied diglyme diluent in the polymer network to understand how thermodynamic interactions affect bulk properties. It was observed at high diglyme content, it partly interacts with the sporadic networks, and the majority remains unbounded. Specifically, at 20% PEGDMA two distinct T_g's were observed, followed by a fully interacting network at 40% and beyond. Thus at 40% PEGDMA content, the diglymes were loosely held by the network resulting in high ionic conductivity in contrast to higher content values. Also, at this critical content of PEGDMA, the solid electrolyte membrane did not manifest any electroconvective flows, thus possessing qualities of both solid and liquid electrolytes. Finally, we utilized the optimum cross-linked network as a solid interphase on lithium metal battery to show stable battery cycling due to inhibition of dendritic growth and electrode-electrolyte side reactions.

10.3 Methods

10.3.1 Fabrication of Cross-Linked Polymer Network and Coated Lithium

Poly(ethylene glycol) Dimethacrylate (Mn = 750), Diglyme and Lithium Nitrate were purchased from Sigma Aldrich. All chemicals were thoroughly dried before usage. The PEGDMA and diglyme were mixed in different ratios as required; however, the $LiNO_3$ content was maintained at $Li{:}EO = 0.1$ for all the samples. The mixture was thoroughly mixed to obtain a uniform solution. After addition of 4 wt.% of a photoinitiator methyl benzoylformate (MBF), the solution was casted on a desired substrate and exposed to UV light (VMR UVAC 115 V ~ 60 Hz 254/365 nm) for 20 min. After the reaction, the membranes were utilized as is for characterizations.

The solid polymer interphase was formed using the same procedure; however, the reaction was carried out on a flat piece of lithium metal anode in an argon-filled glove box.

10.3.2 Material Characterization

The molecular structuring in the glassy electrolytes were studied using attenuated total reflectance – Fourier transform infrared spectroscopy (ATR-FTIR) on a Nicolet iS10 FTIR spectrometer (Thermo Fisher Scientific) equipped with a deuterated triglycine sulfate (DTGS) detector and a SMART iTR diamond ATR accessory. Melting transitions were then investigated using differential scanning calorimetry on a DSC Q2000 (TA Instruments) at a scan rate of 10 °C/min.

10.3.3 Electrochemical Characterization

Ionic transport in the bulk and at the interface in this system was studied using conductivity and impedance measurements using a Novocontrol N40 broadband spectrometer fitted with a Quarto temperature control system. The samples were sandwiched between two gold-plated blocking electrodes. The *I–V* analyses were done using staircase voltammetry where each voltage steps comprise of 20 s using Maccor battery testers.

2030 coin-type cells were assembled in a glovebox (MBraun Labmaster) with nickel cobalt manganese oxide (NCM) cathode (2 mA/cm^2) as the cathode and lithium foil (Alfa Aesar) as the anode. The polymer-coated lithium metal was paired with the NCM cathode, and the in a liquid electrolyte comprised of LiBOB (Oakwood Chemicals), LiTFSI (Sigma-Aldrich) and $LiPF_6$ (Sigma-Aldrich) salts in an Ethylene Carbonate/Dimethyl Carbonate (Sigma Aldrich) mixture.

The coulombic efficiency tests were performed in a cell configuration of lithium anode with or without the solid polymer coating paired against a stainless steel counter electrode. The electrolyte utilized was 1 M EC:DMC $LiPF_6$. In this measurement a fixed amount of lithium was plated onto the stainless steel electrode and stripped back, such that the ratio of stripped and plated lithium determined the coulombic efficiency for each cycle.

The direct visualization experiment was done using two lithium rod-type electrodes in a tube-like visualization cell [27].

Appendix: Supplementary Information

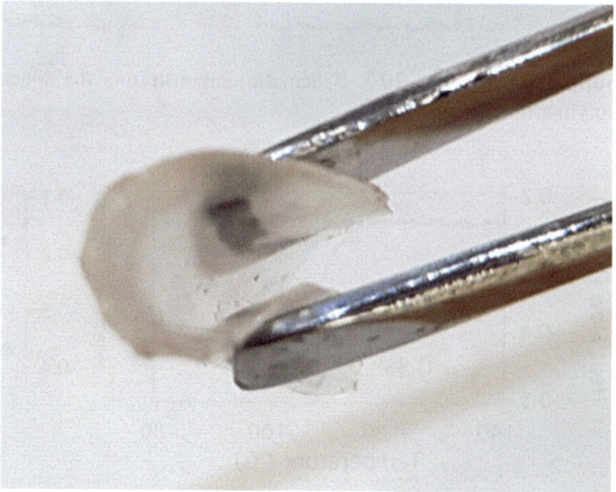

Supplementary Fig. 10.1 Photograph of the flexible cross-linked membrane

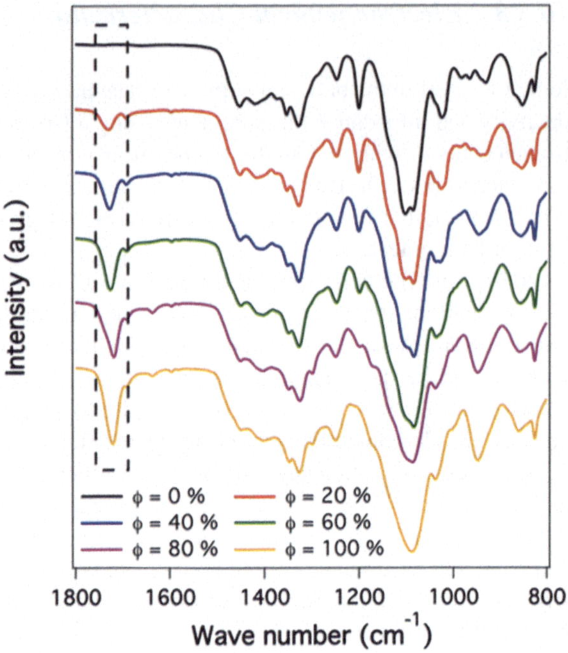

Supplementary Fig. 10.2 FTIR analysis of the cross-linked membrane at various PEGDMA content that shows C=O bond at (1700 cm^{-1}) shifting to lower intensity values as the volume percent PEGDMA is increased in solution

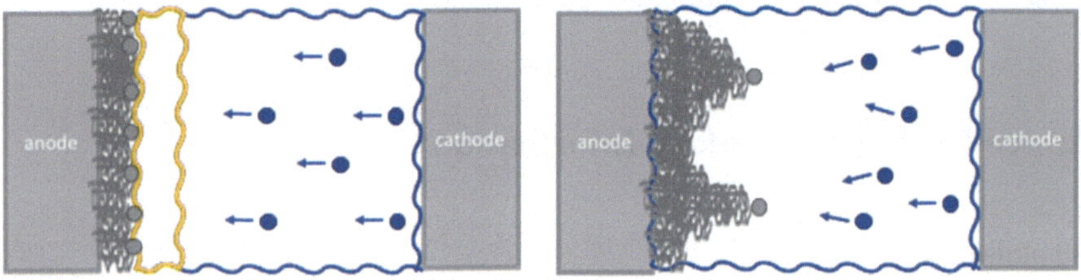

Supplementary Fig. 10.3 Schematic demonstrating the concept of stabilization using a solid polymer interphase

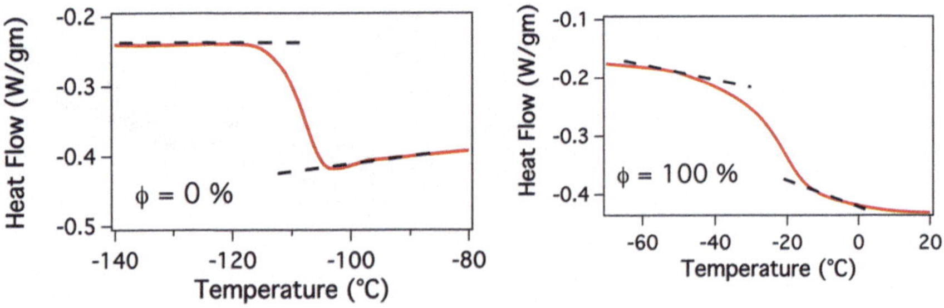

Supplementary Fig. 10.4 Thermograms obtained from differential scanning calorimetry for pure diglyme (Φ = 0%) and pure PEGDMA network (Φ = 100%). The dotted lines mark the step-change in the heat flow

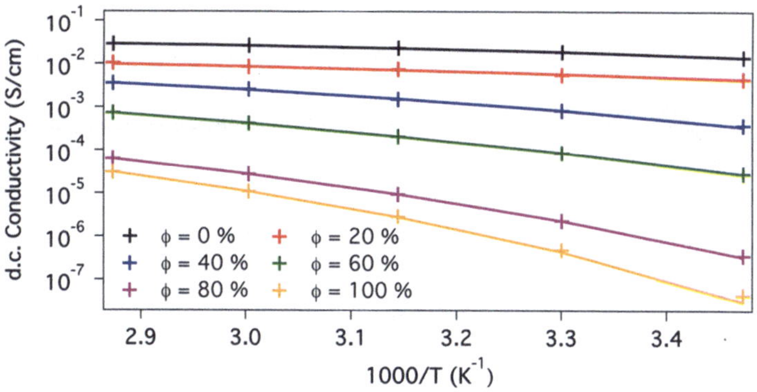

Supplementary Fig. 10.5 d.c. conductivity as a function of inverse absolute temperature. Measured values are shown as markers, and the data is fitted to Vogel-Fulcher-Tamman function

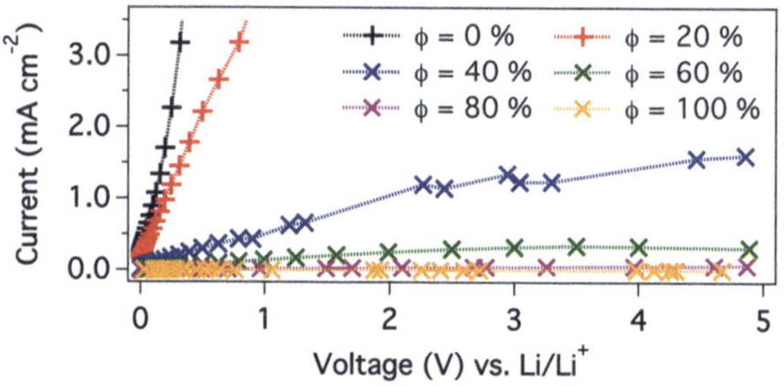

Supplementary Fig. 10.6 Current as a function of voltage. Divergence in current was seen for $\Phi = 0\%$ and $\Phi = 20\%$, signaling the presence of electroconvection. For $\Phi = 40\%$ and beyond, the current reached a limiting value at higher voltages

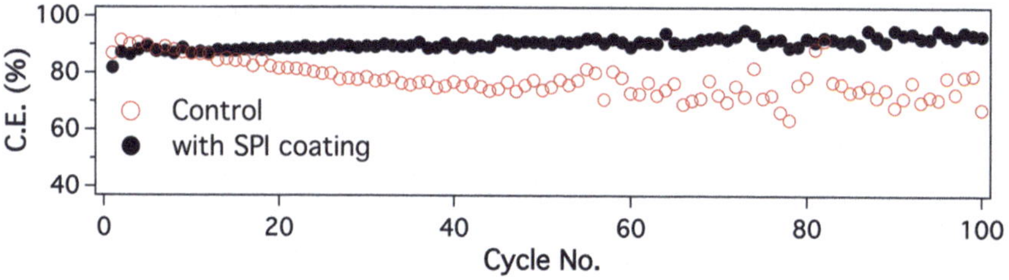

Supplementary Fig. 10.7 Coulombic efficiency measurement in Li||stainless steel coin cell with and without the solid polymer interphase at a current density of 1 mA cm^{-2} and capacity of 1 mAh cm^{-2}, using the 1 M $LiPF_6$ in EC/DMC electrolyte

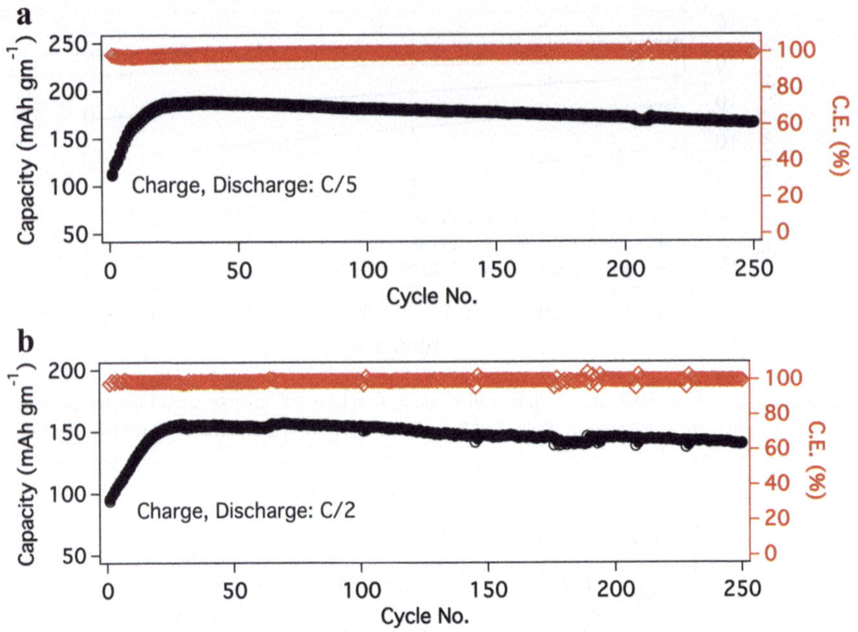

Supplementary Fig. 10.8 Cycling performances for Li||NMC cell operated at a C-rate of (**a**) C/5 and (**b**) C/2. Here the lithium metal electrode was coated with the solid polymer interphase that comprises of the polymer network and diglyme, with PEGDMA content of 40%. The thickness of the polymer coating was ~100 µm. The capacity of cathode is 2 mAh/cm^2, and the electrolyte used here is 0.6 M LiTFSI, 0.4 M LiBOB, and 0.05 $LiPF_6$ in EC/DMC (1:1 by wt.)

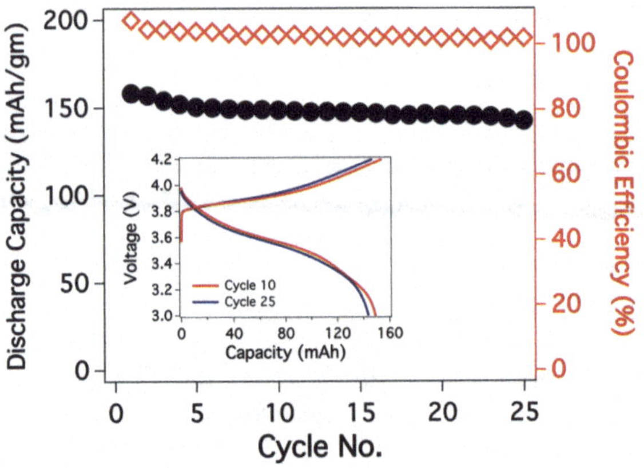

Supplementary Fig. 10.9 Cycling performances for Li||NMC cell operated at a C-rate of C/5. Here the electrolyte utilized was an all-solid-state polymer electrolyte that comprises of the polymer network and diglyme, with PEGDMA content of 40% and with the salt $LiNO_3$ (Li:EO = 0.10) and 0.4 M LiBOB as additive. The thickness of the solid polymer electrolyte was ~400 µm. The capacity of cathode is 2 mAh/cm^2. The cathode surface was wetted by liquid electrolyte of diglyme-$LiNO_3$ (Li:EO = 0.1) and 0.4 M LiBOB

References

1. Wei, S., Choudhury, S., Tu, Z., Zhang, K., Archer, L.A.: Electrochemical interphases for high-energy storage using reactive metal anodes. Acc. Chem. Res. **51**, 80–88 (2018)
2. Cheng, X., et al.: A review of solid electrolyte interphases on lithium metal anode. Adv. Sci. **3**, 1–20 (2016)
3. Lin, D., Liu, Y., Cui, Y.: Reviving the lithium metal anode for high-energy batteries. Nat. Nanotechnol. **12**, 194–206 (2017)
4. Cheng, X.-B., Zhang, R., Zhao, C.-Z., Zhang, Q.: Toward safe lithium metal anode in rechargeable batteries: a review. Chem. Rev. **117**, 10403–10473 (2017)
5. Armand, M., Tarascon, J.-M.: Building better batteries. Nature. **451**, 652–657 (2008)
6. Ma, L., Hendrickson, K.E., Wei, S., Archer, L.A.: Nanomaterials: science and applications in the lithium–sulfur battery. Nano Today. **10**, 315–338 (2015)
7. Aurbach, D., McCloskey, B.D., Nazar, L.F., Bruce, P.G.: Advances in understanding mechanisms underpinning lithium–air batteries. Nat. Energy. **1**, 16128 (2016)
8. Choudhury, S., et al.: Confining electrodeposition of metals in structured electrolytes. Proc. Natl. Acad. Sci. **115**, 6620 (2018)
9. Bai, P., Li, J., Brushett, F.R., Bazant, M.Z.: Transition of lithium growth mechanisms in liquid electrolytes. Energy Environ. Sci. **9**, 3221–3229 (2016)
10. Zhang, S.S.: A review on electrolyte additives for lithium-ion batteries. J. Power Sources. **162**, 1379–1394 (2006)
11. Suo, L., et al.: Fluorine-donating electrolytes enable highly reversible 5-V-class Li metal batteries. Proc. Natl. Acad. Sci. **115**, 1156 (2018)
12. Choudhury, S., Archer, L.A.: Lithium fluoride additives for stable cycling of lithium batteries at high current densities. Adv. Electron. Mater. **2**, 1–6 (2015). https://doi.org/10.1002/aelm.201500246
13. Lu, Y., Tu, Z., Archer, L.A.: Stable lithium electrodeposition in liquid and nanoporous solid electrolytes. Nat. Mater. **13**, 961–969 (2014)
14. Tikekar, M.D., Choudhury, S., Tu, Z., Archer, L.A.: Design principles for electrolytes and interfaces for stable lithium-metal batteries. Nat. Energy. **1**, 16114 (2016)
15. Tikekar, M.D., Archer, L.A., Koch, D.L.: Stabilizing electrodeposition in elastic solid electrolytes containing immobilized anions. Sci. Adv. **2**, e1600320 (2016)
16. Liu, W., Lin, D., Pei, A., Cui, Y.: Stabilizing lithium metal anodes by uniform Li-ion flux distribution in nanochannel confinement. J. Am. Chem. Soc. **138**, 15443–15450 (2016)
17. Zheng, G., et al.: Interconnected hollow carbon nanospheres for stable lithium metal anodes. Nat. Nanotechnol. **9**, 618–623 (2014)
18. Choudhury, S., Mangal, R., Agrawal, A., Archer, L.A.: A highly reversible room-temperature lithium metal battery based on crosslinked hairy nanoparticles. Nat. Commun. **6**, 10101 (2015)
19. Kozen, A.C., et al.: Next-generation lithium metal anode engineering via atomic layer deposition. ACS Nano. **9**, 5884–5892 (2015)
20. Khurana, R., Schaefer, J.L., Archer, L.A., Coates, G.W.: Suppression of lithium dendrite growth using cross-linked polyethylene/poly(ethylene oxide) electrolytes: a new approach for practical lithium-metal polymer batteries. J. Am. Chem. Soc. **136**, 7395–7402 (2014)
21. Bouchet, R., et al.: Single-ion BAB triblock copolymers as efficient electrolytes for lithium-metal batteries. Nat. Mater. **12**, 452–457 (2013)
22. Wei, S., et al.: Stabilizing electrochemical interfaces in viscoelastic liquid electrolytes. Sci. Adv. **4**, eaao6243 (2018)
23. Jonscher, A.K.: The "universal" dielectric response. Nature. **267**, 673–679 (1977)
24. Funke, K., Cramer, C.: Conductivity spectroscopy. Curr. Opin. Solid State Mater. Sci. **2**, 483–490 (1997)

25. Gurevitch, I., et al.: Nanocomposites of titanium dioxide and polystyrene-poly(ethylene oxide) block copolymer as solid-state electrolytes for lithium metal batteries. J. Electrochem. Soc. **160**, A1611–A1617 (2013)
26. Stone, G.M., et al.: Resolution of the modulus versus adhesion dilemma in solid polymer electrolytes for rechargeable lithium metal batteries. J. Electrochem. Soc. **159**, A222–A227 (2012)
27. Choudhury, S., et al.: Electroless formation of hybrid lithium anodes for fast interfacial ion transport. Angew. Chem. Int. Ed. **56**, 13070–13077 (2017)

Chapter 11
Stabilizing Polymer Electrolytes in High-Voltage Lithium Batteries

11.1 Introduction

Small-molecule linear and cyclic ethers/glymes and their carbonate esters formed by reaction with carbon dioxide have emerged as the most important family of electrolytes for lithium batteries. These molecules are attractive for a variety of reasons, including their low viscosity and ability to coordinate with lithium ions, producing higher concentrations of mobile charge carriers than one would anticipate from classical theory, based on their dielectric constants alone [1–6]. Macromolecular analogs, most notably polyethylene glycol dimethyl ether (PEGDME), have been reported to offer additional beneficial effects, including orders of magnitude higher mechanical modulus, low volatility and low flammability, making them attractive candidates for solid-state or flexible lithium batteries in a variety of form factors [7–11]. A substantial body of work focused on charge carrier transport mechanisms in polyethers has shown that lithium-ion mobility is coordinated with molecular motions and that charge carrier transport occurs predominantly in the amorphous phase of the materials where molecular mobility is highest [12–16]. A less studied, but as important trait of ethers is the ease with which they can be electropolymerized at the reducing potentials at the lithium battery anode, as well as at the oxidizing potentials of the cathode. Almost nothing is known about how these processes can be regulated to produce self-limiting interphases and how fast ion transport at such interphases might be used to stabilize deposition at the Li anode.

Reduction of small-molecule ethers and carbonate esters at a lithium battery anode produces less mobile polymeric species by ring-opening and/or anionic chemistries [2, 4, 17–19]. In favorable situations (e.g., at the graphitic carbon anode of state-of-the-art lithium-ion batteries), the reactions are self-limiting and produce a thin coating of a low molar mass polymer-rich phase (*interphase*) at the electrode surface. This so-called solid-electrolyte interface (SEI) limits molecular access to the electrode surface and prevents continuous loss of electrolyte. The SEI is known to be crucial for stable, long-term battery operation, but almost nothing is

S. Choudhury, *Rational Design of Nanostructured Polymer Electrolytes and Solid–Liquid Interphases for Lithium Batteries*, Springer Theses,
https://doi.org/10.1007/978-3-030-28943-0_11

known about how the tools of polymer chemistry can be used to harness it to achieve a similar electrochemical function at more unstable (chemical and morphological) alkali metal anodes. In cells that use lithium metal as anode, spontaneously formed interphases are in fact rarely self-limiting. Numerous studies have begun to appear that center on materials synthesis strategies for creation of specially designed self-limiting interfaces on such anodes using sacrificial, easily reduced species added to an electrolyte [20–23], or application of ion-permeable coatings formed ex situ [24–26].

At the intercalating composite cathodes (e.g., NMC, LMO, LCO) of greatest contemporary interest for lithium cells, electrolyte-electrode interfaces are not restricted to planes. Designing self-limiting interphases able to reduce/prevent electrolyte oxidation is therefore far more complex. Because ethers are particularly vulnerable to oxidative attack, a concession is the field is that ether- and polyether-based electrolytes cannot be used in practical electrochemical cells that employ high-voltage cathodes [27]. As a consequence, solid-state ceramic electrolytes have emerged in recent years as the most promising candidates for all solid-state lithium batteries.

Here, we consider the chemical processes responsible for uncontrolled interphase polymer chain growth at the anode and oxidative degradation of ethers at the cathode of a lithium cell and on that basis show that electrolytes based on ethers can be designed to overcome conventional limitations. We show in particular that inhibition of anionic polymerization of electrolytes based on chain transfer agents (CTAs) offer unusually high levels of interphase stability at a lithium metal anode. We further report that anionic species able to limit transport of polymer intermediates at the cathode are an integral component in designing self-limiting cathode electrolyte interfaces (CEIs) able to stabilize glymes at highly oxidizing electrode potentials. Taken together with recent work showing that polyethers in a variety of cross-linked configurations are able to inhibit rough, dendritic electrodeposition at a lithium metal anode during battery recharge, the results reported herein provide a path towards safe, cost-effective solid-state and flexible batteries based on polymer electrolytes [7, 28, 29].

The electrolyte used in the study is comprised of Bis(2-methoxyethyl) ether (diglyme) and lithium nitrate ($LiNO_3$) salt. The choice of $LiNO_3$ is based on the fact that it is cheaper compared to other salts, and it is known to form a stable interfacial layer on the anode [30]. Diglyme is chosen as the simplest oligo-ether that offers the combination of a high boiling point (162 °C) and appreciable ion transport rate at ambient temperature to be of interest as an electrolyte for the lithium metal battery. The chemical structure of the electrolyte, including the ease with which the molecule can be electropolymerized at the cathode or anode of an electrochemical cell, is shared with all ether-based liquid and solid polymer electrolytes, which means that the interfacial polymerization, oxidative breakdown, and transport characteristics of diglyme at electrodes are to a reasonable approximation representative of a much broader class of polymer electrolyte candidates. The $LiNO_3$ concentration in diglyme is systematically adjusted by varying the ratio, *r*, of Li^+ cations to ether oxygen (EO) molecules in the electrolyte.

11.2 Results and Discussions

Supplementary Fig. 11.1a reports the effect of r on the temperature-dependent electrolyte conductivity. The conductivity values at room temperature are seen exceed 1 mS/cm for all materials used in the study, but there are appreciable variations at sub-zero temperatures. It is clear from the results that diglyme-$LiNO_3$ electrolytes with $r = 0.1$ exhibit the highest conductivity compared across the range of measurement temperatures employed in the study. It is also notable that even at temperature of −30 °C, the conductivity of this electrolyte is >1 mS/cm, which makes it suitable for low-temperature battery operation without any compromise in power density. The continuous lines in Supplementary Fig. 11.1a show that the Vogel-Fulcher-Tammann (VFT) model, $\sigma = A\ \exp(-E_a/R(T - T_0))$. Here, E_a is related to the free volume required to move the ions, and T_0 is related to the glass transition temperature of the polymer (typically found to be in order of T_g-50). E_a, obtained from this analysis, provides a measure of the facility with which ions move in an electrolyte plotted and are reported as a function of r in Supplementary Fig. 11.1b. E_a is seen to increase monotonically with r, similar to the glass transition temperature (T_g), also plotted in Supplementary Fig. 11.1b. This result indicates that there are high levels of molecular association between the diglyme and the salt and is consistent with the idea that as the salt concentration is increased, diglyme molecules move in a more coupled manner. On the basis of these results, we utilize an electrolyte with $r = 0.1$ for all subsequent studies.

Glyme- or ether-based electrolytes are known to undergo anionic polymerization at the surface of alkali metals, particularly at the highly reducing potentials at the anode. The resultant polymer-rich interphases are desirable because they passivate the electrode against parasitic chemical reactions with the electrolyte. Glymes are for this reason among the most preferred electrolytes for electrochemical cells in which alkali metals are to be used as anodes [1, 19, 30]. Unfortunately, left unchecked, the polymers formed may grow to such high molecular weights that Li^+ transport to the electrode is severely retarded. Alkali metals are thought to initiate polymerization by cleaving a proton from the side-chain of a glyme molecule as shown in Fig. 11.1a. The polymer chain grows by an addition process wherein the active anionic reactive center collides with another glyme molecule, extending the length of the chain. Because electrostatic interactions between active centers prevent collisions between growing chains and centers can be stabilized by Li ions in solution, the growth can in principle progress indefinitely to produce extremely large, poorly conductive polymer chains or until all available glyme molecules are integrated into the growing center. In either event, ion mobility in the electrolyte bulk falls and interfacial resistance rises, producing premature failure of the cell by voltage runaway.

We hypothesize that an electrolyte that addresses this fundamental, termination-free characteristic of anionic addition polymerization could limit chain growth to produce self-limited SEI on a metallic electrode. To test this idea, we employ the molecule Tris(hexafluoroisopropyl)phosphate (HFiP) that is known to readily

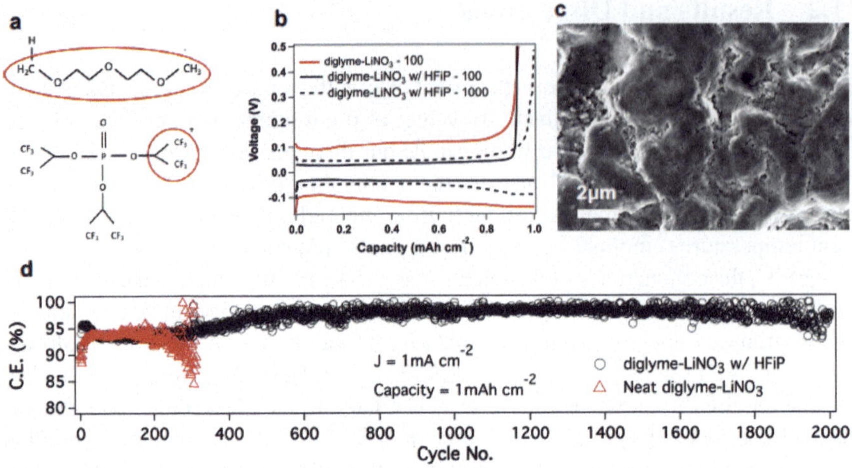

Fig. 11.1 Enabling stable electrodeposition of lithium metal: (**a**) Schematic showing the possible cleavage sites for diglyme and HFiP molecules such that the uncontrolled polymerization of diglyme is quenched by the $CH(CF_3)_2^+$ radical. (**b**) Voltage profile for the electroplating and stripping of lithium metal at the same current density. The different numbers represent cycle no. (**c**) Scanning electron microscopy image of stainless steel substrate after lithium deposition for 6 h at the current density of 1 mA/cm^2. (**d**) Coulombic efficiency measurements in a Lillstainless steel battery at a current density of 1 mA/cm^2 and capacity of 1 mAh/cm^2. The black circles represent the diglyme-$LiNO_3$ electrolyte with the HFiP additive, and red triangles are for neat electrolyte

degenerate to form multiple $CH(CF_3)_2^+$ species per molecule [31, 32]. The large number of electron withdrawing groups near the cationic fragments should enable rapid and efficient quenching of anionic polymerization of glyme molecules by the chain transfer mechanism depicted in Fig. 11.1a. As a proof of concept, we performed a simple analysis wherein lithium metal was dipped in diglyme-$LiNO_3$ electrolyte with and without HFiP additive aged for a month. Supplementary Fig. 11.2 shows the image vial bottles, where it is seen that the electrolyte without the CTA turns yellow due to uncontrolled polymerization of the diglyme molecules, while the lithium is blackened due to surface reactions. In comparison, the HFiP additive stabilizes not only the diglyme solution but also the lithium surface maintains its pristine form. Furthermore, surface of lithium from both vial bottles were analyzed using X-ray Photoelectron Spectroscopy (XPS) and reported in Supplementary Figs. 11.3 and 11.4. The F-1s XPS for the case with HFiP additive has a single peak at 688.9 eV representing –CF3 bond, which is further confirmed from the C-1s XPS from the peak at 293.3 eV, while it is absent in the C-1s for the lithium extracted from additive-free lithium [33]. The absence of a metal-fluoride binding energy peak is a confirmation that the –CF3 groups do not decompose in the presence of the lithium metal electrode, ruling out an alternative stabilizing mechanism reported in our previous work [22, 23].

The effectiveness of the approach to create self-limited interphases on a Li anode was evaluated in an asymmetric electrochemical cell comprised of lithium metal and stainless steel electrodes. By comparing the electric current generated when a specific amount of Li is stripped from the Li electrode and deposited onto the stainless steel electrode, with the current required for the reverse process, the coulombic efficiency (CE) of the cell can be determined. Figure 11.1b shows the voltage profile during a typical measurement in cells with and without the HFiP chain transfer agent. It can be seen that although for the 100th cycle the CE values for the two electrolytes are the same, the overpotential for stripping and plating Li are vastly reduced by the chain transfer agent, consistent with expectations for the CTA's ability to terminate polymer chain growth. The consequence of these effects is quite clearly seen in Fig. 11.1d and Supplementary Fig. 11.5, which report the CE for electrolyte with and without the CTA, at current densities of 1 mA/cm^2 and 0.25 mA/cm^2, respectively, with each half cycle comprising of 1 h. This means that approximately 5 and 1.25 μm of the 450 μm Li electrode is stripped and plated during each cycle, respectively. It is seen that the CE is maintained at a value >98% for 2000 plate-strip cycles, even without efforts to optimize the composition of the CTA in the electrolyte or its efficiency in terminating addition polymerization! This level of stability has to our knowledge not been observed in a lithium metal cell using a liquid electrolyte. The benefits of the CTA are obvious when results for electrolytes with and without this species are compared (Fig. 11.1d). It is observed that whereas the control diglyme-$LiNO_3$ electrolyte with/without the chain transfer agent have similar CE for the initial 200 cycles, upon longer-term cycling large fluctuations appear in the latter that are absent in the former. Similar behavior is observed at the lower current density of 0.25 mA/cm^2; however the fluctuations in CE are seen after 500 cycles. We further characterize the electrodeposition by SEM analysis of the stainless steel electrode after electrodeposition of 6 mAh/cm^2 (ca 30 μm of Li) at 1 mA/cm^2 (Fig. 11.1c) and at 0.25 mA/cm^2 (Supplementary Fig. 11.6), in the glyme electrolyte containing HFiP additive. It is seen that the deposits are compact at both these current densities; also the coverage of the smooth deposits span over several microns indicating the large-scale uniformity.

The fluctuations in CE observed in the control electrolytes are associated with the sporadic electrical connections with electrically disconnected fragments of lithium ("orphaned lithium") formed during the electrodeposition process and are indicative of the irreversibility of the process. These findings are confirmed by postmortem analysis of the electrode surface after cycling the Li||stainless steel cell with and without HFiP additive at current density of 1 mA/cm^2 for 100 cycles, followed by depositing lithium of capacity 1 mAh/cm^2 on the stainless steel electrode. The SEM images of the electrodeposited stainless steel reported in Supplementary Fig. 11.7 indicate that in contrast to open, dendritic, or needlelike deposits observed in the control electrolyte, the CTA containing electrolyte resulted in compact structures. This difference underscores the consequence of faster diffusion of lithium ions and low charge transfer resistance for the anodic reaction: $Li^+ + e^- \rightarrow Li$ at interphases where polymerization of the glyme is constrained.

The continuous polymerization of the diglyme molecules without the HFiP chain transfer agent can lead to increased battery resistance over cycling which can be verified by impedance spectroscopy measurements of batteries. Supplementary

Fig. 11.8 reports the Nyquist plots of the cells containing the electrolyte of diglyme-$LiNO_3$ with and without the HFiP additive. The cells were comprised of lithium metal and stainless steel disk as electrodes and were cycled 100 times at 1 mA/cm^2 current density and 1 mAh/cm^2 capacity with plating in the last step. After fitting the impedance spectra with the appropriate circuit model (shown in the inset of Supplementary Fig. 11.8), it was seen that the interfacial resistance for the cell using control electrolyte was 77.5 Ω while that of the CTA containing cell was 50.9 Ω. Thus, it can be argued that the chain transfer agent enables longer-term stable cycling by preventing electrolyte degradation. We also analyzed the surface of lithium metal extracted from a Li||stainless steel cell with the diglyme-$LiNO_3$-HFiP electrolyte after 100 cycles using XPS (reported in Supplementary Fig. 11.9). It was seen that the majority of the F-1s spectra comprised of the peak at 688.9 eV corresponding to the –CF3; it also shows presence of a peak at 684 eV that can be ascertained to the presence of LiF species. Several previous works on electrode-electrolyte interfaces have demonstrated that LiF stabilizes electrodeposition of metallic lithium [23, 34].

The success of a CTA in limiting polymer growth under the reducing potentials at the Li anode leads us to hypothesize that an analogous approach might be used to enable all ether-based electrolytes to be operated at higher potentials, where oxidative breakdown of the electrolytes is a well-known and long-standing barrier to ether-based electrolytes. Because the cathodes of greatest contemporary interest are porous materials and the active polymer centers once initiated can in principle react with electrolyte solvent able to diffuse from any other location in the cell, a localized strategy that limits active center diffusion away from the electrode-electrolyte interface and lowers solvent migration to the active center is evidently needed. Here, we chose to study interphases formed by the semi-crystalline anionic polymer electrolyte, Lithion (see Fig. 11.2a). This choice is motivated by three primary considerations. First, we discovered that solutions of Lithion in aprotic carbonate ester solvents possess sufficiently low viscosity that the polymer can be transported by liquid carriers into the pores of preformed cathodes. Second, the immobilized anions on Lithion can be thought to provide a barrier to oxidation reactions of the negatively charged species and lewis bases in the electrolyte. We have previously explored electrokinetic attributes of this barrier and on that basis shown that the negative charge adopted by Lithion in solution provides an effective electrostatic shield that limits transport of negatively charged species at planar electrodes, yielding lithium transference numbers approaching unity in liquid electrolytes [35]. Finally, the coexistence if hydrophobic and hydrophilic domains in Lithion means that at appropriate thicknesses, it should be possible to retard molecular solvent transport, without compromising anion mobility.

To evaluate this concept, we performed linear scan voltammetry in a three-electrode cell using Ag/AgCl as the reference electrode and Lithion-coated stainless steel as the working and counter electrodes. A variety of liquid electrolytes were tested ranging from aqueous to oligomers, and the electrochemical oxidation potential was compared to the case without the Lithion coating. The inset of Fig. 11.2b reports the oxidative window of 1 M $LiNO_3$-water as electrolyte. It is seen that the

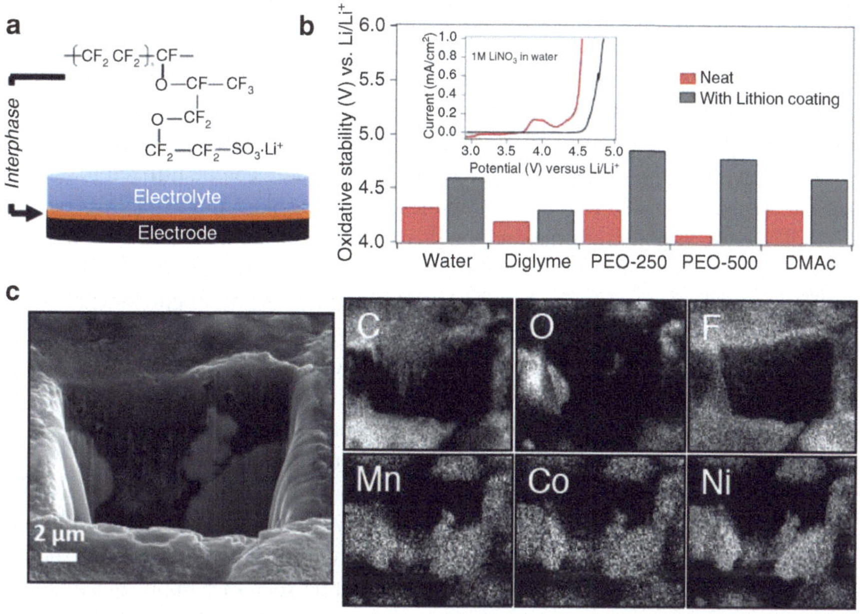

Fig. 11.2 Designing a cathodic interface based on immobilized anions: (**a**) Schematic showing the structure of lithiated nafion (Lithion) utilized to form the artificial interface. (**b**) Bar chart compares the oxidative stability of different electrolytes with (black) and without (red) the Lithion coating on the electrode. The measurements were done in three-electrode setup with Ag/AgCl as reference and stainless steel as counter and working electrodes. The scan rate was 10 mV/s. The electrolytes investigated are 1 M $LiNO_3$ in water, $r = 0.1$ $LiNO_3$ in diglyme, $r = 0.05$ $LiNO_3$ in PEO-250, $r = 0.05$ $LiNO_3$ in PEO-500, and 1 M $LiNO_3$ in dimethylacetamide. The inset shows the linear san voltammetry for the 1 M $LiNO_3$-water electrolyte. All the voltages are shifted with respect to Li/Li^+. (**c**) Cryo-SEM image of the cross-section for a Lithion-coated Nickel Manganese Cobalt Oxide (NCM) electrode obtained by focused ion beam milling. The images on the right represent the EDX mapping of different atoms present in the cross-section

Lithion coating elevates the degradation potential of water by ~0.3 V even in a dilute concentration, in contrast to several recent reports involving high-concentration (water-in-salt) aqueous electrolytes. This finding is important from the fundamental perspective of the working mechanism of supersaturated electrolytes, where it is argued that the solvent molecules bind with the ionic content to kinetically inhibit the electrochemical reactions at the electrode. Here, we demonstrate that enabling an immobilized layer of salt at the electrode surface can serve the same purpose of inhibiting electrolyte degradation at the interface. It is hypothesized that the immobilized ionic species strongly desolvate lithium ions in the interfacial region due to electrostatic interactions; thus the reactive solvent molecules are shielded away from high potentials. Furthermore, it is important to note that this methodology is not limited by solubility limit of salt in the bulk electrolyte and cost-effective from industrial point-of-view. We performed similar voltammetry analysis for nonaqueous electrolytes of dimethylacetamide, diglyme, PEG250, and

PEG500 with $LiNO_3$ salt and reported the results in Supplementary Fig. 11.10. In all the reported electrolytes, an improvement in the oxidative potential was observed, as summarized in Fig. 11.2b. The universality of this effect further supports our claim that we are able to isolate the interfacial degradation mechanism from the bulk electrolyte.

We believe that the generalized finding mentioned here can be utilized to design stable interphases in different electrochemical systems (like Li-O_2, Li||NMC batteries) that operate at potentials beyond the thermodynamic stability limits of the electrolytes and at the same time do not have much room of electrolyte chemistry modifications. In this work, we demonstrate this studied electrochemical characteristics of a cell comprised of a lithium metal anode, NMC cathode, and the base electrolyte (diglyme-$LiNO_3$-HFiP). Specifically, we use drop-cast method to form a coating on the NMC electrode disk. The thickness of the Lithion layer was analyzed using scanning electron microscopy at cryo-temperatures shown in Supplementary Fig. 11.11. It was observed that the Lithion drop-cast method yielded a thickness range of 90 to 30 μm, the edges being relatively thicker than the central regions. We further investigated the Lithion coating on the NMC active particles by a focused ion beam milling of the cross-section and EDX mapping of all the atoms as shown in Fig. 11.2c and Supplementary Fig. 11.12. It can be seen that the C and F atoms are highly populated around the individual metal atoms of Ni, Mn, and Co, which implies that the Lithion layer not only laminates the surface but also conformally surrounds the NMC particles.

The baseline battery performance with the diglyme-$LiNO_3$-HFiP electrolyte without the Lithion coating is demonstrated in Fig. 11.3a, where it is seen that the voltage profile exhibits a prolonged charging step in the first cycle and erratic fluctuations in the second cycle above 3.8 V vs. Li/Li^+. The discharge step does not show such fluctuations; however a high overpotential is observed in the second cycle indicative of high battery resistance due to the oxidative degradation of glyme electrolytes. These results can be compared with observations provided in Fig. 11.3b where the the NMC electrode was coated with the mentioned Lithion layer. As seen from voltage profiles for the first and second cycles in Fig. 11.3b, the Li||NMC cells do not show the prolonged charging characteristics observed for the controls. Furthermore, the battery comprising of a thin metallic lithium (50 μm) as anode shows stable cycling for over 100 cycles as seen in Supplementary Fig. 11.13. In addition, we cycled the Lithion-coated NMC cathodes in lithium metal batteries with varying charging potentials up to 4.4 V, as shown in Supplementary Fig. 11.14. Although previous works based on glyme electrolytes have demonstrated stable cycling with lithium iron phosphate cathode [29], the facile coating technique is able to augment the stability limits for cycling a high-voltage NMC cathode.

We further investigated the high-voltage stabilization using electrochemical floating experiments. In this experiment, Li||NMC cells with and without the Lithion coating are charged at voltages ranging from 3.6 to 4.3 V in a step-wise ramp and the voltage maintained at a targeted value for a period of 24 h, as shown in Fig. 11.3c. The leakage current obtained at each voltage is recorded and can be used to directly assess the importance of electrochemical degradation of electrolytes in the fully

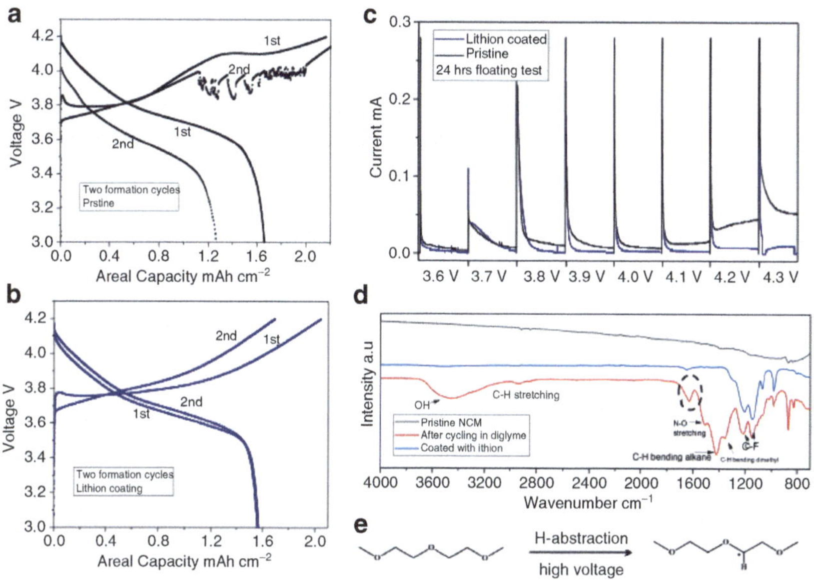

Fig. 11.3 Immobilized anions on cathode prevent glyme oxidation: (**a**) Voltage profile of a lithium||NCM cell using the base electrolyte of diglyme-$LiNO_3$-HFiP at C/10 rate. (**b**) Voltage profile of Li||NCM cell using the same base electrolyte, however the cathode is coated with a layer of Lithion, operated at C/10. (**c**) Floating point experiment in a Li||NCM cell, where the voltage is fixed at different values ranging from 3.6 to 4.3 V for a period of 24 h and the current response is measured. The black curves represent results for uncoated NCM and blue is for Lithion-coated NCM. (**d**) Intensity profile obtained from Fourier Transform Infrared Spectroscopy (FTIR) for pristine (uncycled) NCM and NCM cathode extracted from a Li||NCM cell cycled twice at C/10 with and without the Lithion coating; (**e**) Schematic showing the proposed mechanism for the proton extraction from the diglyme molecule due to oxidation at high voltages

charged state. The results show that the leak current is always higher for the control cells (i.e., without the Lithion electrode treatment) than for those that utilize a Lithion-coated NMC electrode. In addition, it is seen that the leakage current for the neat NMC cell starts to exceed the modified NMC based cell at a faster rate beyond 4 V, which is also consistent with the low coulombic efficiency in the Li||NMC half-cell cycling. Fourier transform infrared spectroscopic analysis was used to provide deeper insight into the mechanism(s) through which the glyme molecules fail at high potentials and how Lithion coatings increase electrolyte stability in the cathode. The NMC cathodes after the formation cycles, with and without the Lithion layer, were analyzed, and the intensity profiles are plotted in Fig. 11.3d. Although most of the chemical bonds are present in both cases of with and without Lithion layer, the major difference is the occurrence of the peak at 1600 cm^{-1} without the Lithion coating that corresponds to the formation of unsaturated C=C bonds due to the H-abstraction from the diglyme molecules as shown in Fig. 11.3e. The deprotonation of the ether oxide molecules during the high-voltage charging can lead to self-polymerization and formation of side products on the electrode surface resulting in high overpotentials and battery fading. Similar, C=C bond formation was also found in the electrolyte remains at the battery separator surface, shown in Supplementary Fig. 11.15 that is absent in pristine diglyme solution.

The effectiveness of the Lithion cathode coatings suggests that other approaches that lead to in situ formation of anionic polymer coatings throughout the cathode would be a more straightforward strategy for enabling ether-based electrolytes in lithium cells employing high-voltage cathodes. To evaluate this concept, we use the lithium salt bis(oxalate)borate (LiBOB) salt as and electrolyte additive in the base electrolyte and, by means of hybrid density functional theory (DFT), computationally study the interphases the salt forms at various electrode potentials. The BOB anion is of interest because it has been reported in previous studies to readily form either an open, dianion by breaking a B–O bond, or can furnish dissociation products [36–38]. The reactions of these intermediate species with diglyme would generate distinct coupling products. We calculated the reaction free energies for the formation of a series of neutral and anionic O–C, C–C, and B–C coupling products from the diglyme and BOB dianion. These transformations proceed through the release of CO_2 molecules. Unique coupling products considered here and the respective free energy changes are presented in Fig. 11.4a. The calculations indicate that the formation of negatively charged species is thermodynamically more favorable than the respective neutral analogs. Among the anionic dimers, the C–C coupling product (a, Fig. 11.4a) formed by the release of a CO_2 molecule is thermodynamically most favorable ($\Delta G = -0.64$ eV).

Starting from the negatively charged dimer, one could envision its subsequent reactions with diglyme and BOB^{2-}, which will generate oligomers, polymers, or a supramolecular assembly at the electrode-electrolyte interface. We have calculated the reaction free energies for the step-wise generation of neutral or negatively charged dimer, trimer, tetramer, and pentamer from BOB^{2-} and diglyme (Fig. 11.4b). These calculations reveal that the formation of neutral or negatively charged trimer and higher aggregates is thermodynamically unfavorable. The formation of anionic and neutral forms of trimer from the dimer is endothermic by 1–4 eV, whereas the generation of higher-order coupling products is highly unfavorable ($\Delta G > 10$ eV). At higher voltages, the trimers could still form; however it is very unlikely that further polymerization will occur. The higher oligomers with multiple charges may not be stable as they would readily dissociate to smaller charged dimers or trimers. We theoretically calculate the redox potentials of glyme molecule and its oligomers with BOB molecules, presented in Fig. 11.4c using the computational methodology described in the Supplementary Information. It is seen that the glyme-BOB oligomers are charged and at the same time electrochemically stable even at high potentials. It can be argued that these initially generated oligomers would form a network at the cathode via strong non-covalent interactions, furnishing a charged supramolecular assembly (as shown in the schematic of Fig. 11.5a). This might be the reason for the prevention of further oxidation of diglyme at the cathode.

To experimentally interrogate the cathode-electrolyte interphase (CEI) formed in the presence of a LiBOB salt, we cycled a Li||NMC battery with LiBOB additive in diglyme-$LiNO_3$-HFiP (base) electrolyte twice at C/10 and extracted the

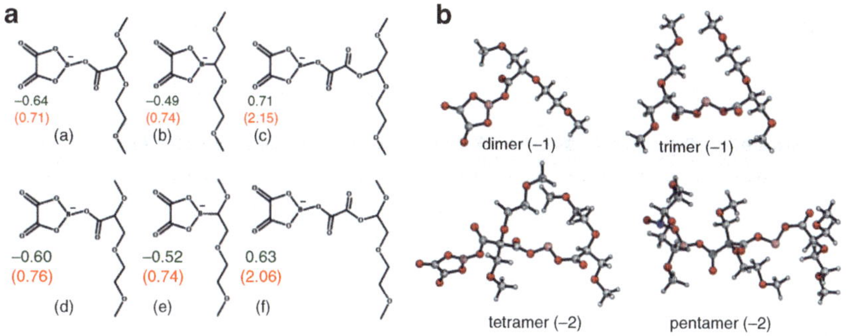

c

Calculated redox potential calculated with respect to that of Li/Li$^+$ couple for diglyme and its oligomers with BOB molecules.

system	Ox/Red charge state	Redox potential (eV)
Diglyme	0/−1	4.14
Dimer	−1/0	−6.02
Trimer	−1/0	−5.77
Tetramer	−1/0	−6.07

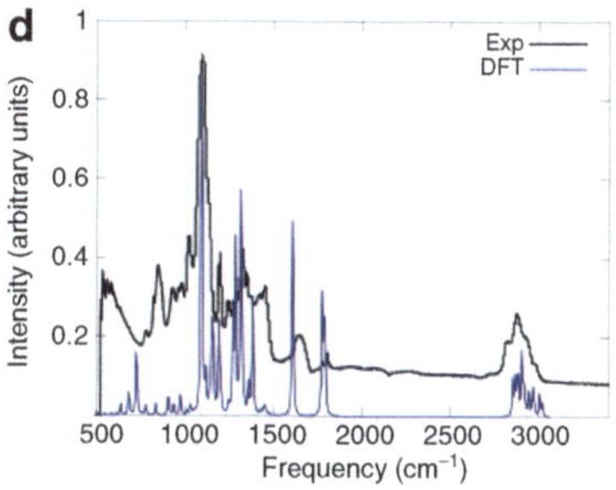

Fig. 11.4 In situ formation of anionic aggregates at cathode interface: (**a**) Structures of plausible coupling products of BOB^{2-} and diglyme. Calculated reaction free energies (in eV) for the formation of anionic (green color) and neutral (red color) dimers are presented. (**b**) Optimized geometries for the dimer and higher-order coupling products of BOB and diglyme. Respective charge is shown in the parenthesis. (**c**) Table showing calculated redox potentials for diglyme and its oligomers with BOB molecules. Oxidation/reduction potentials are calculated with respect to that of Li/Li$^+$ couple. A positive or negative sign is used represent reduction and oxidation potentials, respectively; (**d**) Infrared (IR) spectra comparing the intensity profiles obtained from experiment and DFT calculations. The experimental profile was obtained from a NCM cathode harvested from a Li||NCM cell comprising of diglyme-$LiNO_3$-HFiP electrolyte with 0.4 M LiBOB salt additive and the battery was cycled twice at C/10

NMC cathode for FTIR analysis and compared it with the computational IR-spectra of the predicted oligomeric species, as shown in Fig. 11.4d. We find that there is a good agreement between the peak locations of the simulations and experiments. The differences in the relative intensities can be ascertained to the presence of additional species in the experiment as the electrolytes comprise of additional components (salt, additive, surface impurities) that are not considered in the DFT calculations. The verification of our hypothesis that LiBOB additive in diglyme-based electrolyte enhances the oxidation potential was done using three-electrode linear scan voltammetry and electrochemical floating point test reported in Fig. 11.5b and Supplementary Fig. 11.16, respectively. It is seen that the degradation potential of diglyme electrolyte is enhanced by ~0.3 V when LiBOB was used as additive (Fig. 11.5b). Since, the measurement was done in a three-electrode cell with stainless steel as working electrode, our argument that the LiBOB in situ forms the CEI in the liquid-phase is further strengthened.

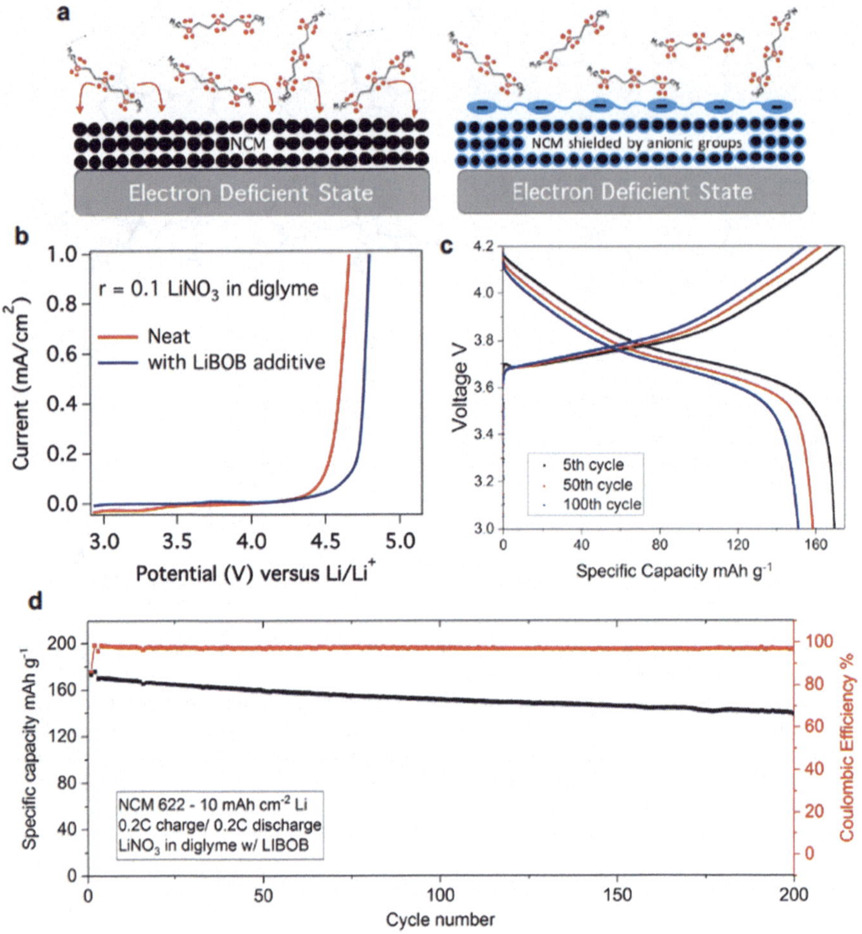

Fig. 11.5 Enabling stable cycling of high-voltage lithium battery: (**a**) Schematic showing the proposed mechanism according to which the oxidation of glymes is inhibited by a layer of immobilized anions. (**b**) Potential-current diagram obtained from linear scan voltammetry in a three-electrode setup comprising of Ag/AgCl reference electrode and stanless steel as working and counter-electrodes. The scan rate was 10 mV/s. The electrolytes used here were diglyme-$LiNO_3$-HFiP, with (blue) and without (red) 0.4 M LiBOB salt additive. (**c**) Voltage profile for 5th, 50th, and 100th cycles of Li||NCM cycling using 0.4 M LiBOB additive. (**d**) Discharge capacity retention and coulombic efficiency over 200 cycles in Li||NCM cell diglyme-$LiNO_3$-HFiP electrolyte with 0.4 M LiBOB additive. Here, the lithium used is 50 μm thick; thus the anode to cathode capacity ratio is 5

The floating experiments with and without LiBOB additive was done using the same protocol as mentioned for the Lithion case, where we charged a Li||NMC battery at varying voltages for 24 h each. The results (Supplementary Fig. 11.16) show that the leak current for the cells containing LiBOB additive is lower than the control cells at all voltages. We also characterized the surface of lithium metal anode using XPS after two initial cycles of charge and discharge at C/10 rate (reported and discussed in Supplementary Fig. 11.17).

Finally, we cycle a lithium metal battery with the NMC cathode and a thin lithium anode (50 μm) such that the anode to cathode capacity ratio is 5:1 and using the base electrolyte (diglyme-$LiNO_3$-HFiP) and LiBOB as additive. The voltage profiles for the 5th, 50th, and 100th cycles are shown in Fig. 11.5c and the cycle life in Fig. 11.5d. It is seen that coulombic efficiency of the cells is high (>98%) and that the discharge capacity is retained to more than 80% for at least 200 cycles at a rate of C/5. A similar performance is also observed at a relatively higher rate of discharge (C/2) reported in Supplementary Fig. 11.18. It is remarkable that by rational design of the cathodic interphases, we have been successful in enabling stable operation of glyme electrolytes in high-voltage batteries for several hundreds of cycles, while they spontaneously fail without the interfacial modification. The concept of high-voltage stabilization using anionic interfaces is not limited in oligomeric liquid electrolytes but also in other classes of polymer electrolytes including gels and cross-linked nanocomposites. We further designed gel electrolytes comprising of 1 wt.% PEG (100k) in diglyme electrolyte (image shown in Supplementary Fig. 11.19) and operated Li||NMC cells with and without LiBOB salt additive at ambient conditions with a rate of C/5 as shown in Supplementary Fig. 11.20. We find that unlike control liquid electrolyte, the gel electrolyte is able to show charge-discharge profiles even without LiBOB for first few cycles. However, beyond 20 cycles, there is sharp drop in the capacity followed by a noisy charge profile due to the breakdown of the gel electrolyte at high voltages similar to its liquid counterparts. The limited stabilization of the control gel electrolytes can be attributed to transport barrier of the viscous components preventing spontaneous breakdown in the initial cycles. In contrast, it is found that LiBOB additive in the gel electrolyte enables stable cycling and high-capacity retention for at least 100 cycles. We also demonstrate the stable cycling behavior of a Li||NMC cell in Supplementary Fig. 11.21 using a recently reported cross-linked nanoparticle membrane [39, 40] as solid-electrolyte by infusing the base diglyme electrolyte with LiBOB additive.

In conclusion, we have shown that cationic chain-transfer agents can be used to terminate anionic polymerization of ether-/glyme-based electrolytes at a lithium metal electrode, producing self-limiting interfaces, high coulombic efficiency, and extend the lifetime of the anode (to over 4000 h) in asymmetric lithium||stainless steel cells. Building on these observations, we show that a long-standing barrier to deployment of glyme electrolytes can be removed using either ex or in situ generated interphases in the cathode that limit transport and reduce reactivity of active polymer centers by what we hypothesize to be an electrostatic shielding mechanism. Specifically, we show that a cathode electrolyte interphase (CEI) that hosts immobilized anions tethered to a polymeric backbone can act as a barrier for the oxidation reaction. Extending this concept to create an in situ generated interphase composed of anionic polymer aggregates at the cathode result in significantly enhanced lifetime of a high-voltage lithium metal battery. We believe, this work opens a new pathway for conventional, solid-state, and flexible lithium metal batteries based on ether- and polyether-based electrolytes.

11.3 Methods

11.3.1 Computational Details

All structures are optimized in the gas-phase using wB97X-D [41, 42] functional and 6-311G(d,p) [43] basis sets implemented in the Gaussian suite of programs [44]. Vibrational frequencies are calculated at the same level of theory to ensure that the optimized geometry represents a true minimum; i.e., no negative frequencies are found. Further, single point calculations are performed on these structures by employing a polarizable continuum model (PCM) to mimic the effects of diglyme [45]. We used a dielectric constant of 7.23 for diglyme. A value of 1.63 eV is assumed for the electron solvation free energy [46].

11.3.2 Experimental Details

The detailed description of the synthesis procedure is provided in the Supplementary Information.

Appendix: Supplementary Information

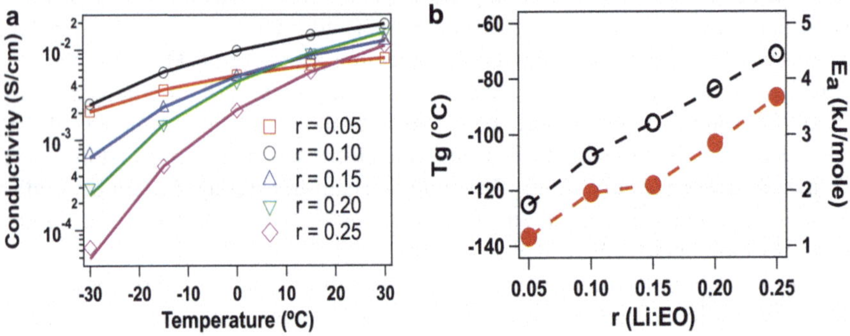

Supplementary Fig. 11.1 (**a**) d.c. conductivity as a function of temperature for different Li:EO (*r*) ratios. The points represent experimental values and lines are the CFT fitting. (**b**) Variation in glass transition temperature obtained from differential scanning calorimetry measurements in the left axis and activation energy obtained from the VFT fitting of conductivity data

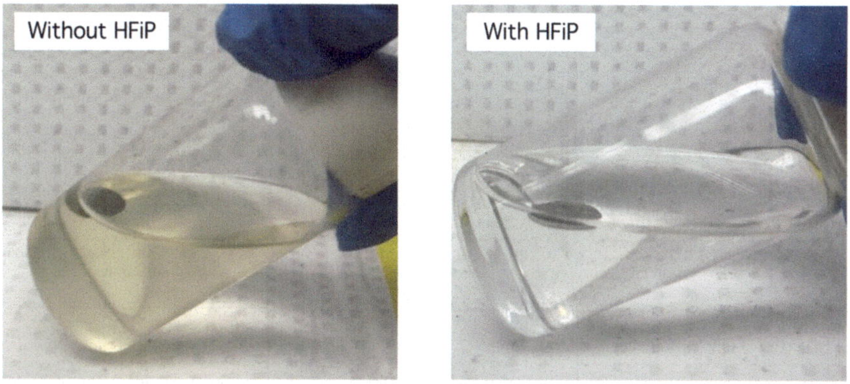

Supplementary Fig. 11.2 A piece of lithium metal was added to the diglyme-$LiNO_3$ electrolyte without 1 wt.% HFiP (left) and with 1 wt.% HFiP additive (right). The electrolyte solutions were aged for 1 month in a vial bottle. It is seen that the electrolyte without HFiP additive turns yellow and also the lithium surface becomes blackened, presumably, due to the polymerization of the glyme molecules

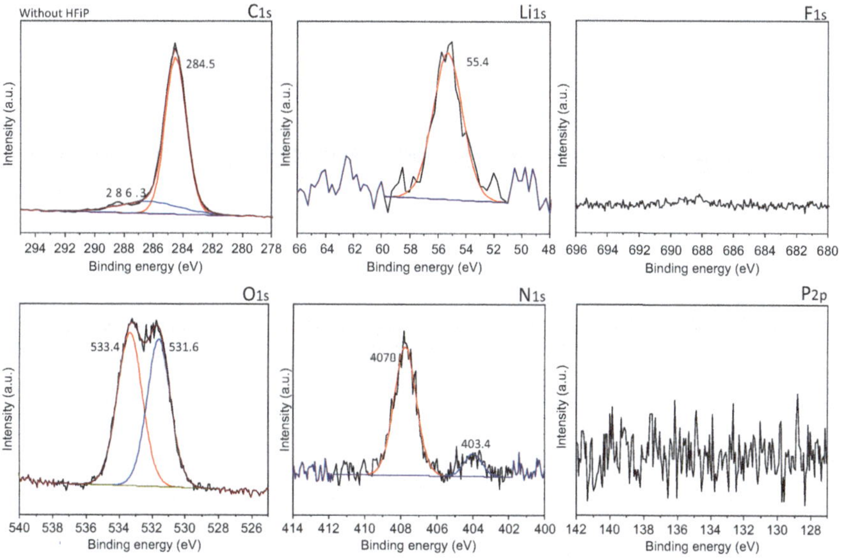

Supplementary Fig. 11.3 Binding energies of different atoms on the surface of lithium obtained from X-ray Photoelectron Spectroscopy. The lithium metal was dipped in an electrolyte solution of diglyme-$LiNO_3$ without any addition of HFIP

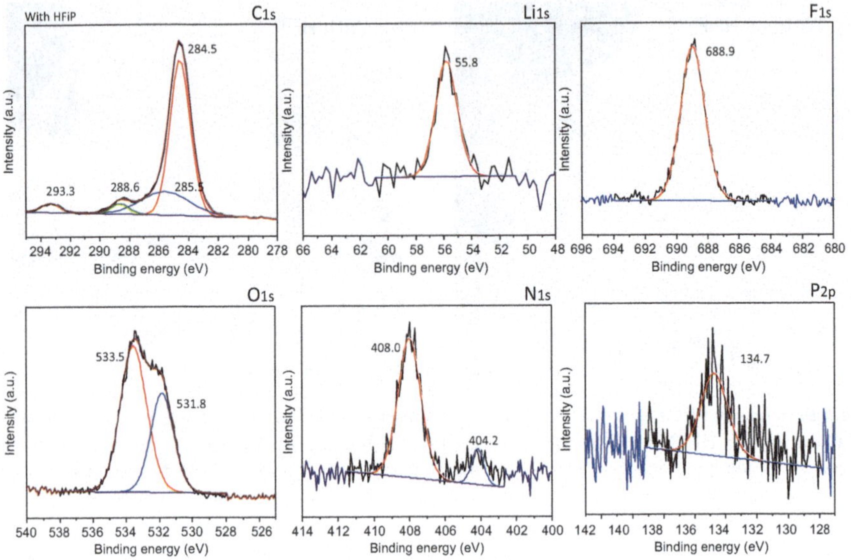

Supplementary Fig. 11.4 Binding energies of different atoms on the surface of lithium obtained from X-ray Photoelectron Spectroscopy. The lithium metal was dipped in an electrolyte solution of diglyme-$LiNO_3$ with 1 wt.% HFIP additive

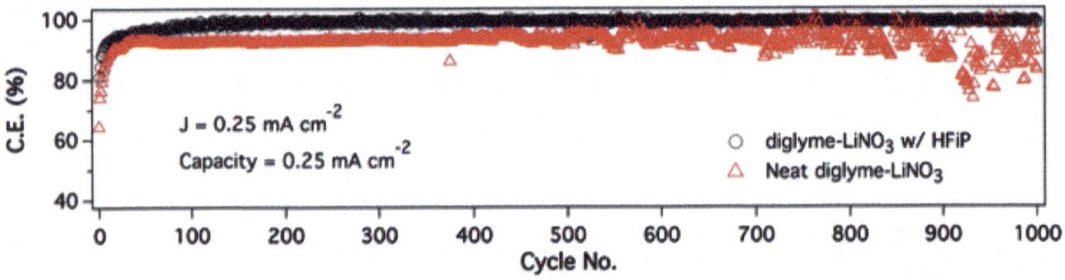

Supplementary Fig. 11.5 Coulombic efficiency measurements in a Li‖stainless steel battery at a current density of 0.25 mA/cm^2 and capacity of 0.25 mAh/cm^2. The black circles represent the diglyme-$LiNO_3$ electrolyte with the HFiP additive, and red triangles are for neat electrolyte

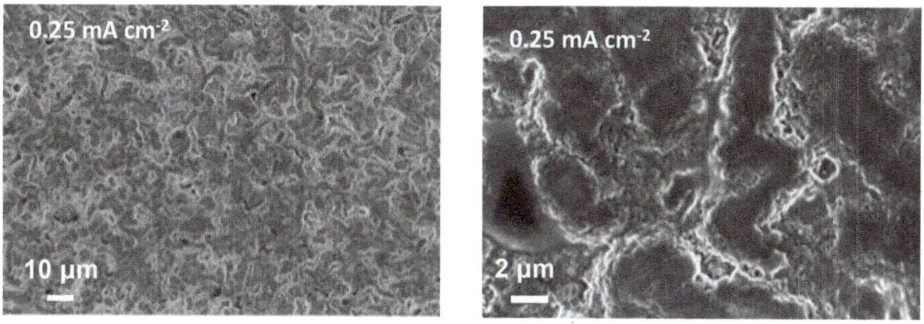

Supplementary Fig. 11.6 Scanning electron microscopy image of stainless steel substrate after lithium deposition for 24 h at the current density of 0.25 mA/cm^2, using the electrolyte diglyme-$LiNO_3$ with 1 wt.% HFiP

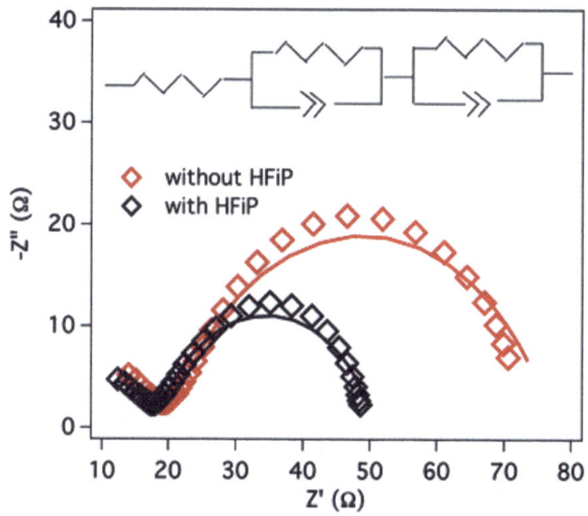

Supplementary Fig. 11.7 Nyquist diagrams obtained from impedance spectroscopy for lithium vs. stainless steel cell that was cycled 100 times at a current density of 1 mA/cm^2 and capacity of 1 mAh/cm^2 before depositing lithium onto the stainless steel electrode. The red symbols are for electrolyte of diglyme-$LiNO_3$, while the black are for the same electrolyte with 1 wt.% HFiP additive. The inset shows the circuit model utilized to fit the data

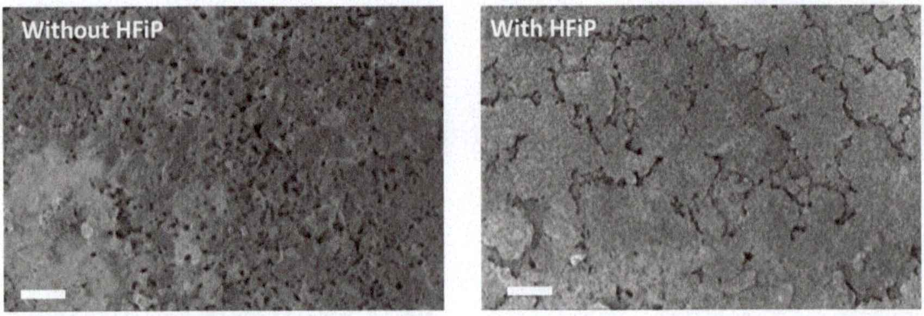

Supplementary Fig. 11.8 Scanning electron microscopy image of stainless steel substrate after 100 cycles of plating and stripping lithium for 1 h at the current density of 1 mA/cm^2. In the last step, lithium metal was plated onto stainless steel electrode. The left image is for the diglyme-$LiNO_3$ electrolyte, without any additive, and the right is for same electrolyte with 1 wt.% HFiP additive. The scale bar in both images represents 20 μm

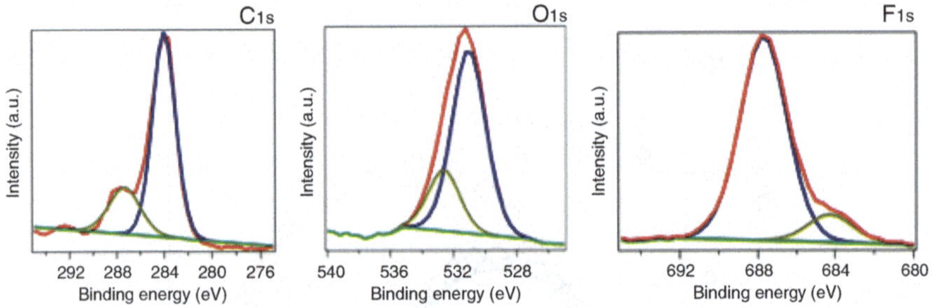

Supplementary Fig. 11.9 Binding energies of different atoms obtained from X-ray Photoelectron Spectroscopy measurements for the lithium surface, extracted from a cell of after cycling 100 times at a current density of 1 mA/cm^2 and capacity of 1 mAh/cm^2 in a cell with configuration of Li||stainless steel electrolyte comprised of diglyme-$LiNO_3$ with 1 wt.% HFiP additive

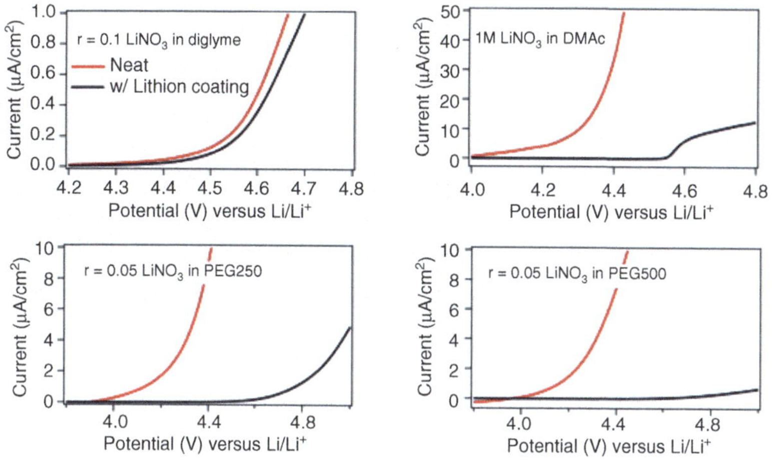

Supplementary Fig. 11.10 Linear scan voltammetry in a three-electrode cell with Ag/AgCl electrode as reference electrodes, while stainless steel used as both reference and counter electrodes. The scan rate utilized was 10 mV/s, and the potentials were shifted with reference to Li/Li^+. The red curves represent cases where the stainless steel was coated with Lithion layer and black represent pristine stainless steel electrodes

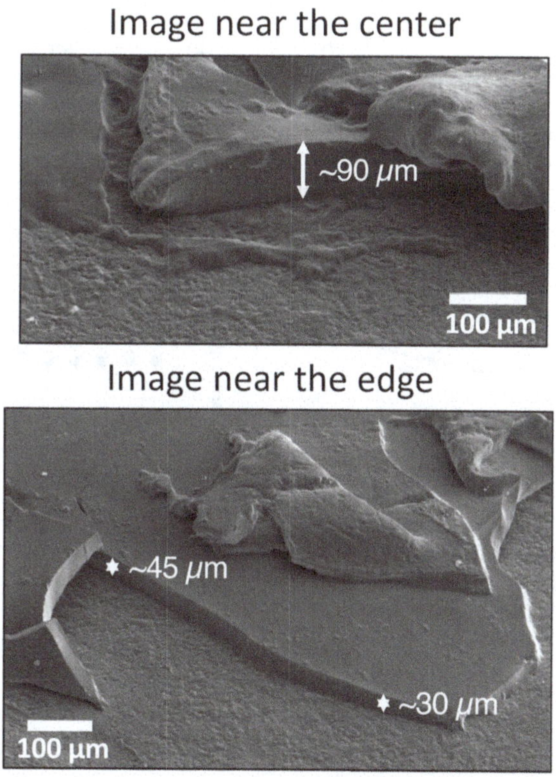

Supplementary Fig. 11.11 SEM images of the Lithion-coated NCM surfaces at cryo-temperatures. A Lithion layer was present on the NCM cathode (cracked during preparation). A thickness gradient was present, from ~90 µm thick near the center to ~30 µm near the edge

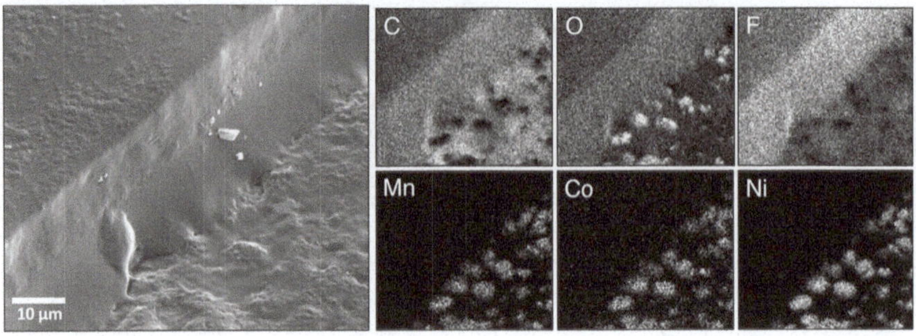

Supplementary Fig. 11.12 EDX of the edge of the cracked Lithion layer showed no nickel (or other metals) in the Lithion

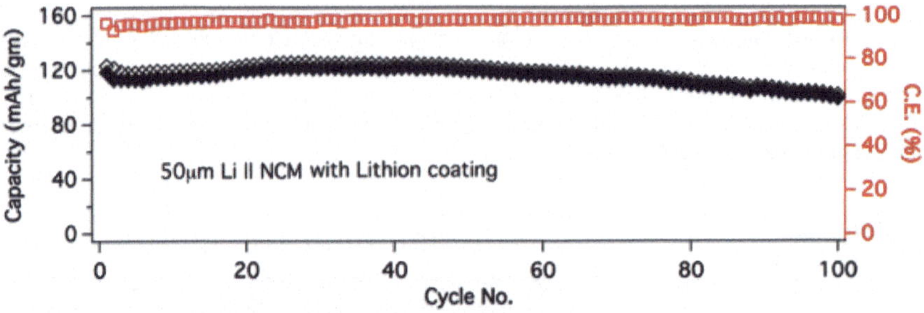

Supplementary Fig. 11.13 Cycling of a thin lithium versus NCM battery, where the capacity of the lithium anode was 10 mAh/cm^2 and that of cathode was 2 mAh/cm^2. Here, the NCM cathode was coated with a layer of Lithion layer. The current density is 0.4 mA/cm^2 (C/5)

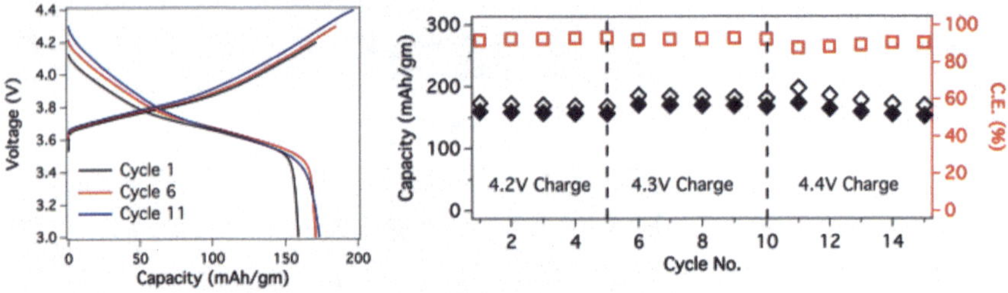

Supplementary Fig. 11.14 Cycling of a thin lithium versus NCM battery, where the capacity of the lithium anode was 10 mAh/cm^2 and that of cathode was 2 mAh/cm^2. Here, the NCM cathode was coated with a layer of Lithion layer. The current density is 0.2 mA/cm^2 (C/10)

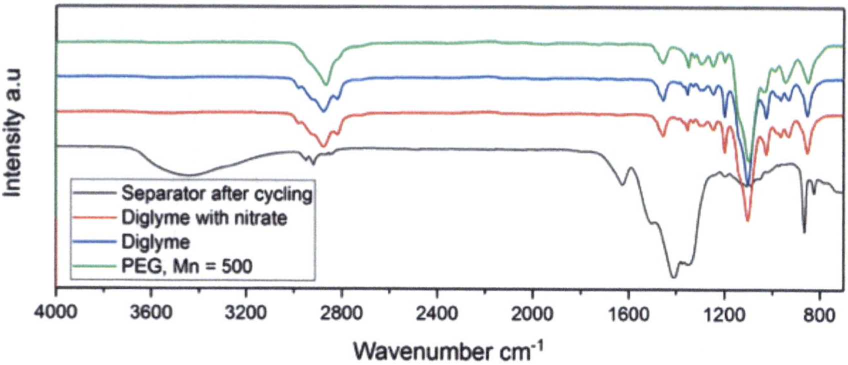

Supplementary Fig. 11.15 FTIR spectra for diglyme electrolyte infused separator extracted from a Li||NCM cell without any Lithion coating, after it was cycled twice at C/10. It is compared with neat diglyme solvent, diglyme with $LiNO_3$, and PEG500 liquid

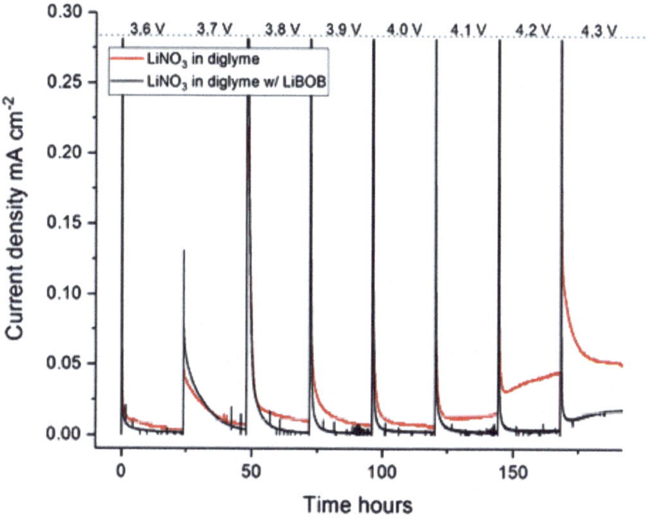

Supplementary Fig. 11.16 Floating experiments using Li||NCM cell, where the cells are charged at different voltages for 24 h each and the leak current is recorded to determine the side reactions at the operated voltages. The red line is for the diglyme-$LiNO_3$-HFiP electrolyte, while the black line is result for the same electrolyte with 0.4 M LiBOB salt as additive

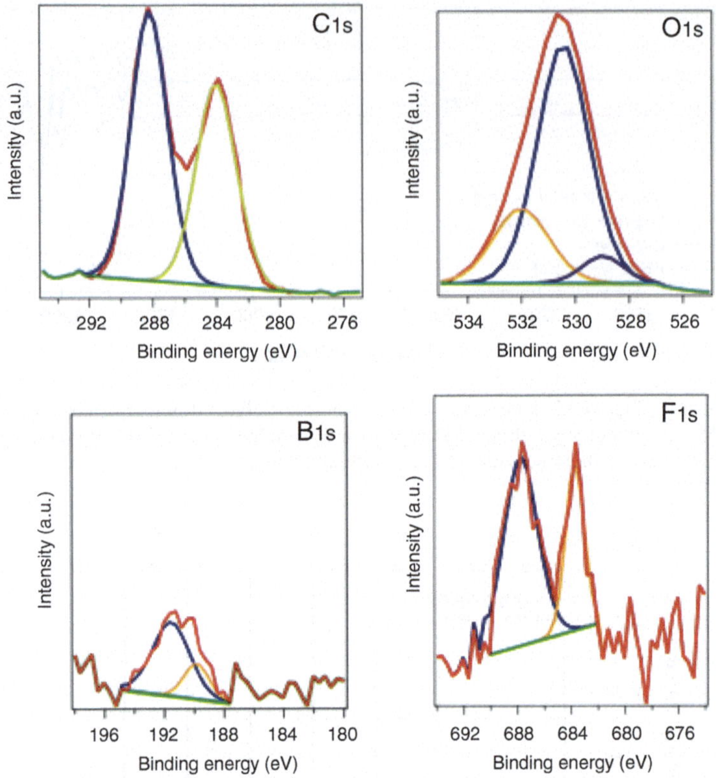

Supplementary Fig. 11.17 XPS spectra from obtained from the surface of lithium metal anode harvested from a Li||NCM cell cycled twice at a C/10 rate using the electrolyte diglyme-$LiNO_3$-HFiP with 0.4 M LiBOB additive. Here, the fluorine spectra indicates that there is presence of both LiF and –CF3 content from the HFiP, while the LiBOB plays a role of forming boro-oxolate compounds in the anodic interfacial layer due to low potential reduction

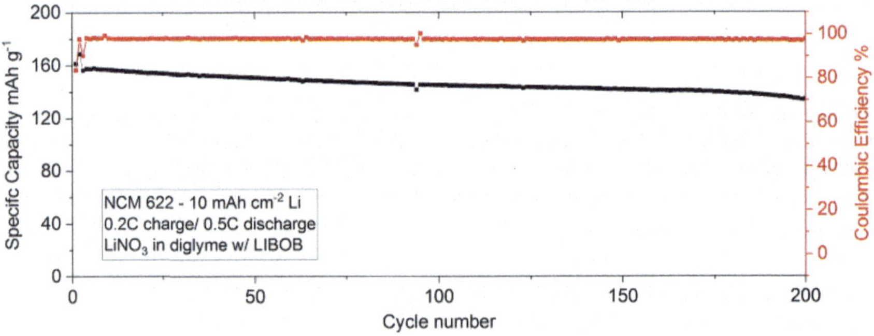

Supplementary Fig. 11.18 Li||NCM cycling results using a rate of 0.2 C charge and 0.5 C discharge. The cathode loading is 2 mAh/cm², and the lithium metal anode is 50 μm thick that corresponds to 10 mAh/cm² capacity. The electrolyte used here is diglyme-$LiNO_3$-HFiP with 0.4 M LiBOB salt additive

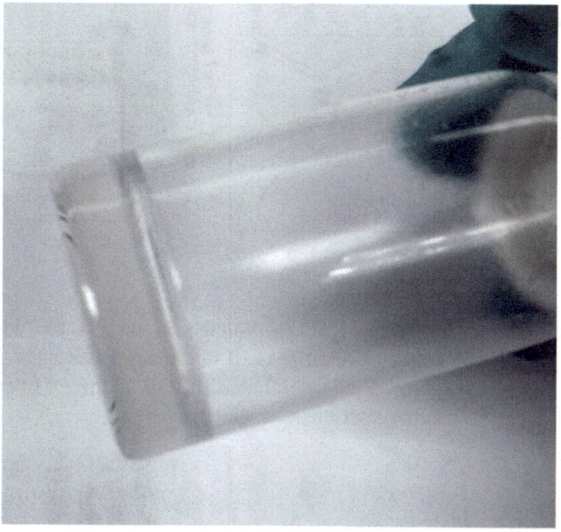

Supplementary Fig. 11.19 Image showing the gel electrolyte used for cycling at room temperature. The composition is 1 wt.% 100k PEG in diglyme with $LiNO_3$ and HFiP

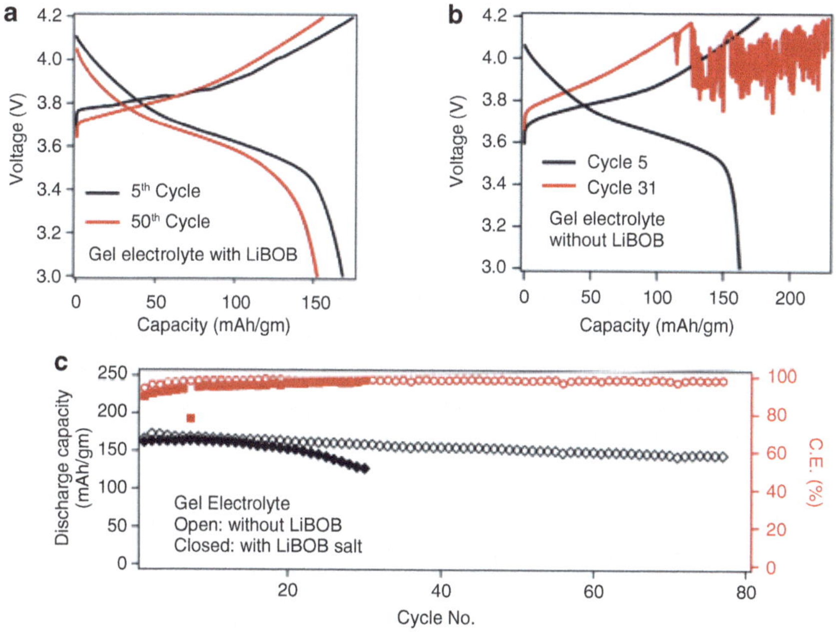

Supplementary Fig. 11.20 Voltage profile of Li||NCM cell with a thin (50 μm) lithium and 2 mAh/cm^2 cathode cycled at C/5 rate using the gel electrolyte comprising of 1 wt.% PEG 100k in diglyme-$LiNO_3$-HFiP, (**a**) with LiBOB salt additive and (**b**) without LiBOB. (**c**) Cycling performance of the gel electrolyte with (filled symbols) and without LiBOB salt (unfilled symbols)

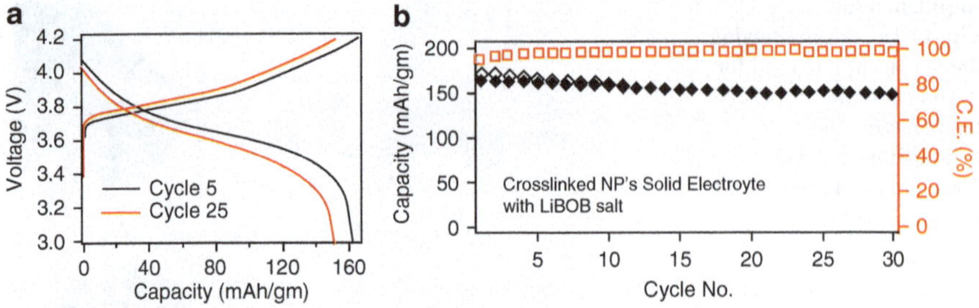

Supplementary Fig. 11.21 (**a**) Voltage profile of Li||NCM cell with a thin (50 μm) lithium and 2 mAh/cm^2 cathode cycled at C/10 rate using cross-linked hairy nanoparticles soaked with the electrolyte diglyme-$LiNO_3$-HFiP and LiBOB additive, (**b**) cycling profile showing the coulombic efficiency and charge/discharge capacity

Methods

Computational Details

Scheme Used to Calculate the Redox Potentials

By using the thermodynamic cycle shown in Scheme 11.1, Gibbs free energy change in solution phase ($\Delta G_{sol}{}^{EA}$) during oxidation process could be estimated from Eq. (11.1):

$$\Delta G_{sol}{}^{EA} = \Delta G_{G}{}^{EA} + \Delta G_{sol}\left(M^{-}\right) - \Delta G_{sol}\left(M\right) - \Delta G_{sol}\left(e^{-}\right) \tag{11.1}$$

Then, the oxidation potential for a given molecule (*M*) is calculated as

$$E^{0} = \frac{-\Delta G_{sol}{}^{EA}}{F} \tag{11.2}$$

where *F* is the Faraday constant.

Practically, it is difficult to calculate the solvation free energy of an electron, $\Delta G_{sol}(e^{-})$. Therefore, we have calculated the relative oxidation potential with respect to Li/Li^{+} electrodes (−3.04 eV) using Eq. (11.3):

$$E^{0} = \frac{-\Delta G_{G}{}^{EA} + \Delta G_{sol}\left(M^{-}\right) - \Delta G_{sol}\left(M\right)}{F + 3.04} \tag{11.3}$$

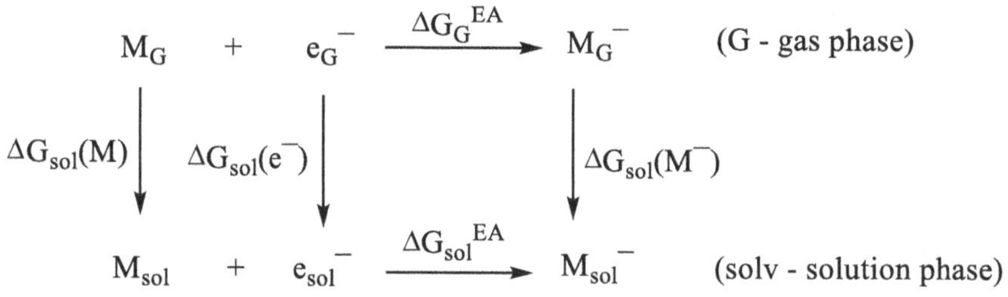

Scheme 11.1 Thermodynamic cycle used to calculate the oxidation/reduction potential of diglyme and its oligomers formed via reactions with BOB

Experimental Details

Materials

Lithium disks were obtained from MTI Corporation. Diglyme and lithium nitrate were all purchased from Sigma Aldrich. Tris(hexafluoroisopropyl) phosphate was obtained from Synquest Laboratories. Celgard 3501 separator was obtained from Celgard Inc. Lithion solution (LITHion™ dispersion, ~10 wt.% in isopropanol) was purchased from Ion Power Inc. The Lithion is composed of a nafion-type perfluorinated polymer having the sulfonic acid groups (EW ~ 1100) ion exchanged by lithium ions. Nickel Manganese Cobalt Oxide (NCM) cathodes were obtained from Electrodes and More Co. All the chemicals were used as received in after rigorous drying in a ~0 ppm water level and <0.1 ppm oxygen glove box.

Coating of NCM Electrode with Lithion Solution

NCM electrodes were punched out using a hole-punch of diameter 3/8″. On a flat bench-top, the NCM cathodes were laid and ~20 μl of Lithion solution was dropped to evenly cover the entire surface. Thereafter the electrodes were dried in open air for 6 h, followed by rigorous drying in a vacuum oven at a temperature of 60 °C for 24 h.

Synthesis of Gel and Cross-Linked Nanoparticles Electrolyte

The gel electrolyte was prepared by dissolving 1 wt.% of PEG-100 kDa (Sigma Aldrich) in an electrolyte solution of diglyme-$LiNO_3$-HFiP (with and without 0.4 M LiBOB salt additive) and thereafter heating the solution to 60 °C overnight. Thereafter the gel electrolyte was brought to room temperature before usage. It was used with a 3501 celgard separator for lithium battery cycling.

The cross-linked solid electrolyte was prepared using the same procedure reported in our earlier work [39, 40]. After thoroughly drying, the membrane was soaked in the diglyme-$LiNO_3$-HFiP-LiBOB solution for a period of 2 days before using in the battery. No separator was used in these batteries.

Dielectric Spectroscopy

The ionic conductivities of the electrolytes were measured at room temperature the desired electrolytes between two gold-plated copper disks using a Novocontrol Broadband Dielectric spectrometer with a frequency range of 10^{-3} to 10^{6} Hz. The electrolyte was sandwiched between the disks using a Teflon o-ring. The DC conductivities were obtained from the plateau of real part of the conductivity versus frequency curve. The Dielectric Spectroscopy instrument was calibrated initially using a 1 M KCl standard solution.

Scanning Electron Microscopy

Surface analysis of electrodeposited stainless steel was done using SEM with the LEO155FESEM instrument. The sample was prepared by depositing 6 mAh cm^{-2} in battery comprising of lithium vs. stainless steel comprising of diglyme-$LiNO_3$-HFiP electrolyte and celgard separator.

X-Ray Photoelectron Spectroscopy

XPS was conducted using Surface Science Instruments SSX-100 with operating pressure of ~2×10^{-9} torr. Monochromatic Al K-α X-rays (1486.6 eV) with beam diameter of 1 mm were used. Photoelectrons were collected at an emission angle of 55°. A hemispherical analyzer determined electron kinetic energy, using pass energy of 150 V for wide survey scans and 50 V for high-resolution scans. Samples were ion-etched using 4 kV Ar ions, which were rastered over an area of 2.25 × 4 mm with total ion beam current of 2 mA, to remove adventitious carbon. Spectra were referenced to adventitious C 1s at 284.5 eV. CasaXPS software was used for XPS data analysis with Shelby backgrounds. The lithium and NCM cathode samples were lightly washed in pure diglyme before XPS measurements. Also, the samples were transferred in an air-tight argon-filled puck from the glove box to the XPS chamber. Hence, there is minimal or no exposure to air.

Floating-Point Experiment

Floating-point experiments were performed in a cell comprising of lithium vs. NCM using various electrolytes reported in the main text. The batteries were charged at constant current of 0.4 mA cm^{-2} up to different voltages from 3.6 to 4.3 V and then held at a constant voltage for 24 h and the values of the leak current at various voltages were measured.

Fourier Transform Infrared Spectroscopy

The NCM electrodes were harvested after constant voltage charge at 3.8 V for 24 h in a battery comprising of lithium anode and NCM cathode using the electrolyte of diglyme-$LiNO_3$-HFiP with and without LiBOB additive. After drying for 24 h in the glove-box antechamber, ATR-FTIR was used in the wavelength range of 800 and 4000 cm^{-1}.

Three-Electrode Voltammetry

Three-electrode cell was prepared in a vial-type cell that comprised of a Ag/AgCl electrode (prior soaked in standard 1 M KCl brine) as the reference electrode and stainless steel disk (2 mm) as the working electrode at room temperature. The scan rate utilized was 10 mV/s.

Battery Performance

2032 type Li||stainless steel coin cells with and without HFiP additive in diglyme-$LiNO_3$ electrolyte were prepared inside an argon-filled glove box. The amount of electrolyte used for all battery testing was 60 μl. The cells were evaluated using galvanostatic cycling in a Neware CT-3008 battery tester. Coulombic efficiency test was performed in Li||stainless steel cell with different current densities with one each cycle comprising of 1 h. Half-cell test was performed in Li||NCM at different C-rates after initial two formation cycles of C/10. The cathode loading was 2 mAh/cm^2, and all the Li||NCM experiments were performed using a thin lithium (50 μm) as anode. Unless stated in the figure, the voltage ranges were chosen from 4.2 to 3 V. All the coin cells were crimped to a pressure of ~2500 psi. Except for the results using the cross-linked nanoparticles electrolytes, all the battery comprised of a 3501 celgard separator. Unless stated otherwise, the $LiNO_3$ content in the battery was $r = 0.5$ (molar ratio between EO and Li ions). In all the battery measurements with the HFiP chain-transfer agent, the amount added in the electrolyte was 1 wt.%. In the measurements using LiBOB salt additive, the amount utilized was 0.4 M.

References

1. Qian, J., et al.: High rate and stable cycling of lithium metal anode. Nat. Commun. **6**, 6362 (2015)
2. Cheng, X.-B., Zhang, R., Zhao, C.-Z., Zhang, Q.: Toward safe lithium metal anode in rechargeable batteries: a review. Chem. Rev. **117**, 10403–10473 (2017)
3. Cheng, X., et al.: A review of solid electrolyte interphases on lithium metal anode. Adv. Sci. **3**, 1–20 (2016)
4. Tikekar, M.D., Choudhury, S., Tu, Z., Archer, L.A.: Design principles for electrolytes and interfaces for stable lithium-metal batteries. Nat. Energy. **1**, 16114 (2016)

5. Wei, S., Choudhury, S., Tu, Z., Zhang, K., Archer, L.A.: Electrochemical interphases for high-energy storage using reactive metal anodes. Acc. Chem. Res. **51**, 80 (2017). https://doi.org/10.1021/acs.accounts.7b00484
6. Lin, D., Liu, Y., Cui, Y.: Reviving the lithium metal anode for high-energy batteries. Nat. Nanotechnol. **12**, 194–206 (2017)
7. Long, L., Wang, S., Xiao, M., Meng, Y.: Polymer electrolytes for lithium polymer batteries. J. Mater. Chem. A. **4**, 10038–10069 (2016)
8. Porcarelli, L., Gerbaldi, C., Bella, F., Nair, J.R.: Super soft all-ethylene oxide polymer electrolyte for safe all-solid lithium batteries. Sci. Rep. **6**, 19892 (2016)
9. Khurana, R., Schaefer, J.L., Archer, L.A., Coates, G.W.: Suppression of lithium dendrite growth using cross-linked polyethylene/poly(ethylene oxide) electrolytes: a new approach for practical lithium-metal polymer batteries. J. Am. Chem. Soc. **136**, 7395–7402 (2014)
10. Bouchet, R., et al.: Single-ion BAB triblock copolymers as efficient electrolytes for lithium-metal batteries. Nat. Mater. **12**, 452–457 (2013)
11. Agrawal, A., Choudhury, S., Archer, L.A.: A highly conductive, non-flammable polymer–nanoparticle hybrid electrolyte. RSC Adv. **5**, 20800–20809 (2015)
12. Srivastava, S., Schaefer, J.L., Yang, Z., Tu, Z., Archer, L.A.: 25Th anniversary article: polymer-particle composites: phase stability and applications in electrochemical energy storage. Adv. Mater. **26**, 201–234 (2014)
13. Rojas, A.A., et al.: Effect of lithium-ion concentration on morphology and ion transport in single-ion-conducting block copolymer electrolytes. Macromolecules. **48**, 6589–6595 (2015)
14. Chintapalli, M., et al.: Effect of grain size on the ionic conductivity of a block copolymer electrolyte. Macromolecules. **47**, 5424–5431 (2014)
15. Schaefer, J.L., Yanga, D.A., Archer, L.A.: High lithium transference number electrolytes via creation of 3-dimensional, charged, nanoporous networks from dense functionalized nanoparticle composites. Chem. Mater. **25**, 834–839 (2013)
16. Schaefer, J.L., Moganty, S.S., Yanga, D.A., Archer, L.A.: Nanoporous hybrid electrolytes. J. Mater. Chem. **21**, 10094 (2011)
17. (Youngman) Chusid, O, Ein Ely, E., Aurbach, D., Babai, M., Carmeli, Y.: Electrochemical and spectroscopic studies of carbon electrodes in lithium battery electrolyte systems. J. Power Sources. **43**, 47–64 (1993)
18. Aurbach, D.: Review of selected electrode–solution interactions which determine the performance of Li and Li ion batteries. J. Power Sources. **89**, 206–218 (2000)
19. Miao, R., et al.: Novel dual-salts electrolyte solution for dendrite-free lithium-metal based rechargeable batteries with high cycle reversibility. J. Power Sources. **271**, 291–297 (2014)
20. Zhang, X., Cheng, X., Chen, X., Yan, C., Zhang, Q.: Fluoroethylene carbonate additives to render uniform Li deposits in lithium metal batteries. Adv. Funct. Mater. **27**, 1605989 (2017)
21. Choudhury, S., et al.: Electroless formation of hybrid lithium anodes for fast interfacial ion transport. Angew. Chem. Int. Ed. **56**, 13070–13077 (2017)
22. Choudhury, S., Archer, L.A.: Lithium fluoride additives for stable cycling of lithium batteries at high current densities. Adv. Electron. Mater. **2**, 1500246 (2015)
23. Lu, Y., Tu, Z., Archer, L.A.: Stable lithium electrodeposition in liquid and nanoporous solid electrolytes. Nat. Mater. **13**, 961–969 (2014)
24. Zheng, G., et al.: Interconnected hollow carbon nanospheres for stable lithium metal anodes. Nat. Nanotechnol. **9**, 618–623 (2014)
25. Li, N.-W., Yin, Y.-X., Yang, C.-P., Guo, Y.-G.: An artificial solid electrolyte interphase layer for stable lithium metal anodes. Adv. Mater. **28**, 1853–1858 (2016)
26. Tu, Z., et al.: Fast ion transport at solid–solid interfaces in hybrid battery anodes. Nat. Energy. **3**, 310 (2018). https://doi.org/10.1038/s41560-018-0096-1
27. Suo, L., et al.: Fluorine-donating electrolytes enable highly reversible 5-V-class Li metal batteries. Proc. Natl. Acad. Sci. **115**, 1156 (2018)
28. Stephan, A.M.: Review on gel polymer electrolytes for lithium batteries. Eur. Polym. J. **42**, 21–42 (2006)

29. Zhang, W., et al.: Design principles of functional polymer separators for high-energy, metal-based batteries. Small. **14**, 1703001 (2018). https://doi.org/10.1002/smll.201703001
30. Li, W., et al.: The synergetic effect of lithium polysulfide and lithium nitrate to prevent lithium dendrite growth. Nat. Commun. **6**, 7436 (2015)
31. Tan, S., et al.: Tris(hexafluoro-iso-propyl)phosphate as an SEI-forming additive on improving the electrochemical performance of the Li[Li0.2Mn0.56Ni0.16Co0.08]O2 cathode material. J. Electrochem. Soc. **160**, A285–A292 (2013)
32. von Cresce, A., Xu, K.: Electrolyte additive in support of 5 V Li ion chemistry. J. Electrochem. Soc. **158**, A337–A342 (2011)
33. Verma, P., Maire, P., Novak, P.: A review of the features and analyses of the solid electrolyte interphase in Li-ion batteries. Electrochim. Acta. **55**, 6332–6341 (2010)
34. Choudhury, S., Archer, L.A.: Lithium fluoride additives for stable cycling of lithium batteries at high current densities. Adv. Electron. Mater. **2**, 1–6 (2016). https://doi.org/10.1002/aelm.201500246
35. Tu, Z., et al.: Designing artificial solid-electrolyte interphases for single-ion and high-efficiency transport in batteries. Joule. **1**, 394–406 (2017). https://doi.org/10.1016/j.joule.2017.06.002
36. Amine, K., et al.: Mechanism of capacity fade of MCMB/$Li_{1.1}[Ni_{1/3}Mn_{1/3}Co_{1/3}]_{0.9}O_2$ cell at elevated temperature and additives to improve its cycle life. J. Mater. Chem. **21**, 17754–17759 (2011)
37. Xiao, A., Yang, L., Lucht, B.L., Kang, S.-H., Abraham, D.P.: Examining the solid electrolyte interphase on binder-free graphite electrodes. J. Electrochem. Soc. **156**, A318–A327 (2009)
38. Zhu, Y., Li, Y., Bettge, M., Abraham, D.P.: Electrolyte additive combinations that enhance performance of high-capacity $Li_{1.2}Ni_{0.15}Mn_{0.55}Co_{0.1}O_2$–graphite cells. Electrochim. Acta. **110**, 191–199 (2013)
39. Choudhury, S., Mangal, R., Agrawal, A., Archer, L.A.: A highly reversible room-temperature lithium metal battery based on crosslinked hairy nanoparticles. Nat. Commun. **6**, 10101 (2015)
40. Choudhury, S., Vu, D., Warren, A., Tikekar, M.D., Tu, Z., Archer, L.A.: Confining electrodeposition of metals in structured electrolytes. Proc. Natl. Acad. Sci. **115**, 6620–6625 (2018)
41. Chai, J.-D., Head-Gordon, M.: Long-range corrected hybrid density functionals with damped atom–atom dispersion corrections. Phys. Chem. Chem. Phys. **10**, 6615–6620 (2008)
42. Chai, J.-D., Head-Gordon, M.: Systematic optimization of long-range corrected hybrid density functionals. J. Chem. Phys. **128**, 84106 (2008)
43. Hariharan, P.C., Pople, J.A.: The influence of polarization functions on molecular orbital hydrogenation energies. Theor. Chim. Acta. **28**, 213–222 (1973)
44. Frisch, M.J., Trucks, G.W., Schlegel, H.B., Scuseria, G.E., Robb, M.A., Cheeseman, J.R., Scalmani, G., Barone, V., Mennucci, B., Petersson, G.A., Nakatsuji, H., Caricato, M., Li, X., Hratchian, H.P., Izmaylov, A.F., Bloino, J., Zheng, G., Sonnenberg, J.L., Hada, M., et al.: Gaussian 09. Gaussian, Inc., Wallingford (2016)
45. Tomasi, J., Mennucci, B., Cammi, R.: Quantum mechanical continuum solvation models. Chem. Rev. **105**, 2999–3094 (2005)
46. Zhan, C.-G., Dixon, D.A.: The nature and absolute hydration free energy of the solvated electron in water. J. Phys. Chem. B. **107**, 4403–4417 (2003)

Biographical Sketch

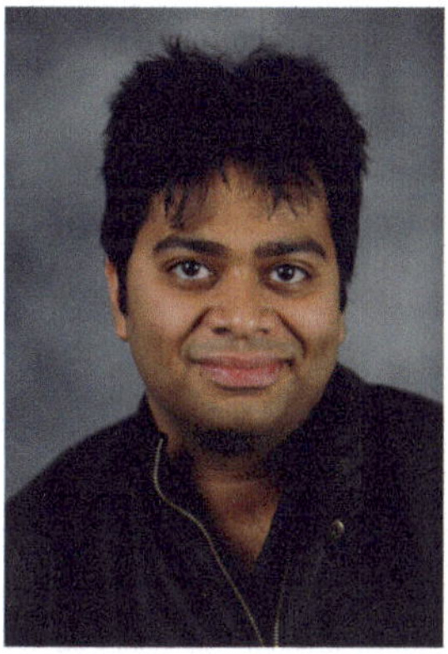

Snehashis Choudhury was born in a small town of Brajrajnagar in the state of Odisha in India. He went to high school in Kolkata, India. As a child, his favorite topics were Mathematics, Chemistry, and Economics. Following his high school, he decided to pursue Engineering over his second choice of Economics. He went to the National Institute of Technology Calicut to pursue Bachelor of Technology in Chemical Engineering in the year 2009. His 4-year experience in NIT Calicut was one of the best times of his life, where he made several great friends and also learned about independent research. He joined Cornell University for his Master of Engineering in the School of Chemical and Biomolecular Engineering in the year 2013. He worked in the group of Professor Lynden A. Archer and thereafter continued as a PhD student starting 2014. In his PhD, he started his work on designing covalently grafted hairy nanoparticles to understand their structure and dynamics. In his second year, he developed an interest in understanding the instabilities and thereafter stabilizing of metal-based batteries. He found that custom-designed hairy nanoparticles are excellent candidates in serving as solid or gel electrolytes to inhibit battery short circuits in lithium metal batteries. However, a glaring failure mechanism he discovered in these metal batteries was the unwanted side reactions at the battery interfaces causing slow fading in the capacity. He took this challenge head on by designing artificial interfaces to inhibit these side reactions, thereby enabling stable cycling of a solid polymer electrolyte in a high-energy metal battery. He will always be grateful to all the collaborators and group members for their constant support not only in research but also in

S. Choudhury, *Rational Design of Nanostructured Polymer Electrolytes and Solid–Liquid Interphases for Lithium Batteries*, Springer Theses,
https://doi.org/10.1007/978-3-030-28943-0

other endeavors. Following his PhD, he joined as a postdoctoral researcher in the lab of Professor Zhenan Bao in the Department of Chemical Engineering of Stanford University, where his research focusses on developing new synthetic routes of self-healable and stimuli-responsive ionic polymer networks for application in electrochemical devices.

The manufacturer's authorised representative in the EU is Springer Nature Customer Service Centre GmbH, Europaplatz 3, 69115 Heidelberg, Germany. If you have any concerns regarding our products, please contact ProductSafety@springernature.com

Printed and bound by CPI Group (UK) Ltd, Croydon, CR0 4YY
05/10/2026
02228130-0003